B. P. Pratten

Contagious Diseases of Domesticated Animals

Continuation of investigation

B. P. Pratten

Contagious Diseases of Domesticated Animals
Continuation of investigation

ISBN/EAN: 9783337241193

Printed in Europe, USA, Canada, Australia, Japan

Cover: Foto ©berggeist007 / pixelio.de

More available books at **www.hansebooks.com**

DEPARTMENT OF AGRICULTURE.

SPECIAL REPORT—No. 22.

CONTAGIOUS DISEASES

OF

DOMESTICATED ANIMALS.

CONTINUATION OF INVESTIGATION

BY

DEPARTMENT OF AGRICULTURE.

WASHINGTON:

GOVERNMENT PRINTING OFFICE.

1880.

CONGRESS OF THE UNITED STATES,
IN THE HOUSE OF REPRESENTATIVES, *May* 6, 1880.

The following resolution, originating in the House of Representatives, has this day been agreed to:

Resolved by the House of Representatives (the Senate concurring therein), That there be printed one hundred thousand copies of Special Report number twenty-two of the Commissioner of Agriculture, containing the reports of the Veterinary Surgeons appointed to investigate diseases of swine, and infectious and contagious diseases incident to other classes of domesticated animals, of which sixty thousand copies shall be printed for the use of Members of the House, twenty-five thousand copies for the use of Members of the Senate, and fifteen thousand copies for the use of the Commissioner of Agriculture.

Attest: GEO. M. ADAMS,
Clerk.

By—   GREEN ADAMS,

2   *Chief Clerk.*

TABLE OF CONTENTS.

	Page.
INVESTIGATION OF SWINE PLAGUE—Introductory	7
Report of Dr. H. J. Detmers	13
Report of Dr. James Law	68
Swine Plague in the Southwest	81
Review of Special Report No. 12	89
Swine Plague in Florida	94
INVESTIGATION OF SOUTHERN CATTLE FEVER:	
Report of Dr. D. E. Salmon	98
Southern Cattle "Distemper"	142
CONTAGIOUS LUNG PLAGUE OF CATTLE:	
Review of Dr. Law's monograph work	143
Report of Dr. Chas. P. Lyman	163
MISCELLANEOUS:	
Pleuro-Pneumonia, or Bovine Lung Plague	179
Cattle Plague, or Rinderpest	185
Glanders and Farcy in horses	202
Experiments in inoculation of Charbon	209
Experiments in "Chicken Cholera"	211
Catarrh in sheep	214
Statistics of domesticated animals in Europe	261
CORRESPONDENCE:	
Relating to sanitary regulations and preventive measures	216
Relating to prevalence of diseases among farm animals	232

3

LIST OF ILLUSTRATIONS.

Page.

SWINE PLAGUE:

Plate I. Lung of experimental heifer, showing hepatization 67

Plate II. External surface of left lobe of lung of diseased pig 67

Plate III. Nos. 1, 2, 3, 4, and 5, thin transversal sections of lung stained in Kleinenburg's solution. No. 6, same, but not stained 67

Chart illustrating microscopic investigations 67

CONTAGIOUS LUNG PLAGUE OF CATTLE:

Plate I. Section of diseased lung; thin end shows black hepatization; thick end interlobular infiltration. Several blocked vessels are shown.. 162

Plate II. Section of healthy lung (cow) showing the abundance of interlobular, cellular, or connective tissue, of a bright rose color. Average weight of either right or left lung, 3¼ pounds 162

Plate III. Section of left lung of a cow in an advanced stage of contagious pleuro-pneumonia, showing the characteristic marbled appearance formed by the exudation and consolidation of lymph into the interlobular, cellular tissue. Weight of left lung, 37 pounds; right lung, 5 pounds 162

MAP. Showing States and counties in which contagious pleuro-pneumonia exists 179

RINDERPEST:

Plate I. Skin of udder on the sixth or seventh day, showing, in addition to the usual eruption, patches of redness on the teats.................... 196

Plate II. Skin of udder, showing eruption in more advanced stages of the plague.. 196

Plate III. Lips and gums, showing apthous condition 196

Plate IV. Roof of mouth, showing excoriations 196

Plate V. Tongue and throat, showing thickening of epthelium, with excoriations and congestions...................................... 196

Plate VI. Surface of lung, showing interlobular emphysema, extending, in some places, into the sub-pleural tissue............................ 196

Plate VII. Portion of fourth stomach of cow, about eighth day of cattle plague, showing patches of ecchymosis and deep ulcers................. 196

Plate VIII. Rectum and anus, showing deep congestion 196

INVESTIGATION OF SWINE PLAGUE.

INTRODUCTORY.

In the further investigation of swine plague Dr. Detmers finds but few additional symptoms worthy of mention. As his observations extended through the winter he was enabled to observe the disease closely during those months when it is neither so general nor so fatal as during other seasons of the year. During the winter months, therefore, and in the early spring, he found bleeding from the nose and symptoms of respiratory disorders quite frequent, but there seemed to be fewer indications of gastric disorders than he had observed during the summer and fall months. Neither was the prognosis, as a rule, so hopeless in the winter and spring as during the summer and early autumn. This he regards as attributable to the fact that during the former seasons the seat of the morbid process is limited more frequently to the respiratory organs and to the pulmonal tissue, and is not found so often in the intestinal canal. Thirty additional *post-mortem* examinations failed to reveal any new morbid changes worthy of special mention. A few variations, and in some cases an unusual combination of morbid changes, were observed, which will be found accurately described in the text of Dr. Detmers' report. The absence of worms or entozoa in 75 per cent. of the whole number of animals dissected prior to December 1, and their entire absence in every animal examined between that date and the 15th of the same month, would seem to prove conclusively, Dr. Detmers thinks, that the morbid changes characteristic of swine plague cannot be attributed to the work of entozoa. He thinks, however, that the presence of worms in large numbers, occurring in weak, poorly-kept, and neglected animals, may cause considerable mischief, and sometimes occasion death, but in such cases the cause should not be attributed to swine plague.

Former experiments seemed to fully demonstrate the fact that swine plague is an infectious and contagious malady, and that it is easily communicable from one animal to another by means of direct inoculation, and by the introduction into the digestive organs of the infectious principle by means of food and drinking water; second, that an exceedingly small quantity of the virus or infectious principle is sufficient to produce the disease; third, that the period of incubation does not exceed fifteen days, and lasts on an average from six to seven days; fourth,

7

that the small microphytes (*bacilli*) found in all the morbid products, in the blood and other fluids, and in all excretions of the animals, would seem to constitute the infectious principle of the malady. This being the case Dr. Detmers instituted a series of experiments in order to determine, if possible, first, whether the infectious principle consists solely in the *bacilli* and their germs; second, whether an animal that has recovered from the disease has gained immunity from a second attack; and, third, whether the affection can be transmitted to other classes of domesticated animals. The result of these experiments proved, first, that an inoculation with *bacilli* and bacillus-germs cultivated in so innocent a fluid as milk will produce the disease with just as much certainty as an inoculation with pulmonal exudation from a diseased or dead hog; second, that an animal that has been afflicted with the plague has not lost its susceptibility, but may contract the disease again, though probably in a milder form.

In order to test the susceptibility of other animals to the contagion, Dr. Detmers inoculated two heifers with the virus of swine plague, and the elevation of temperature and the subsequent *post-mortem* examination would seem to indicate that the disease prevailed, at least in one case (heifer No. 2), to a considerable degree of intensity. The autopsy revealed distinctly limited (circumscribed) hepatization at several points in both lobes of the lungs, each single patch comprising only a few lobules, but these were distinct and well defined. The most extensive hepatization was found along a larger bronchus in the posterior part of the left lobe. The hepatized parts or patches amounted to about 4 or 5 per cent. of the whole pulmonal tissue. The mucous membrane of the bronchiæ was found to be slightly swelled; a small quantity of serum was found in the pericardium and in the chest, and a few ounces also in the abdominal cavity. The lymphatic glands of the chest, and those belonging to the mesenterium, were enlarged, and some of them, especially the latter, to a considerable extent. The other organs exhibited no abnormal changes. The result of this experiment indicates that while cattle are not as susceptible to this plague as swine, yet it may be transmitted to them, in a mild form, by direct inoculation.

Many illustrations are given of the manner in which the contagious principle is transmitted from herd to herd and from farm to farm. Believing that the seeds of the disease consist in the *bacilli* and their germs, as fully described in his former report, Dr. Detmers is of the opinion that these microphytes can be conveyed from one place to another, not only in the morbid products of the disease, such as the tissues, fluids, and excretions of affected and dead animals, but also by adhering to and contaminating inanimate objects, both fluid and solid, and independently of any vehicle, through the air to a distance of a mile if the conditions are favorable, and in the water of running streams. The last mode is one of the most prolific sources of infection, as these microphytes propagate and multiply in water, especially if it should be contaminated by a mixture of organic matter.   After citing many cases in illustration of the various ways in which the contagion is spread, Dr. Detmers is of the opinion that the following facts have been established:

First. The plague is not easily communicated unless the infectious principle is introduced either into the digestive apparatus with the food or with the water for drinking, or directly into the blood through wounds, sores, scratches, or external lesions.

Second. That the carcass of an animal that has died of the plague will communicate the disease to healthy swine if eaten before it is thoroughly putrified.

Third. That even severe frost is not sufficient to destroy the infectious principle if the same is protected against external influences by some organic substance.   Former experiments by Dr. Law also demonstrated this fact.

Fourth. That the plague is readily and frequently communicated to healthy animals by means of the water used for drinking, especially if the same should be contaminated by the carcass of an animal that has died of the disease, or by the excrements, urine, nasal discharges, saliva, &c., of animals afflicted with the malady.

Fifth. That in localities where the plague is prevailing, every wound, scratch, or sore on the surface of the body constitutes a port of entry for the infectious principle.

Sixth. That old straw stacks and other decaying porous bodies may preserve the infectious principle for months, and in many cases even for a whole year.

Seventh. That the infectious principle enters the animal organism and communicates the disease more readily through external sores and lesions than through the digestive canal.

Many experiments were instituted for the purpose of determining upon a system of preventive measures and of testing the value of certain remedial agents.   As to the result of these experiments and the conclusions deduced therefrom, the reader is referred to the detailed report of Dr. Detmers.

When Dr. Law closed his former report he had just commenced some important experiments for the purpose of determining the susceptibility of other animals to swine plague.   These experiments resulted, as he had previously foreshadowed, in the successful inoculation of sheep and rats, and in the transmission of the disease from these animals back to swine in a virulent and intensified form.   His first experiment was undertaken for the purpose of determining at what period or stage of the malady it is most easily and certainly transmissible from one animal to another by cohabitation.   In his previous report an experiment of this kind was furnished, and a deduction made that the disease was most virulent when at its height, inasmuch as the exposed pig seemed to resist the contagion from an animal in process of convalescence, but within twelve days fell a victim to the disease when placed alongside of a pig in which the malady was rapidly advancing. In the accompanying report the necropsy of this pig is given, from which it will be observed that it was afflicted with the plague in an intensified form.

The autopsy of an infected lamb, noticed by Dr. Law in his former report, is also given in full. The intestinal irritation and catarrh, shown in the tenderness of the anus and the mucus discharges accompanying the feces, together with the elevated temperature and large lymphatic glands, presented much in common with the affection in the pig. The marked eruption in the ears might be accepted as representing the skin lesions, so common in swine suffering with the plague. The more characteristic lesions revealed by the *post-mortem* examination were the purple mottling of the liver, kidneys, and heart, the grayish consolidation of portions of the lungs, and the deep pigmentation of the lymphatic glands in general.

The next experiment was that of a merino sheep infected by inoculation. The record and the results of the autopsy are very similar to those furnished by the lamb. Here, again, the principal changes consisted in purple mottling of the liver and heart, and the deep pigmentation of the lymphatic glands. Dr. Law is of the opinion that the yellowish-brown coloration of the kidneys in this case implied antecedent changes, probably of the nature of inflammation or extravasation.

From virus taken from these animals Dr. Law successfully inoculated a pig. The pig was inoculated twice, at an interval of fifteen days, with mucus from the anus of the infected sheep, and with scabs from the ear of the lamb. Enlarged lymphatic glands were observable before the last inoculation, and six days after there was a febrile temperature and the more violent manifestations of the malady. The following characteristic lesions were revealed by the necropsy, viz: The intestines contained patches of congestion; the follicles were enlarged, and the rectum ulcerated; purple discolorations were present in the liver, kidneys, and heart; the lymphatic glands were enlarged and congested by a deep red, in some cases almost black. While the evidence of the presence of the disease in this case was quite positive, in order to confirm it beyond question, Dr. Law inoculated a second pig from virus taken from the first. This inoculation was successful. While followed by but a slight elevation of temperature, all the other characteristic symptoms of the disease were well marked. The *post-mortem* examination was made on the twelfth day after the inoculation, when the usual lesions of the plague were found. The red and black blotches on the skin were extensive, the ears blue, the intestines extensively congested, with enlarged follicles in the cæcum and colon, and blood extravasations and ulcers in the rectum. Purple discolorations and petechiæ were numerous on the liver, kidneys, and heart, and finally the lymphatic glands in general were in part congested of a deep red, and in part pigmented as the result of a previous congestion.

Another successful inoculation of a pig was made from virus from an infected lamb. The pig was inoculated from material taken from the swelling in the axilla (near the seat of inoculation) of the lamb. The inoculation produced fever, with the general malaise, moping, peevish grunt, inapetence, and the cutaneous blotches of swine plague. The animal was killed on the eleventh day after inoculation, when the autopsy was at once made. The skin showed a number of red and purple blotches, and was covered with the black unctuous exudation so frequently observed in this disease. The bowels contained patches of congestion, the crecum and colon were enlarged, and the follicles and the rectum were ulcerated. The liver, kidneys, and heart contained the usual purple blotches. Finally, the lymphatic glands in the abdomen were enlarged and congested of a deep red or black, and those in the chest and guttural region were darkly pigmented. This was regarded as a most unequivocal case, and fully confirmed the position heretofore taken by Dr. Law, viz., that the virus of swine plague may be transmitted through the sheep and conveyed back to the pig with active and deadly effect.

This experiment was followed by the inoculation of a pig with virus from infected rat and lamb, and also inoculations with virus from infected pig, rat, and lamb. By reference to the first experiment it will be seen that while the pig showed but little elevation of temperature, there was a purple cutaneous eruption of the skin on the fifth day and enlarged glands on the twelfth, when it was inoculated with bloody mucus from the anus of the infected lamb. After this the symptoms became much more severe, and when killed, twenty-two days after, the animal showed unmistakable lesions of the disease.

The next subject was that of a healthy female Suffolk pig. This was inoculated with albumen which had been charged with a drop of blood containing bacteria drawn from an infected pig. For fifteen days nothing more was shown than a few purple spots and patches on the rump, tail, and hocks. The subject was then reinoculated with the congested intestine of the rat which had died two days after inoculation. The intestine had been frozen over night. For thirteen days more the same equivocal symptoms continued. A third inoculation was now practiced, this time with bloody mucus from the anus of the lamb. Twenty-two days after this inoculation the pig was sacrificed, but beyond some pigmentation of the lymphatic glands presented no distinct lesions that could be held characteristic of the specific fever.

On February 5, 1879, Dr. Law inoculated a rat with virulent matter that had been preserved for seventy-eight days closely packed in dry wheat bran. The rodent was preserved for thirteen days, when it was killed and immediately dissected. The symptoms during life and the lesions found after death were so closely in keeping with those of swine plague, that there seemed no reasonable grounds for doubting the entire success of the inoculation. In order to confirm this, however, and place the matter beyond doubt, the following experiment was undertaken: On

February 19, 1879, a healthy pig was inoculated with the congested lymphatic glands and lungs of the above-mentioned rat. On the sixth day after inoculation there was much malaise, with redness of the skin and the appearance of the black unctuous exudation on the ears and legs. These went on increasing, and black spots and patches, ineffaceable by pressure, appeared on the inside of the thighs and hocks. The subject was destroyed on the twentieth day, and showed the usual symptoms of the disease. The lesions were as unequivocal as in any case where inoculation was made from a sick pig direct, and would seem to prove conclusively that the rat is capable of containing this disease and of conveying it back to the pig.

The above is a brief *résumé* of the results of the more important experiments undertaken and completed by Dr. Law after the closing of his report last season. In view of the recent discoveries of M. Pasteur and other eminent scientists in inoculations for charbon and anthrax, a brief mention of which is made elsewhere in this report, Dr. Law is now engaged in a like series of experiments to determine if diluted inoculations will not produce swine plague in a mild form and in such manner as to guarantee immunity from a second attack.

INVESTIGATION OF SWINE PLAGUE.

SECOND REPORT OF DR. H. J. DETMERS, V. S.

Hon. WM. G. LE DUC,
 Commissioner of Agriculture, Washington, D. C.:

SIR: After I sent you my supplemental report, dated December 1, 1878, you ordered me to go to work again and complete the investigation of swine plague, begun under your directions on August 1, 1878. In compliance with this order I at once started for a field of operations, and arrived at Dixon, Lee County, Illinois, on December 28. After a brief survey I established my experimental station at Gap Grove, a small village six miles west of this place, and at that time the center of an infected district. I remained there until February 8, 1879, when you ordered me to the Union Stock Yards of Chicago, to inspect cattle. My investigation, in consequence, was thereby interrupted, until on the 16th of May, 1879, when you requested me to resume my former work.

The results of my work in investigating swine plague from December 15 to February 7, and from May 18 to July 4, will be found briefly reported in the following pages: For convenience and to avoid unnecessary repetition, I shall arrange the various chapters in the same order as in my first report, and shall exclude, as much as possible, everything already stated in the latter. The following chapters, therefore, may be considered as supplementary to those of my first report.

1. DEFINITION OF SWINE PLAGUE.

Nothing new needs to be added, except that swine plague, although a disease *sui generis*, peculiar to swine, can be transferred by inoculation, and undoubtedly, also, by means of infected food and water, to other animals, such as rats, rabbits and sheep (Prof. J. Law), cattle and dogs.

2. SYMPTOMS.

I have visited thirty-two different herds, and examined a large number of diseased animals, but have very little to add to my former report. The essential differences observed are as follows: During the winter and in the spring bleeding from the nose, difficulty of breathing, and symptoms of respiratory disorders in general seem to be more, and symptoms of gastric disorders less frequent than in the summer and in the fall.

3. PROGNOSIS.

Though always unfavorable, the same, as a rule, is not quite so hopeless in the winter and spring as in the summer and early autumn, probably because in the former seasons the seat of the morbid process is limited more frequently to the respiratory organs and to the pulmonal tissue, and is not found so often in the intestinal canal. Still the difference, partially due no doubt to some other causes or conditions, to be explained further below, is not a very great one, especially if it is taken into consideration that swine plague is always more fatal to very young pigs than to older animals or full-grown hogs; and that more pigs are born in the spring than at any other season of the year. Consequently, the average age of the pigs diseased with swine plague is much less in the summer than in the winter.

4. MORBID CHANGES.

Since December 15 numerous examinations (thirty-one is the exact number) have been made, but no new morbid changes not met with before have been discovered, and the combinations in which the various morbid changes presented themselves did not essentially differ from those already recorded in my first report. It will, therefore, be sufficient, in order to avoid unnecessary repetition, and to give at the same time a complete description of the various combinations of morbid changes which have come under my observation, to describe only those few cases which presented such variations as may possibly serve to throw more light upon the nature of the morbid process, by pointing out some of the probable causes of the great diversity of morbid changes found in different animals. Swine, just as well as other domesticated animals, and perhaps even more than other animals, on account of being omnivorous, and having therefore more opportunity to pick up worm-brood, are subject to being inhabited and preyed upon by various species of parasitic worms or entozoa, especially at certain seasons of the year. These worms, of course, occur just as well in those animals that are afflicted with swine plague as in those that have never been exposed to any infection; consequently they are found quite often on *post mortem* examinations.

As mentioned in my first report, some species of entozoa—*Strongylus paradoxus* (in the bronchial tubes), *Tricocephalus crenatus* (in the cæcum), and a few others—were found in thirteen animals, or at least 25 per cent. of the *post mortem* examinations made last summer and fall, prior to December 1; but no entozoa whatever could be found, notwithstanding I searched for them in nearly every examination made since December 15. The absence of worms or entozoa in 75 per cent. of the whole number of animals dissected prior to December 1, and their entire absence in every animal examined after death since December 15, proves conclusively, if anything, that the morbid changes characteristic of swine plague cannot be the work of worms or entozoa. I will not deny that the latter, if present in large numbers, or occurring in weak, starving, and neglected animals, may be able to cause considerable mischief, and even death; but such cases must not be mistaken for swine plague. Worms or entozoa are found very often in healthy hogs and pigs independent of the morbid process of swine plague, and have no connection whatever with that disease. I therefore simply mentioned their occurrence in my first report, and did not deem it necessary to dwell on their natural history or on the damage which they may be able to do.

The most essential difference between the morbid features presented at the *post mortem* examinations previous to December 1 and those found in the animals examined in the winter and spring, consists in a more frequent affection of the large intestines (cæcum and colon) in the summer and fall, while in the winter and spring the principal seat of the morbid process was almost invariably in the organs of the chest, but especially in the pulmonal tissue. In the summer and fall, or previous to December 1, ulcerous tumors in either one or both of the large intestines, cæcum and colon, were found in about 90 per cent. of the whole number of cases examined, while in the winter and spring they did not exist in more than about 50 per cent.

This difference it seems to me is not accidental, but admits of an explanation. At any rate, the predominating affection of the organs of the chest, and especially the extensive embolism and exudation in the lungs, observed invariably in every case in which the large intestines were free from ulcerous tumors, may be traced to distinct causes, acting principally during the winter. Swine, especially in the cold season of the year, on entering their lair and going to sleep in the evening, are in the habit of crowding close together, of lying on top of each other, and of frequently passing the night in very close quarters. Such crowding into a narrow space cannot fail to heat their bodies, to vitiate the atmosphere, and to accelerate the respiration. Consequently it will prevent a proper decarbonization of the blood and retard its circulation in the pulmonal capillaries, and cause more or less congestion of the lungs, and prepare those organs for just such morbid changes as are effected by the *bacilli* and their germs. In the morning, after the animals have been heated during the night, and are rising from their lair in search of food, the air, especially in the winter, is usually cold and chilly, and, but a moment ago reeking and steaming with perspiration, they become chilled and commence to shiver. Such a sudden change of temperature necessarily causes a disturbance of the functions of the lungs and of the skin, contracts the expanded capillary vessels of the latter, and thereby compels the blood to rush to the heart and to the interior parts of the body. All this cannot fail to predispose, especially the lungs and heart, to become the principal seat of the morbid process of swine plague, if the infectious principle, the *bacilli* and their germs, have entered the organism. Moreover, it appears to be probable that an imperfect decarbonization of the blood promotes the tendency of the bacillus-germs and partially developed *bacilli* to agglutinate to each other, and to form those irregularly shaped clusters which clog the capillaries, and cause in that way extravasation of blood and extensive exudations. Some other influences may be acting, but those mentioned seem to be the principal ones. The greater frequency of morbid changes in the large intestines and in the digestive canal in general, in the summer and fall, has probably an equally good cause.

Swine as a rule lead a more independent life in the summer and fall than during the winter. In the summer and fall a great many have access to a pasture, others are allowed to roam at large, and get their food and water wherever it is convenient, and almost all receive at least some green food. Further, the *bacilli* and their germs propagate more rapidly in the summer and early fall than in any other season of the year, and the heavy dews, common from July or August till November, carry down in the night and morning those germs that may have risen in the air during the day and deposit them upon the surface of the earth, but especially upon the grass and herbage of field and pasture, and in the water of pools, brooks, &c; consequently, there can be no doubt that an introduction of the disease-germs into the animal organism with the food and water for drinking constitutes a more frequent means of infection in the summer than in the winter. Now, according to my observations, as mentioned in my first report, the morbid process seems to have a special tendency to attack the digestive canal, and to produce ulcerous tumors in the large intestines, if the infectious principle has been introduced with the food or with the water for drinking; but this is often, though not always, confined to the organs of the chest and to the lymphatic system if the disease-germs have entered the organism through a scratch or wound.

A FEW SPECIAL CASES.

1. *Post mortem* examination of a barrow, nine months old, belonging to H. Miller, Prairieville, Lee County, Illinois. Date, December 29.
Externally.—No *rigor mortis*, and carcass yet warm. Skin reddened and purplish on lower surface of body, between the legs, behind the ears, and on the neck.
Internally.—Lymphatic glands enlarged; in the lungs gray hepatization and numerous embolic hearths, looking like tubercles, comprising over one-third of the whole pulmonal tissue; more than one pint of serum in the chest, and over four ounces in the pericardium, capillaries, and larger blood-vessels of the heart, but especially of the auricles, very much injected and tinged with dark-colored blood, so as to give the auricles an almost uniform black-brown appearance; extravasated blood and large quantities of gelatinous exudation imbedded in the tissue (walls) of the auricles; no ulcerous tumors in the intestinal canal.
2. *Post mortem* examination of a pig nine months old, weighing 250 pounds, and belonging to Mrs. Harms, Gap Grove, Lee County, Illinois. The animal had just died. Date, December 30.
Externally.—Blood oozing from the nostrils.
Internally.—Pulmonal and costal pleura nearly everywhere more or less firmly united; brown hepatization extending over three-fourths of the lungs; some parts almost gangrenous, and containing patches of extravasated blood; capillary blood-vessels of the heart, but especially of the auricles, very much injected, and tinged with dark-colored blood; left auricle perfectly black; blood everywhere dark-colored; lymphatic glands everywhere enlarged; some incipient ulcerous tumors in cæcum, and erosions in the mucous membrane of the blind end of the same intestine. The carcass was in very good flesh, and rather plethoric.
3. *Post mortem* examination of another pig of same age, belonging to same party. Date, January 8.
Externally,—No changes worth mentioning.
Internally.—Lymphatic glands somewhat enlarged; adhesion between pulmonal and costal pleura; one-half of the pulmonal tissue hepatized, numerous embolic hearths and extravasations of blood in the hepatized parts of lungs, and in tissue of the heart and of the posterior aorta; some serum in pericardium; auricles of the heart spotted with numerous specks of extravasated blood; flesh (fat and lean) everywhere very yellow; liver more or less sclerotic, and contents of gall-bladder very dark-colored and semi-solid; spleen dark-colored, and spotted with small rust-colored and elevated spots; small specks and patches of extravasated blood in mucous membrane of stomach; several ulcerous tumors—some very large, others small—in colon; erosions and specks of extravasated blood, but no ulcerous tumors, in mucous membrane of cæcum.
4. *Post mortem* examination of a pig belonging to Mr. Swigart, Palmyra, Lee County, Illinois. Date, January 14. Pig was killed by bleeding. While alive, was bleeding from the nose, and breathing with difficulty.
Morbid changes.—All lymphatic glands very much enlarged; pulmonal

pleura of left lobe of lungs adhering to costal pleura; red and brown hepatization and numerous specks of extravasated blood in lower half and posterior part of left lobe; in right lobe, three-fourths of the whole tissue hepatized and containing numerous specks of extravasated blood in lower half and posterior part of left lobe; in right lobe three-fourths of the whole tissue hepatized and containing numerous specks of extravasated blood, and also an abundance of fresh and partially coagulated exudation in the pulmonal tissue, and on the surface of the pulmonal pleura. The lungs, but especially the right lobe, presented a very marked appearance—gray, brown, and red hepatization—and fresh exudation in adjoining lobules alternating with each other. Some serum in pericardium; blood in the heart and everywhere else, dark-colored, as in all animals in which the pulmonal tissue constitutes the principal seat of the morbid process; one ulcerous tumor in colon; cæcum healthy.

5. *Post mortem* examination of another pig, nine months old, belonging to Mrs. Harms. Carcass in first-rate condition as to flesh; weight 220 pounds. Date, January 18.

Morbid changes.—All lymphatic glands enlarged; pleura of right lobe of lungs partially coalesced to costal pleura; fully two-thirds of the tissue of both lobes of lungs hepatized—red and brown hepatization—and containing innumerable small patches of extravasated blood and embolic hearths; some serum in thoracic cavity and in pericardium; capillary vessels of the heart, but particularly of the auricles, very much injected and tinged with dark-colored blood; numerous small red spots (extravasated blood), as large as a millet-seed or smaller, in the serous membrane of the small intestines (jejunum and duodenum); numerous large and, some of them, confluent ulcerous tumors penetrating into the external or serous membrane in cæcum; several smaller or medium-sized ulcerous tumors in colon; liver yellowish; contents of gall-bladder semi-fluid and granular.

6. *Post mortem* examination of a pig belonging to John Lord, Palmyra, Lee County, Illinois. The pig was killed by bleeding to obtain material for experimental purposes. Date, January 21.

Morbid changes.—Besides those usually found in the organs of the chest and in the lymphatic system were, a profuse proliferous growth of connective tissue and epithelium cells in process of decay, presenting a profuse ulcerous tumor on the mucous membrane of the stomach at its large curvature, but no ulcerous tumors in any other intestine.

7. *Post mortem* examination of two hogs belonging to G. Sartories, near Gap Grove, Lee County, Illinois. Both animals had recovered from an attack of swine plague over two months ago, and were butchered for pork. Date, January 22. One of the hogs, No. 7, dressed 180 pounds, and the other one, No. 8, dressed 260 pounds. Both were of about the same age, and over a year old.

Morbid changes in No. 7.—Pulmonal and costal pleura connected everywhere by means of a loose and very meshy connective tissue, which could be torn without using great force; remnants of partially absorbed hepatization in anterior lobes of the lungs; firm and inseparable adhesion (union) almost everywhere between the external surface of the heart and the internal surface of the pericardium. (The animal, before it was killed, exhibited asthmatic symptoms.) No other morbid changes could be found, except enlargement of most of the lymphatic glands situated in the thoracic and abdominal cavities.

Morbid changes in No. 8.—All lymphatic glands considerably enlarged; remnants of hepatization, but yet very distinct, and inclosing two nodules (one of the size of a small cherry, and the other the size of a large pea), sequestered by an envelope of firm and solid connective tissue in the anterior parts of the lungs; no adhesion between the pleuras, and no other morbid changes. (In the chapter headed "Contagion" I shall have to refer again to these two animals.)

9, 10. *Post mortem* examination of two hogs belonging to John Lamken, near Prairieville, Lee County, Illinois. The same had been afflicted with swine plague, but had recovered over two months ago, and were butchered for pork. Both were of about the same age and size, and dressed each about 275 pounds. Date, February 3. No morbid changes in hog No. 9, except small remnants of hepatization in the lungs; and in hog No. 10 no morbid changes, except a small and unimportant scar in the cæcum, indicating the former existence of an ulcerous tumor. Both hogs undoubtedly had only a very mild attack of swine plague, otherwise more important morbid changes would have been left behind. If any exudation or hepatization had existed in the lungs of hog No. 10—and it would be very strange if it had not, because I never found it entirely wanting in any other of the numerous *post mortem* examinations I had an occasion to make (about one hundred since August 1)—it must have been perfectly absorbed, because no trace could be discovered.

11. *Post mortem* examination of a pig belonging to Mr. Horace McKay, twelve miles north of Champaign, Champaign County, Illinois. A small, evidently stunted animal, which had been sick for two months, had a large tumor of the size of a man's fist on the left side of the anterior part of its nose. The temperature of the animal, which was killed by bleeding, was 104.3° F. Date, June 17. *Morbid changes:* The tumor was hard, mainly composed of a dense connective tissue, similar to that of an intestinal ulcerous tumor, and undoubtedly originated in the mucous membrane of the nasal cavity. Its interior concave surface, communicating with the latter, was coated with a thick layer of a dirty-white or gray-yellowish detritus, the same as coats the surface of the ulcerous tumors which occur in the large intestines. Its external surface was convex or semi-globular, dark-colored, and almost smooth.

Internal morbid changes.—Two-thirds of both lobes of the lungs hepatized; some almost clear or slightly straw-colored serum in the pericardium and in the chest; froth in the bronchial tubes; all lymphatic glands enlarged, but no other morbid changes, except a small quantity of serum in the abdominal cavity. The tumor, according to Mr. McKay, made its appearance two months ago, at the beginning of the disease.

12. *Post mortem* examination of a small pig four months old, belonging to M. Philippi, ten miles north of Champaign, Champaign County, Illinois. This animal was very emaciated, and may not have weighed over ten pounds. Its temperature was 106° F. It had been sick four weeks, and was killed by bleeding. Date, June 24.

Morbid changes.—Small quantities of serum in the pericardium and in the thoracic and abdominal cavities; lymphatic glands enlarged; capillary vessels of the auricles of the heart gorged with dark-colored blood, and hepatization in the lungs, comprising three-fourths of the left and one-half of the right lobe.

5. EXPERIMENTS.

The experiments made previous to December 1, and recorded in my first report, have proved: 1. That swine plague is infectious, and can be communicated or transmitted from diseased hogs or pigs to healthy animals in two different ways: by direct inoculation, and by an introduction of the infectious principle, either with the food or with the water for drinking, into the digestive canal. 2. That an exceedingly small quantity of the infectious principle is sufficient to produce the disease. 3. That the period of incubation, or, more correctly, the stage of colonization, does not exceed fifteen days, and lasts on an average from six to seven days. 4. That small *Schizomycetes*, the *bacilli suis* and their germs, which are found in all the morbid products, in the blood and other fluids, and in all excretions of the diseased animals, constitute, almost beyond a doubt, the infectious principle and the real cause of the disease.

I concluded, when I went to work again on the 15th of December last, to make another series of experiments for the purpose of ascertaining with certainty, if possible: 1, whether the infectious principle consists solely in the *bacilli* and their germs; 2, whether an animal that has had the disease, and has recovered, has lost all further susceptibility, or is yet subject to future attacks; and, 3, whether swine plague can be communicated to other animals besides swine. To enable the reader to draw his own conclusions, I will first briefly relate the experiments made, and then state the conclusions I have arrived at.

On January 9 I bought two healthy pigs (which I shall designate as pigs Nos. 1 and 2) of Mr. H. Lamken, at Prairieville, and put each pig by itself in a clean and comfortable pen, which had not been occupied by any hogs or pigs for a long time. On January 17 I bought another pig, seven or eight months old, of Mr. E. Taddicken, at Prairieville. This pig had recovered from swine plague about two months before, and had become somewhat stunted in consequence of its sickness. but had a very good appetite, and did not exhibit any symptoms of existing fever or of active disease. It was designated as pig No. 3, and was put by itself in the pen occupied by pig No. 2, which latter was put in with pig No. 1.

January 21.—Charged one ounce of fresh milk, just drawn from the cow, with a mere speck of the proliferous growth of the stomach of John Lord's pig, which had been killed by bleeding. The milk thus charged, and contained in a perfectly clean two-ounce vial, closed by a tight-fitting glass stopper, was kept at a constant temperature of 90 to 100° F.

January 22.—None of the experimental pigs, so far, have shown any symptoms of disease. All seem to be in good health. Nos. 1 and 2 are thrifty and growing. Inoculated No. 3, the one that recovered from a previous attack, in the ear, by means of a small inoculation-needle, with a little juice (less than half a drop) pressed out of the proliferous growth of the stomach of John Lord's pig, killed by bleeding for the purpose of obtaining fresh material for inoculation, so as to exclude any possibility of producing pyæmia.

January 23.—Inoculated pigs Nos. 1 and 2 also in the ear, by means of a small inoculation-needle, with the milk charged with *bacilli* and bacillus germs on January 21. Examined under the microscope, the milk, besides its normal constituents, contained numerous *bacilli* and bacillus germs.

January 24.—All three pigs apparently in good health. None of them show any symptoms of a reaction.

January 25.—All experimental pigs apparently healthy.

January 26.—Experimental pigs apparently healthy; all have good appetite.

January 27, 28, and 29.—All three experimental pigs have good appetite. No symptoms of disease.

January 30 —Pigs Nos. 1 and 3 indisposed, but have some appetite. Pig No. 2 apparently healthy.

January 31.—Pigs Nos. 1 and 3 show plain symptoms of disease, are sneezing frequently, and show a tendency to hide in their bedding. Pig No. 2 apparently all right.

February 1.—Experimental pigs Nos. 1 and 3 evidently sick, both sneeze and cough a great deal, and do not seem to have much appetite. Pig No. 2 apparently not affected.

February 2.—Experimental pigs about the same as yesterday.

February 3.—Pig No. 1 hides in its bedding, is emaciated, and has no appetite. Pig No. 3 is sick, but eats some. No. 2 is doubtful.

February 4.—All three pigs about the same as yesterday.

February 5.—Pig No. 1 has no appetite whatever, and is very poor. Nos. 2 and 3 about the same as the day before.

February 6.—Pig No. 1 about the same as yesterday. No. 3 appears to be slightly improving. No. 2 sneezes and shows other symptoms of a mild attack.

February 7.—Pig No. 1 eats a little. In Nos. 2 and 3 no visible changes.

As I was called away to Chicago, I had to leave the pigs to their fate; but in order to learn what would become of them, I left them with Mr. H. Lamken, with the understanding that he was to pay for pigs Nos. 2 and 3, should they be alive three weeks after date. Pig No. 1 was considered as not being worth anything. In due time Mr. Lamken sent me the money and a note, in which he stated that pigs Nos. 2 and 3 were alive and improving, and pig No. 1 of no account, but still alive.

I intended to subject the causal connection of the *bacilli* and their germs with swine plague to one more (negative) test by inoculating healthy animals with morbid fluids (exudations) of diseased animals after they had been freed from bacillus-germs, and filtered for that purpose some pulmonal exudation and blood serum through sixteen papers, but did not succeed. The last filtrate examined under the microscope still contained a large number of bacillus-germs or globular bacteria.

The experiments related above prove two things : First, that an inoculation with *bacilli* and bacillus-germs, cultivated in an innocent fluid, such as fresh milk, can and will produce the disease with just as much certainty as an inoculation with pulmonal exudation, or with any other bearer of the infectious principle taken directly from a dead or diseased hog. Second, that an animal that has been afflicted with swine plague, and has recovered, has not lost its susceptibility, but may contract the same disease again, though probably in a milder form. The latter fact has received further confirmation by a statement of Mr. Reichard, an intelligent farmer and reliable observer, residing near Prairieville, who informed me that one of his hogs had been sick with swine plague three times, but had (partially) recovered after each attack, and was still living, but of not much value. Such cases would probably occur oftener, if it was not for the malignancy of the disease; the first attack has generally a fatal termination, and the usually very short life of the hog.

When called away to Chicago I was about to commence a series of experiments with cattle, for the purpose of deciding whether swine plague can be communicated to these animals, the same as of sheep and rabbits, which, I had seen stated, had been successfully inoculated by Professor James Law, at Ithaca, N. Y. Considering that question at any rate as of great practical and scientific importance, something happened while I was employed in the Union stock-yard of Chicago as inspector of cattle which made it still more desirable to settle the question as soon as possible. While there I had to inspect, from February 10 to May 16, over 300,000 head of cattle. Among that vast number I found only one animal exhibiting symptoms decidedly suspicious of contagious pleuro-pneumonia, or lung fever. The animal in question was a yearling heifer, and had come in, together with another one, in a car-load of hogs from Sublette, Lee County, Ill. It was shipped by its owner—so I learned afterwards—because it had been ailing for some time, and was not doing well. In order to decide whether those suspicious symptoms exhibited during life were those of pleura-pneumonia, or of some other respiratory disorder, I bought the heifer and had it killed by bleeding for *post-mortem* examination. The morbid changes were as follows: The lungs filled the whole thoracic cavity so completely as to show on their surface plain impressions of the ribs. Their surface was uneven to the touch, and on further examination distinctly limited hepatization, such as is characteristic of contagious bovine pleuro-pneumonia, or lung plague of cattle, presented itself. It was most developed in the left lobe, and particularly in its anterior part, but quite large and distinctly limited patches of hepatized lobules, some gray, and some red or brown, presented themselves also when the left lobe was cut into, in its central and posterior portion. Externally the central and posterior part of the left lobe, if looked at superficially, seemed to be healthy, because the lobules next to the pleura were not affected. The right lobe, too, contained several patches of hepatization, but was on the whole, much less affected than the left lobe. I cut off some of the worst hepatized parts, and put them in a bucketful of clean water; they went to the bottom like a rock. Only one small portion of the pleura, say about three inches in diameter, and coating a portion of lung in which the hepatization extended to the surface, was coated with a slight layer of exudation. Most of the lymphatic glands in the chest and in the abdominal cavity appeared to be enlarged. No other morbid changes were found.

As hepatization in the lungs of cattle is, to say the least, an exceedingly rare occurrence except in contagious pleuro-pneumonia—in a practice of over twenty years I have never seen it except in that disease, neither have other experienced practitioners whom I have consulted (I will only name Dr. J. C. Meyer, sen., of Cincinnati, and Dr. F. W. Prentice, of Champaign, and refer to Professor Gerlach's work on Veterinary Jurisprudence)—and as Prof. James Law, of Ithaca, N. Y., had succeeded in communicating swine plague, a disease also characterized by distinctly limited hepatization in the lungs, to other animals than swine by means of inoculation, the question arose: Can swine plague be transmitted also to cattle, and, if so, what is the case in question? Is it contagious bovine pleuro-pneumonia, or is it swine plague transmitted to cattle? It was clear to my mind that if it was contagious pleuro-pneumonia, several cases, or at least more than one case, would be existing at the place where the heifer had come from; and if swine plague, some lasting and intimate contact or association with diseased hogs must have taken place. I communicated my views to John B. Sherman, superintendent of the Union stock-yard, and to Nelson Morris, the largest cattle-dealer and exporter in Chicago, and, on consultation, it was concluded, in order to obtain certainty, to send Dr. F. W. Prentice, Professor of Veterinary Science in the Illinois Industrial University, Champaign, Ill., at once to Sublette, where the heifer had come from, to make a thorough and searching investigation. Until his return, and the contrary had been proven, the worst of the two possibilities had to be accepted, as it was not known that swine plague could be communicated to cattle. Dr. Prentice made a thorough investigation, but failed to find any bovine pleuro-pneumonia, or any trace of its existence; he learned, however, that the heifer in question had been raised in the hog-lot, among the swine, by the same farmer who shipped her to the stock-yard, and I know that in Sublette and immediate vicinity an immense number of hogs and pigs had died of swine plague in the latter part of last fall and the early part of last winter. The absence of any contagious pleuro-pneumonia, and the fact that the heifer in question had been born and raised on the same farm from which it had been shipped, were sufficient proofs that we had not to deal with the bovine lung plague. Dr. Prentice and myself were therefore able to contradict, on his return, certain perverted statements which had been published in several papers. Still, although fully convinced that we had not to do with a case of contagious bovine pleuro-pneumonia, we had not sufficient proof to authorize us to pronounce the morbid changes in question the product of transmitted swine plague.

When, in compliance with your order, I resumed the investigation of swine plague in May, which had been interrupted in February, it was one of my first attempts to ascertain by experiment whether swine plague can be communicated to cattle or not. On May 26 I bought two healthy heifers, one a common scrub, and about eight or nine months old, and the other a half-breed Jersey, about four months old. The latter, designated as heifer No. 1, was kept in a good pasture on the same farm on which it had been raised, and received, besides grass, some milk while being experimented with. The former, designated as heifer No. 2, being old enough to eat hay, was kept in a good stable in the city of Champaign, and was fed with good hay, oats, chopped feed, and water. Both animals were inoculated in the ear—received each two punctures— by means of a small inoculation-needle, No. 1 with less than a quarter of a drop of blood, and No. 2 with a similar quantity of serum pressed out of an ulcerous tumor situated in the scrotum of a recently castrated pig, sick with swine plague. Up to June 5, neither of the heifers showed any symptom of disease, but it may be remarked that heifer No. 1, being in a large pasture over two miles from town, could not be visited and examined every day; but heifer No. 2, being in a stable in town, and therefore always approachable, was examined at least twice every day.

June 5.—Heifer No. 2 appears to be less lively; its muzzle is dry and warm, and the temperature (in rectum) 102.5° F. Heifer No. 1 perfectly healthy.

June 6.—Heifer No. 2, appetite changeable; muzzle dry; temperature 102.6° F.

June 7.—Heifer No. 2, muzzle moist; otherwise no change; temperature, 102.4° F.

June 8.—Heifer No. 2, muzzle moist; appetite good. (Broke thermometer, and therefore failed to ascertain temperature.) Heifer No. 1 evidently all right in every respect.

June 9.—Inoculated heifer No. 2 at 5 o'clock p. m., by means of a hypodermic syringe with half a drachm of pulmonal exudation, obtained from the lungs of a pig belonging to Mr. Coffee, in Champaign. The pig was examined immediately after death, and presented all those morbid changes which are characteristic of swine plague. The injection was made just behind the shoulder-blade into the subcutaneous connective tissue. Heifer No. 1 was inoculated by the same means with one drachm of the same material. The injection was made into the loose connective tissue of the dewlap. The exudation used was perfectly free from any putrid smell, and contained, examined under the microscope, numerous bacillus-germs, and some *bacilli*.

June 10.—No visible reaction in either animal.

June 11.—Heifer No. 2, no change, except a very slight swelling attacked by a few flies at the place of inoculation.

June 12.—Heifer No. 2, changeable appetite.

June 13.—Heifer No. 2, no morbid symptoms whatever; temperature, 102° F.

June 14.—Both heifers apparently in first-rate health.

June 15.—Heifer No. 2 shows signs of illness; breathes fifty-six times a minute; muzzle dry and abnormally warm; appetite slow and irregular; eyes somewhat dull.

June 16.—Heifer No. 2 shows at times plain indications of illness. and at times seems to be all right; coughs some; dung rather hard, dark-colored, and coated with sticky mucus. Temperature, 103.4° F.

June 17.—Heifer No. 2 evidently sick; muzzle dry and hot; appetite irregular and changeable; dung hard and dark-colored as yesterday; respiration accelerated. The animal acts rather dull, and shows a tendency to lie down. Temperature, 103.6° F.

June 18.—Heifer No. 2, in the forenoon the same as yesterday, except the temperature, which was as low as 102° F. In the afternoon apparent improvement; muzzle moist, but temperature 103° F.

Heifer No. 1, till date, has not exhibited any conspicuous symptoms of disease; at least none has been observed by Mr. Moore, who keeps the animal in his pasture, and is perfectly familiar with all the various symptoms of swine plague. Examined the animal at 9 o'clock, a. m., and found the muzzle dry and abnormally warm; the breathing accelerated, and the temperature, taken without any struggling or resistance, 103.5° F.

June 19.—Heifer No. 2, no essential change; muzzle sometimes moist, sometimes dry; temperature, 103.4° F.

June 20.—Heifer No. 2 about the same; dung of the consistency of stiff dough, and blackish in color (the food consists of very good hay, some oats, and occasionally some bran or chop-feed); temperature, 103.5° F.

June 21.—Heifer No. 2, no essential change; breathing a little more accelerated, but the eye somewhat brighter; temperature, 103.6° F.

June 22.—Heifer No. 2, no essential change; temperature, 104° F.

June 23.—Heifer No. 2, temperature in the morning, 104.4° F.; in the evening, 103° F.

June 24.—Heifer No. 2, temperature in the morning, 103° F. Inoculated the same in the evening once more by means of a hypodermic syringe with ten drops of the pulmonal exudation of Mr. Philippi's pig.

June 25.—Heifer No. 2 less lively, more dumpish than on preceding day; temperature, 103.6° F.

June 26.—Heifer No. 2, no essential change; temperature, 104° F.

June 27.—Heifer No. 2, about the same; temperature, 103.7° F.

June 28.—Heifer No. 2 appears to be more lively; appetite improved; temperature, 103° F.

June 29.—Heifer No. 2 eats and drinks well; muzzle moist; temperature, 103.6° F.

June 30.—Heifer No. 2, muzzle hot and dry in the morning. Took at 1 o'clock, p. m., a few drops of blood from a vein of its left ear, which, examined under the microscope, contained a few moving *bacilli* and several clusters of bacillus-germs (see drawings). The temperature, taken at the same time, was only 102° F. At 6 o'clock p. m., respiration fifty-six breaths in a minute, and temperature 104.4° F.

Heifer No. 1, examined in the afternoon, appeared to be all right.

According to Mr. Moore, it had acted dumpish and been out of appetite for a few days, but had recovered. So it may be concluded that heifer No. 1 has had a very mild attack, but its vigorous constitution has enabled it to overcome the effects of the infectious principle.

July 1.—Heifer No. 2, muzzle hot and dry; temperature 103.8° F.

July 2.—Heifer No. 2, muzzle moist; respiration accelerated; the animal breathes over sixty times a minute; auscultation reveals a slight rubbing sound, and increased bronchial breathing; temperature 104.6° F. In the evening temperature down to 103° F.

July 3.—Heifer No. 2, temperature at 9 o'clock a. m., 103.6° F. (It may be remarked here that heifer No. 2, during the whole experiment, had a very quiet, clean, and moderately dark stall, 5×10, where she was not at all, or but very little, molested by flies, where the air was always pure, and where nothing occurred liable to raise the temperature of the body above normal; on the contrary, where the conditions were rather such as to keep the temperature at the lowest point, because the animal was tied, had no exercise, and was naturally of a very quiet and docile disposition. It scarcely ever offered any resistance while being examined.)

At 9.15 o'clock, a. m., heifer No. 2 was killed by bleeding by a professional butcher.

Morbid changes found at the post-mortem examination.—Distinctly limited (circumscribed) hepatization at several places in both lobes of the lungs, each single patch comprising only a few lobules, but very distinct and well defined. (See photograph of Plate I, of a portion of the anterior part of the left lobe, which shows two small hepatized patches.) The most extensive hepatization was found along a larger bronchus in the posterior part of the left lobe. The hepatized parts or patches amounted to about 4 or 5 per cent. of the whole pulmonal tissue. The mucous membrane of the bronchiæ was found to be slightly swelled; a small quantity of serum was found in the pericardium and in the chest, and a little more, a few ounces, in the abdominal cavity. The lymphatic glands situated in the chest, and those belonging to the mesenterium, were enlarged, some of them, especially the latter, to a considerable extent. All other organs appeared to be perfectly healthly and normal. The blood probably was a shade darker than that of perfectly healthy cattle butchered or killed by bleeding.

The experiment with heifer No. 2 has proved beyond a doubt that swine plague can be communicated to cattle by direct inoculation, though perhaps only in a mild form; 2, that cattle possess less susceptibility than swine, and are not easily infected; and 3, that the principal morbid changes of swine plague, communicated to cattle by inoculation, present themselves as hepatization of the pulmonal tissue, and are essentially the same in cattle as in swine.

Since the possibility of a communication of swine plague from hogs to cattle has thus been proved, and since it has been ascertained by other experiments that swine plague is communicated from hog to hog, not only through wounds and scratches (direct inoculation), but also with equal facility by an introduction of the infectious principle with the food, or with the water for drinking, into the digestive canal, there remains in my opinion, not the least doubt that the heifer killed in February in the Union stock-yard, which was raised in a hog-lot among diseased hogs, and compelled not only to eat and drink with diseased hogs, but probably also to consume food and water soiled and contaminated with the exceedingly infectious excretions of diseased hogs, was diseased with communicated swine plague, aggravated, maybe, by rough treatment and transportation by rail. Nay, more, it was even possible that the cattle (steers) condemned last winter in England as affected with pleuro-pneumonia, and alleged to be American, and even Western cattle, have either not come from the West, or from any of the Western States, in which contagious pleuro-pneumonia has ever been known to exist, or have not been diseased with contagious bovine pleuro-pneumonia, but only with communicated swine plague. On a great many farms in nearly all the Western States, the steers and hogs to be fattened for the market are frequently fed in one and the same feed-lot, and eat the same food and drink of the same water. It is therefore possible that swine plague, since it prevails almost everywhere in the whole stock-producing West, has been communicated in a few instances to steers; that those steers affected with only a very mild attack, too mild to be noticed, passed through the stock-yards in the West and at the sea-coast as unsuspected and healthy animals, and that the originally mild form of communicated swine plague became sufficiently aggravated by transportation, exposure, hardship, and confinement on the Atlantic steamer, to be readily mistaken for bovine lung-plague or contagious pleuro-pneumonia by the time the cattle arrived in England.

6. SWINE PLAGUE IN OTHER ANIMALS.

Professor Law succeeded in communicating swine plague to sheep and rabbits, and Professor Klein successfully inoculated rats, and so there is no doubt that those animals can contract the disease and become the means of its spreading. It may therefore be almost superfluous to mention that I have seen, while acting as inspector of cattle in the Union Stock Yard, several rats evidently diseased with swine plague. Professor Law's experiment in inoculating a dog has not been as successful as he desired, and there is no doubt that dogs possess comparatively little susceptibility, but they are, notwithstanding, able to contract the disease, as will be seen from the following: Mr. David Moore, a farmer residing two miles north of Champaign, is known to be a reliable man and a close observer of all the symptoms of swine plague in its various phases. Last year he lost nearly every hog he had on his place, and this spring he lost fourteen pigs. Late in the fall, so Mr. Moore informed me, his dog, a pointer, feasted on the unburied carcasses of hogs that had died of swine plague. In less than two weeks the dog was taken sick and showed symptoms identical, Mr. Moore says, with those exhibited by his diseased hogs. In about two weeks the dog was emaciated to a mere skeleton. It was over four years old, and Mr. Moore is positive that the disease was communicated swine plague and not common dog distemper, a disease which, by the way, was not prevailing in the neighborhood, and which very seldom attacks dogs over four years old. Of course this was not a case witnessed by myself, but I considered it worth relating, because I know Mr. Moore and cannot doubt his veracity.

7. THE CONTAGION OR INFECTIOUS PRINCIPLE.—ITS SPREADING, ITS PROPAGATION, AND ITS VITALITY.

That the *bacilli suis* and their germs constitute the contagion or the infectious principle and the true cause of the disease has been confirmed not only by the result of my experiments with pigs Nos. 1 and 2, but also by numerous clinical observations. 1. None of the inoculations made since August 1 produced any local reaction except the second inoculation of heifer No. 2, which was followed by a very slight local reaction, consisting in a scarcely perceptible local swelling, easily accounted for by the manner in which the operation was performed. The point of the hypodermic syringe used was very weak and rather dull, and an opening through the skin had to be made with a knife, which caused a wound sufficient to produce such a slight swelling. In heifer No. 1. inoculated on the same day, and with double the amount of the same material, but by means of another hypodermic syringe with a point strong and sharp enough to penetrate the skin, no swelling whatever appeared. If the infectious principle consisted in something of the nature of a virus, or in something that possesses chemical properties, or does not need to propagate and to multiply before it can act, a local reaction would have taken place.

On the other hand, if an animal infected with swine plague receives a wound or an external lesion sufficient to cause congestion and inflammation, the morbid process is almost sure to localize in the congested or inflamed parts. Further, if the infectious principle is introduced into a wound or a lesion with inflamed, swelled, or congested borders—for instance, in a wound caused by ringing or by castration. &c.—the morbid process is sure to develop in the inflamed or congested borders of that wound. All this is easily accounted for if the *bacilli* and their germs constitute the infectious principle, and if the mode and manner in which they obstruct and clog the capillary vessels is taken into consideration; but it is utterly irreconcilable with the non-appearance of any local reaction after an inoculation by means of a wound too slight to cause congestion if the infectious principle possesses the nature of a virus or of a chemical agency.

2. Swine plague, until the last days of December, or until the ground becomes covered with snow and the weather exceeding cold, was spreading from farm to farm and from place to place, but as soon as the temperature commenced to remain below the freezing point, at noon as well as at night, it at once ceased to spread from one farm or locality to another. At the same time, however, it was also observed that the very cold weather of the last days of December and of the first days of January—at seven o'clock in the morning of the 2d day of January the thermometer indicated at Gap Grove, Lee County, Illinois, a temperature of 28° below zero, and at the same hour on the day following a temperature of 24°—did not materially interfere with the spreading of swine plague from one animal to another in all pens and hog-lots in which the disease had previously made its appearance, and in which the way of feeding and watering the animals was such as to allow a contamination of the food and of the water for drinking with the excrements or other excretions of the diseased hogs, or in which the hogs and pigs, still healthy, had open wounds, sores, or scratches, and had to sleep together with the diseased hogs in the same sleeping place and on the same litter— old straw and manure, for instance. Afterwards, when milder weather had set in, the spreading from one place to another very slowly commenced again.

Now, if the *bacilli* and their germs do not constitute the infectious principle and the cause of the disease; if, on the contrary, the latter consist in some mysterious poison, or an invisible chemical fluidum, the facts and observations just related cannot be explained, because it must be supposed that the low temperature prevailing at the end of the old and the beginning of the new year, would have affected the infectious agency either not at all, or just the same within as without the hog-lot, and, at any rate, would not have prevented the spreading of the plague except by destroying the infectious principle. The latter, however, is not easily destroyed by frost, but only caused to become dormant till the temperature rises again, otherwise the exceedingly cold weather and continuous frost of last winter would have been sufficient to extinguish the disease; and the new outbreaks, or the renewed spreading, which took place when the weather became warmer, not only in one locality but in a great many, would not have been possible. All the facts and observations, however, will become perfectly harmonious, and be fully explained, if the means by which the disease is produced and communicated consists of something corporeal, endowed with vitality and means of propagation; in other words, if the *bacilli* and their germs constitute the infectious principle and the cause of morbid process, as will become more evident by the following results of my investigation:

Last summer and fall it was found that the *bacilli* and their germs, present in immense numbers in the excrements, urine, and all other excretions of the animals diseased with swine plague, were carried upward into the air by the evaporation of the fluid parts or watery constituents of those excretions, and came down again with the dew, the rain, and other precipitates of atmospheric moisture, and were deposited on the surface of everything wetted by the dew or the rain, on the grass and on other food-plants of field and pasture, and in that way were conveyed from one place to another. Such a rising in the air, and such a conveyance of the bacillus germs from one place or locality to another, cannot be accomplished at all, or only to a very limited extent, while everything is frozen or covered with snow, because in that case all the moisture and watery parts, which otherwise might have evaporated, are locked up by frost—have become solid.

3. It was further observed that swine plague spread the most rapidly, and was the most malignant, among herds in which the animals had external wounds, sores, or lesions, caused by recent ringing, castration, &c., and in all those swine-yards or hog-lots in which an old straw-stack served as shelter and sleeping place, wounds, sores, and scratches constitute a port or entrée for the disease-producing germs, and partly rotten and constantly damp old straw-stacks not only catch the organic particles, such as the *bacilli* and their germs, that may be floating in the air, but also shelter and protect them against destructive influences, and favor and promote their development, propagation, and dissemination, first by being warmer, in the winter at least, than the surrounding atmosphere, and secondly, by absorbing and causing to evaporate, in consequence of their porous condition, a great deal of moisture. Clinical observations have convinced me that an old straw-stack may preserve the infectious principle for several months. The above facts, too, if looked upon in a proper light, will go far to show that the infectious principle must be something endowed with vitality and means of propagation.

4. When resuming my investigation in May, I went again to Champaign, Champaign County, Illinois, because I had been informed of the existence of swine-plague in the immediate vicinity of that place. Arriving there I found my information to be correct, but found also that the disease, which had never entirely ceased to exist in that county since July a year ago (1878), was spreading very slowly, and made a temporary stop, or ceased to spread immediately after each heavy or pouring rain, and during the spring most rain-storms in the West are of this character. I found, further, that even its propagation within the herd became visibly slower or stopped altogether for several days after each violent or pouring rain in all such herds as were kept in a pasture

or a hog-lot sufficiently drained to enable the water to flow off; but the spreading was not visibly interrupted in such herds as were kept in a timber-lot or in a pen under roof. So I have necessarily come to the conclusion that each pouring rain brought down the *bacilli* and bacillus germs floating in the air and washed them away at once, not only from the grass and herbage, but also from the surface of the ground. In timber lots, however, it was different; there the force of the rain was broken by the trees and the usually rank vegetation beneath, and there the water does not run off as fast as from a pasture, or from a bare hog lot. Besides, the drainage in the timber, as far as Illinois is concerned at least, is usually very indifferent.

As to the nature of the infectious principle there can be, in my opinion, no more doubt; and in regard to its spreading my recent observations have corroborated the conclusions arrived at last summer and fall. To sum up, swine plague spreads and is communicated to healthy animals: first, by an introduction of *bacilli* and bacillus germs into the digestive canal with the food and water for drinking; and, second, through wounds, sores, and scratches, or by direct inoculation. Whether they can also enter (and communicate the disease) through the whole skin, and through the whole respiratory mucous membrane, free from any lesions whatever, is doubtful, and a question I have not been able to decide. According to what I have been able to see and to observe it is not probable, still it may be possible.

The *bacilli* and their germs' can be conveyed from one place to another not only in and with the morbid products of the disease, and the tissues, fluids, and excretions of the diseased and dead animals by themselves, or by adhering to and contaminating other inanimate things, fluid or solid, but also independent of any other vehicle through the air at a distance of a mile, if circumstances are favorable, and in the water of running streams. They are even able to propagate in water, especially if it is not free from organic admixtures. An incident happened while I was, last winter, at Gap Grove, which is worth relating. On January 27, in the afternoon, I filtered some pulmonal exudation of a pig that had died of swine plague through several papers for the purpose of freeing it from the bacillus germs which it contained. The filtering was done on a small table in a corner of the room, and the apparatus was left standing on that table with the wet papers (4) in the funnel after the filtrate had been removed. In the evening the latter was examined under the microscope on another table in the opposite part of the room, and as my two highest objectives are immersion lenses, I had to use water, and had a tumblerful of clean well-water on my table, just drawn from a deep well. When through with my work, instead of pouring the water out, I placed the tumbler on another table about four feet distant from the filtering apparatus. Next morning I went to Chicago to return on the 30th. In Chicago I procured a new objective, also an immersion lens, and about the first thing I did after my return was to try that objective. Finding everything undisturbed in my room, and the tumbler with water exactly where I placed it, and not intending to examine but a test object, I did not go for fresh water, but used a drop of the water in the tumbler for the immersion. While adjusting the focus, I discovered that the water, which I knew had been absolutely free from organic bodies, was swarming with *bacilli* and bacillus germs of the same kind as those in the pulmonal exudation. I made then a thorough examination of the water not only with the new, but also with the old objectives, and found that every drop taken from above (the surface) contained myriads of *bacilli*, some of them moving very lively, while in a drop taken from near the bottom but comparatively few could be found. The filtering paper left in the funnel wet and full of bacillus germs and *bacilli* was perfectly dry. All the moisture had evaporated; the aqueous vapors had carried the bacillus germs with them into the air, and many of them undoubtedly had been deposited in the tumbler and in the water it contained, and had there developed and propagated. Another solution is not well possible. The next day the water was examined once more, and it was found that the number of the *bacilli* had become still greater. Soon after I dropped a few grains of thymol into the water, and two hours later every bacillus had been destroyed—at least none could be found.

The peculiarities and the "freaks" in the spreading of swine plague are best illustrated by a brief history of the disease and its progress on Henry Miller's farm, one mile north of Prairieville. Late in the fall of 1877, when no swine plague was existing within fifteen or twenty miles of his place, Mr. Miller bought twenty-six shoats in a part of Whiteside County in which swine plague at that time was prevailing and had been prevailing very extensively. Those shoats themselves *appeared* to be healthy, but had been exposed, as was learned afterwards, to the influence of the infectious principle, and it is possible and even probable that one or more of them suffered from a mild attack; at any rate, those shoats introduced the germs of the disease into Mr. Miller's herd, because soon after their coming swine plague made its appearance in a (so-called) sporadic form. Whether one of the new shoats or an animal belonging to the old herd was the first victim Mr. Miller does not remember. A few words concerning Mr. Miller's farm and swine yard will be necessary. His farm consists of 320 acres of undulating prairie, divided by Sugar Creek into two parts, and his swine yard is large, slopes a little towards the creek, and contains several hog sheds and cow sheds, which are covered with old straw. The losses during the winter, or until spring, were not very severe, only now and then a few animals died, but in the spring, after the sows had farrowed, Mr. Miller lost a great many or most of his young pigs, and only a few of his older hogs, something not very strange if it is taken into consideration that the season, a cold winter, had not been favorable to a rapid and vigorous propagation of the infectious principle, and that young pigs not only possess the greatest susceptibility and succumb to the slightest attack, but also have for obvious reasons far more chances to become infected than older hogs. As soon, however, as the heavy spring rains set in the disease ceased to make much progress—at any rate, from May till August but few new cases and few deaths occurred. The pouring rains, it seems, washed away most of the disease germs into the creek, and the current carried them off. But in the early part of August, as soon as the season for heavy dews arrived, the disease almost at once commenced to spread very rapidly, and the swine died very fast. Mr. Miller's whole herd consisted of 240 head, and 237 died; only three survived or remained exempted. At that time no other case of swine plague existed in the whole neighborhood, and, according to the best information I could obtain, there was none within twenty miles. Soon, however, the disease commenced to spread from Mr. Miller's herd to those of his neighbors, first to the herd of his neighbor towards the north—the prevailing wind was from the south—then all around, and finally over the whole township and beyond. In November, 1878, Mr. Miller, when he had only three hogs left, bought again thirty-two head. These, too, very soon became infected, and commenced to die at the rate of one, two, and three a day. On December 29, fourteen had died, two died that day, and most of the

others were sick and died afterwards. The fluctuations in the progress
of the plague in Mr. Miller's herd may seem to be strange at a first view,
but if all circumstances are taken into consideration, they become very
interesting, and contribute very much to a better understanding of the
nature of the disease.

Another case, which shows how easily swine plague may be communicated, may also be worth relating. Pat Murphy lives 1¼ miles south
of Gap Grove. Up to January 2, he had lost five hogs out of a herd of
ten head; seven had been sick, but two had recovered. Mr. Murphy's
place, although on a public road, which, however is but very little used,
is rather secluded. He made the following statement, which scarcely
needs any comment: About ten days or two weeks before his hogs
showed any symptoms of disease, a wagon loaded with several carcasses
of dead hogs on the way to a rendering establishment passed by his
hog lot adjoining the road on the east, and separated from it only by a
fence. Whether Mr. Murphy's hogs became infected by the passing of
the wagon with the dead hogs—the wind was from the west and blew
the emanations of the latter into the hog lot—or not, is a question difficult to decide. One thing, however, is certain, Mr. Murphy's hogs were
the first ones that were taken sick in his immediate neighborhood, and
those of his next neighbor south, Mr. Hadeler's, became affected next.
Mr. Hadeler lost one hundred head, and saved nine. His hogs affected
those of Mr. Lawrence, who lives a little further south, close to the
northern bank of Rock River. From Mr. Lawrence's farm the disease
traveled west half a mile, and invaded Mr. Muller's herd. I was at his
place on January 3, soon after the plague had made its appearance.
Mr. Muller had his herd divided, and kept one part in one yard, and the
other in an adjoining one separated from the former by a board fence.
The disease was prevailing only in one yard, in the one toward the east.
Five animals had died. Owing, probably, to the severe cold, and to the
15 or 18 inches of snow covering the ground and preventing evaporation,
the plague remained confined to the eastern yard, and the animals in
the western yard escaped.

I could cite many more cases illustrating the peculiarities of swine
plague in its spreading or propagation, but those given, I think, may
suffice. The mortality, all other conditions being equal, is always
greater the larger the herd and the younger the animals.

In my first report I stated that the vitality of the *bacilli* and their
germs is not very great, except where circumstances and surroundings
are favorable. This opinion has been confirmed by further observations
and experiments. In all animal substances the *bacilli* and their germs,
are destroyed, or at least disappear, as soon as putrefaction sets in; or,
to be more definite, they begin to disappear in animal fluids and other
animal substances as soon as the putrefaction bacteria make their appearance (see drawings), and cannot be found after the putrefaction bacteria have become numerous. On the other hand, if contained in a
fluid that does not undergo putrefaction, or in which *bacterium termo*
does not appear, the vitality of the *bacillus suis* is a great one. On the
27th of January last I put some filtrated pulmonal exudation (of a pig
that died of swine plague) swarming with bacillus germs, but consisting
of about one-half of water, which had been added by moistening the
filtering papers in a 1-ounce vial with a tight-fitting glass stopper, and
left it untouched until the 12th of April, when I examined it again, and
found numerous *bacilli suis*, some of them moving very lively. The vial
and its contents, meanwhile, had been exposed to a variety of temperature, ranging from the freezing point to nearly 100° F.

On June 10 I took two perfectly clean 4-ounce vials, and put in each three ounces of clean well-water in which no bacteria nor any other living thing could be found. In one vial, marked No. 1, I put half a drop of the fresh pulmonal exudation of a pig that had died of swine plague (Mr. Coffee's), and in the other vial I put one drop of the same pulmonal exudation and three drops of pure carbolic acid. Both vials were immediately closed with new corks, and sealed perfectly air-tight with asphaltum. Both vials were opened and their contents examined on July 24. The water in vial No. 2 was examined first, and contained a few motionless *bacilli* and some clusters of bacillus germs. The water in vial No. 1, which was examined next, contained a few moving and several motionless *bacilli*, numerous germs, single and double, several clusters, and a few (two or three on a slide) well-preserved blood-corpuscles.

As has been stated in the chapter on "Morbid Changes" (cases 7 and 8), I had an opportunity on January 22 to make a *post mortem* examination of two hogs which had been down with swine plague in the early part of November, and had recovered two months ago, and had thus a chance to see to what extent the morbid changes had been reduced by melting and absorption of the morbid products, and retrogressive processes in general. On examining the lungs of one of those hogs (No. 8) microscopically, it was found that the serum and melted exudation, which could be pressed out of the hepatized portions, still contained some *bacilli* and bacillus germs, but no clusters (see drawing), which leads me to suppose that under favorable circumstances an animal that has recovered from swine plague may, after two months, be able to communicate the disease to healthy pigs. Unfortunately just then no healthy pig, not already designed for another purpose, was available; otherwise, I would have put that question to a test. If swine plague can be communicated by an animal two months after recovery—of bovine pleuro-pneumonia it is well known that it can be spread by cattle that have been convalescent for over two months—many, otherwise mysterious, outbreaks of swine-plague may be explained.

8. THE MORBID PROCESS.

Since my first report was written (December 1) numerous microscopic examinations of morbid tissues, morbid products, blood, &c., have been made, and *bacilli suis* in different stages of development have been found in every case (see drawings), but as to the manner in which the morbid changes are produced nothing new has been discovered; consequently I have nothing to add to what has been stated in my first report, except that all my observations tend to show that most, if not all, of the morbid changes—at any rate those in the lungs and in the skin—are brought about by the *bacillus* clusters clogging and obstructing the capillary vessels.

9. PERIOD OF INCUBATION OR STAGE OF COLONIZATION.

Its duration seems to depend somewhat upon the number of the *bacilli* and bacillus germs introduced at once into the system, and also upon the stage of development of those disease-producing germs at the time of introduction. At any rate, the average time which elapses after an inoculation before plain symptons of the plague make their appearance, varies somewhat according to the quantity of infectious material inoculated, and probably also to the resistibility of the animal organism. A

large quantity inoculated at once may cause a temporary reaction on
the second day, while a very small quantity, say one-sixth or one-eighth
of a drop, of pulmonal exudation does not produce any visible effect in
less than five to seven days.

10. MEASURES OF PREVENTION.

As it will not be necessary to repeat what I have said under this heading in my first report, I may be brief, as I have but little to add. The
cheapest and best way to get rid of swine plague is to stamp it out, notwithstanding the disease has been allowed to exist a whole quarter of a
century, and has been permitted to spread over twenty-nine States and
Territories. A radical extermination is the only thing that will be effective, unless it can be proved that a spontaneous development is taking place, or can take place, within the borders of the United States.
Fortunately, the low temperature of the winters in our principal pork-
producing States facilitates a stamping out, if undertaken at the proper
time—in the winter and in the spring—because a low temperature (frost),
and especially snow, interrupt very essentially the propagation of the
disease-germs and the spreading of the disease, and, although not
absolutely destroying or killing the *bacilli* and their germs, cause a great
many of them to perish or to be in a dormant state for some time. ·Be-
sides that, the number of hogs and pigs in existence from the first of
January to the first of April is a comparatively small one, because most
of the hogs have been shipped and butchered, and the young pigs have
not been born. But the measures of extermination or stamping out must
be thorough. Anything undecided, doubting, hesitating, or wavering
and favoring, will be of no avail, but will only tend to prolong the existence of the plague and increase the cost. Still, as long as we have
no stringent legislation that applies to the whole country and will be
obeyed and be enforced everywhere, no results can be expected. In the
first place, a competent and reliable person must be appointed in every
county, or where a great many hogs are raised, and, where the country
is thickly settled, in every township, with authority to institute, superintend, and enforce a strict execution of such measures of extinction and
prevention as may be authorized by law. 2. Every owner of hogs or
pigs must be compelled by law, under sufficient penalty, to inform, say
within twelve hours, the officer above mentioned, of every case of swine
plague in his herd, or in any other herd that may come to his knowledge. 3. Every hog or pig that shows symptoms of swine plague must
immediately be destroyed, and either be buried, four to six feet deep,
or be cremated; and all those hogs and pigs that have been exposed to
infection, or been in contact with diseased or dead hogs or pigs, or have
occupied infected premises, must be kept under quarantine for several
weeks or be killed. 4. All infected premises must be thoroughly cleaned
from half-rotten manure, old straw. hay, &c.; if practicable, be disinfected, and remain unoccupied by swine for at least six weeks or two
months. Infected old straw and hay must be burned. 5. No hog or
pig must be allowed to run at large, or to have access to running water,
if swine plague has made its appearance within ten miles. 6. Every
owner of diseased swine who has given due and timely information of
the outbreak or existence of swine plague to the proper officer, and has
done and is doing everything in his power to assist that officer in exe-
cuting the measures of prevention and extinction, must be entitled to
receive payment for every one of his animals killed on account of the
disease, but not more than, say, half value for those animals found to

be diseased with swine plague at the *post mortem* examinations, and not more than three-fourths of the full value for all those which are killed because supposed to be infected but do not present any visible morbid changes at the *post mortem* examinations. As to the latter, the value of the carcass, if it can be turned to proper use, and is left to the owner, must be deducted from the appraised value of the animal. Owners of diseased swine who have not notified the proper officer of the existence of the disease in due time, or are otherwise guilty of gross neglect, or have done anything to prevent the execution of the measures of extinction and prevention, must not be entitled to any compensation whatever, but must be held responsible for any damage that may result from their carelessness or gross neglect. 7. Railroad companies and other public carriers must be forbidden, under severe penalty, to receive and to load any hog or pig, or number of hogs or pigs, in any, or from any, township or county after notification has been given to their local agent or business manager by the proper officer or inspector that swine plague is existing in such county or township, unless the said officer or inspector gives a special written permit for each hog or pig, or number of hogs and pigs, belonging to the same herd. Such a permit, which ought to be granted where it can be done without any damage whatever, should be given in duplicate, one to be kept on file in the local office of the railroad company, and one to accompany the hog or pig, or number of hogs or pigs, to their destination. 8. Any transportation of hogs or pigs diseased with swine plague, or of carcasses or parts of carcasses of hogs and pigs that have died of swine plague, or were affected with that disease at the time of death, must be prohibited under all circumstances. 9. Any transportation of hogs or pigs in a wagon or any other conveyance, or on the hoof, within, from, or through an infected township or county, must be prohibited, or be allowed only on a special permit from the proper officer or inspector, and under such restrictions and precautions as to exclude any possibility of such a transportation becoming the means of spreading the disease. 10. Any importation of hogs or pigs from foreign countries must be strictly prohibited, or be allowed only under such restrictions and precautions as will make an importation of the infectious principle an impossibility.

The expenses of such thorough means of extinction and prevention as have just been outlined will, of course, be very heavy, and should be paid in part by the county, township, or corporation, and partly by the general government.

Although not called upon to propose any law or legislation, I consider it my duty to lay before you a plan which, if executed, will lead to a prompt and effective suppression and the final extinction of that terrible plague which costs the country every year many millions of dollars, and undermines the prosperity not only of individual farmers, but of whole States. The execution of such forcible measures, I know, will be very expensive, and may meet with very much opposition and resistance, but I know also that in the end it will prove to be by far the cheapest that can be done, if anything is to be done at all. Any other measures of prevention not aiming at complete eradication or stamping out of the plague can, at best, be only partially successful, need constant repetition, and leave the country in continual danger.

As to local measures of prevention, in every case they must consist in a thorough destruction of the infectious principle, or, what is practically the same, in promptly removing the animals to be protected out of the reach or influence of the *bacilli* and their germs. Whether the latter are destroyed by physical agencies or by chemical means, so-called disin-

3 C D

fectants, is immaterial. What I have said in my first report in regard
to keeping not more than two or three animals together in movable pens
constitutes probably the best means of protection, as far as single herds
are concerned. But I admit that such a separation is sometimes im-
practicable, or may be considered as too expensive or too troublesome
by the owner, and it may also happen that an infection has taken place
before the necessary preparations have been made. In such a case a
strict and, if necessary, repeated separation of the healthy animals from
the diseased ones, not only as to pens and yards, but also as to attendance, and a thorough cleaning and disinfection of the infected premises,
constitute the least that may be expected to afford any protection. That
the food and water given to the healthy animals must be clean and uncontaminated with the infectious principle, and that dead animals must
be buried or be cremated at once, may not be necessary to mention
again. As a disinfectant, I would recommend carbolic acid as one of the
cheapest and most convenient, notwithstanding that some others may
be more effective.

A few cases will illustrate what is necessary and what may be expected of simple and local means of prevention, but it must be kept in
mind that in the summer and in the fall, when everything favors a rapid
development, propagation, and dissemination of the disease germs, much
more circumspection and thoroughness is required than in the winter,
when a low temperature and a limited evaporation of moisture retard
the propagation and dissemination of the *bacilli* and their germs, or in
the spring, when heavy rains may wash the latter away. In winter and
spring strict separation and good care are usually sufficient to prevent
a serious spreading of the disease; in the summer and fall the most
scrupulous care will be required in guarding against an introduction of
the infectious principle and in destroying it wherever it may happen to
exist, provided it is contained in, or adheres to, something on or in which
it can be destroyed, either with or without its vehicle.

Mr. H. Fisher lives one and a half miles north of Prairieville, and
half a mile north of H. Miller. He makes swine breeding his principal
business, and his accommodations for his hogs are nearly perfect. His
swine-yard is divided into several divisions, and each division again
into several separate apartments, composed each of a spacious yard and
a good and well-ventilated pen with a wooden roof. Each separate
yard, finally, contains a good trough for water and a wooden platform
for food. Consequently, his herd, when occupying the swine-yard, is
practically divided into many small herds, perfectly independent of
each other. The food (corn) is thrown on the platforms, and the water
for drinking is pumped from a well by a windmill, and conducted
through pipes and hose into the numerous troughs. In the early part
of August, 1878, Mr. Fisher sold two hundred hogs and pigs at auction,
which sale reduced his herd to seventy-eight head, the number of which
it consisted when swine plague invaded his place. When the first case
occurred most of the seventy-eight animals were running out in the
pasture, and there, it must be supposed, most of the animals that were
taken sick became infected; at least but a few new cases of disease
occurred after the hogs were kept up again in their yards and pens.
Although Mr. Fisher did not use any medicines whatever, his total loss
amounted to thirty-three head out of seventy-eight; forty-five head remained exempted (most of them) or recovered (a few), while his nearest
neighbor, Mr. Miller, lost two hundred and thirty-seven animals out of
two hundred and forty. Fisher's sanitary arrangements were good—
nearly perfect—and his herd was divided into small lots, none of them

numbering more than five or six animals, while Mr. Miller's hogs and shoats were all in one herd. Comments will not be necessary.

Mr. F. Brauer, at Gap Grove, had, in the early part of January, one hundred and forty hogs and shoats in two yards, separated by a fence— sixty barrows in one yard and about eighty sows in the other. Mr. Brauer's nearest neighbors west and east live only a little more than a quarter of a mile from his house; the neighbors northwest and south- east are farther away, and due north and south no house is nearer than a mile. Swine plague prevailed or had been prevailing between Sep- tember and January, on every farm adjoining Mr. Brauer's. On his place the two swine-yards, which are side by side and destitute of any old straw stack and of half-rotten piles of old straw or hay, are on high ground sloping toward the east, and are protected toward the west by barns, stables, and sheds. The food consists of corn from a corn-crib, which constitutes a part of the northern fence or inclosure of the yard occupied by the barrows, and the water for drinking is pumped by a windmill from a deep well, and conducted through iron pipes into the troughs. On the morning of January 6, one of the barrows was found dead, and presented at the *post mortem* examination, which was made immediately, just such morbid changes as are characteristic of swine plague. The infectious principle, it is supposed, had been introduced by some horses which were running at large, jumping fences, and in the habit of visiting all the swine-yards and corn-cribs in the whole neighborhood in search of corn. Mr. Brauer, to avoid greater losses after that one barrow had died, sold and shipped immediately forty-six of his barrows, so that only thirteen animals remained in the north- ern yard. The latter was cleaned at once, and disinfected by a liberal sprinkling with diluted carbolic acid once a day, on January 6, 7, and 8. The thirteen barrows in the northern yard and the eighty sows in the southern yard have remained healthy, and no new cases have occurred.

Mr. Swigart, in Palmyra Township, kept his hogs and cattle (steers to be fattened) in a yard which contained two old straw stacks, and was well littered with half-rotten straw and hay. When I visited his place the first time, on the 14th of January, fourteen hogs had died, several were sick, and some apparently yet healthy. The first cases had occurred only a week or two previous. The diseased hogs were all bleeding from the nose. I advised Mr. Swigart to immediately separate the apparently healthy animals from the sick ones by removing them to a non-infected place, and give to each animal twice a day about ten drops of carbolic acid in the water for drinking. This advice was complied with, and none of the animals removed from the infected yard became diseased.

Mr. Dillon, one and three-quarters miles north of Champaign, had lost two pigs diseased with swine plague on June 10, on which day he re- moved his small herd of fourteen head to a non-infected locality. No other deaths had occurred when I left Champaign on July 5.

II. TREATMENT.

In regard to treatment no new discoveries have been made, but my views, expressed in my first report, have been very much confirmed. Good care, clean and uncontaminated food and water, strict separation from diseased animals, and scrupulous cleanliness, so as to prevent the animals from satisfying their vitiated appetite for excrements and urine, and from introducing thereby into their organisms more and more of

the infectious principle, go a good ways in preventing an attack of
swine plague from becoming very malignant and in facilitating a recovery. Medicines seem to be of little avail—at least everything that has
been tried without any prejudice has failed to produce visible good results. Patent nostrums and secret medicines have done more harm
than good. Mr. Hoyt, of Mendota, informed me that one of his neighbors, who had extensively invested in "Eureka Specific," had lost in
proportion more hogs than anybody else in the neighborhood that had
not used any medicines whatever.

If it is intended to stamp out the disease, any treatment of the sick
animals should be prohibited by law, unless a sufficient bond is given
to cover any possible damage that may result, because the treatment of
such a contagious or infectious disease always involves great danger in
so far as it tends to preserve the infectious principle and facilitates the
spreading of the plague. To destroy the cause, or, what is the same,
the infectious or contagious elements, wherever and in whatever shape
and form or substance it may exist, is the only rational way of dealing
with such diseases. Swine plague should and ought to be treated the
same as rinderpest or cattle plague, pleuro-pneumonia or lung plague,
glanders, and farcy. The most thorough and decisive measures are in
the end the cheapest.

Respectfully submitted,

H. J. DETMERS, *V. S.*

CHICAGO, ILL., *July* 25, 1879.

SUPPLEMENTAL REPORT.

SIR: Immediately after you re-employed me, on the 8th of October
last, and instructed me to resume the investigation of swine plague, I
took the necessary steps to obtain reliable information as to where the
disease might be prevailing to such an extent as to afford sufficient
material for my purpose, and soon learned that the disease existed in
several counties in Illinois and Wisconsin, within a radius of two hundred miles from Chicago. For several reasons I chose as a suitable
locality for my investigation the county of Henderson, in the western
part of the State of Illinois, and on the eastern bank of the Mississippi
River, notwithstanding sufficient material might have been found much
nearer my home—for instance, in the county of La Salle. Every county
and every place in this State, in which swine plague is or has been prevailing, contains one or more rendering-tanks, and men who speculate
upon the credulity of the farmer when in distress, and try to sell him
a "sure cure for hog cholera" at an enormous price. I know a large
number of farmers who paid from $30 to $60 for a worthless prescription, and others who paid as much as $100 for worthless medicines,
composed of substances that can be bought in the market for about $5.
These persons—the tank-men and the "sure-cure men"—find it in their
interest to keep the farmer ignorant, to prejudice his mind, and to prevent, if possible, a thorough investigation. So it happens that many
farmers deny the existence of the disease if approached by a stranger,
or are asked questions concerning the health of their hogs. A great
many farmers have also another motive for keeping the existence of
swine plague a secret. They sell their hogs and pigs for whatever they
can get, and ship them to Chicago as soon as the well-known disease
makes its appearance. In Chicago, however, the city board of health

is at present more vigilant than formerly, and condemns a few diseased
hogs almost every day. This has had a good effect, in so far as the
buyers have become a little shyer and more careful, and refuse to buy
every diseased animal that is offered; they have also commenced to inquire where the diseased hogs are shipped from, and where swine plague
is existing. The farmers and country dealers who send them are, therefore, interested in denying and concealing the existence of the disease.
Some farmers, to my certain knowledge, have even stooped so low as
to sell and ship their diseased hogs, not in their own name, but in that
of some irresponsible person, and don't like to hear swine plague mentioned. Consequently, any investigation of the disease is exceedingly
difficult and almost impossible, unless the investigator is either personally known or introduced by a citizen who commands the confidence
of his community. Not being personally acquainted in any of those
counties in which the disease, according to information received, was
prevailing to an extent sufficient for my purpose, I chose a place where
I could procure such an introduction. I happened to be acquainted
with one of the most prominent and influential citizens of Henderson
county, Mr. James Peterson, at Oquawka, who, on corresponding with
him, invited me to his place, stated that he would take great interest in
my investigation, and promised to go with me through the county and
introduce me to the farmers whose herds had become affected. His in-
vitation, of course, was accepted, and as his promise has been fully
redeemed, his kind offer has considerably facilitated my work. One
other reason induced me to select Henderson County. I considered it
of some importance to observe the disease in different localities, different at least as to soil and drainage. In most of the places in which I
carried on my former investigations, the soil is entirely different from
that of Henderson County, which is very sandy, especially along the
Mississippi River. Champaign County, for instance, is almost level,
and the soil is a rich black loam; Lee County, or at least that portion
of it in which I investigated last winter, is somewhat similar, only more
undulating and better drained; Stevenson County, in the neighborhood
of Freeport, is still more undulating, and Fulton County is again somewhat similar to Champaign.

In my former investigations of swine plague, I made it my principal
object to ascertain the nature and the workings of the morbid process,
and the real cause or causes of the disease and its spreading. In resuming my investigation this fall—in October last—I thought it would best
serve the purpose to make it a special object to obtain or to search for
such results as are of an immediate and practical value to the farmer,
pork-producer, and swine-breeder. In other words, to ascertain as far
possible the means or media by which swine plague is actually and principally spread from place to place, from herd to herd, and to learn by
observation and experiment what may be done by the individual farmer
and swine-breeder to protect his herd, and to effectually prevent the
spreading of the plague, or to stop its progress. I made it also an object
to decide, by means of experiment and observation, whether the morbid
process, once developed, can be arrested by a simple medical treatment
—such a one as can be applied by the farmer—or not. Before I state
the results of my present investigation, it may be in order to first make
a few general statements, and to give the facts and observations upon
which those conclusions have been based, so as to enable the reader to
judge without bias, and to form an opinion of his own. I may also be
allowed to state that to obtain these facts and make these observations
I have visited twenty-five different herds of swine in different parts of

Henderson County, and several of them from four to eight times; have made fifteen *post mortem* examinations; subjected to a special treatment six different herds, namely, those of Messrs. Kennedy, Gilchrist, Rice, Morris, Beaty, and Graham; and have experimented on three healthy pigs, specially procured for that purpose. It may further be stated that the disease is, or was, prevailing this fall and winter, or from October 13 till the present, in a much milder form in Henderson County, a few herds excepted, than it was last year at the corresponding season in the counties of Champaign, Stevenson, Fulton, and Lee. At any rate, the prevalence of the disease was not as general, its spreading was not as rapid, and the mortality was not as great as during the same months of last year in the counties named. The morbid process, too, in a majority of cases at least, was found to be limited almost entirely to the organs of the chest (lungs, pleuras, and heart), and to the lymphatic system; while last year serious morbid changes in the intestines, such as ulcerous tumors in the cæcum and colon, presented themselves in about 75 per cent. of all the cases examined, in addition to the morbid changes invariably found in the respiratory organs. This greater leniency of the disease must, of course, be taken into consideration in judging the results of the experiments, and the effect of the measures of prevention and of the medical treatment.

Still, notwithstanding this greater leniency and the frequent absence of conspicuous morbid changes in the intestines, numerous examinations of living animals, fifteen *post mortem* examinations, and repeated microscopic investigations have convinced me that the disease prevailing this fall and winter among the swine in Henderson County is exactly the same swine plague found last year in the counties of Champaign, Stevenson, Fulton, and Lee, only this year's epizootic is milder, and the digestive organs, but especially the colon and cæcum, are less frequently affected, which may account for the decreased malignancy or fewer deaths and the slower spreading, because the infectious principle is always the most concentrated, or, what is the same, the disease-producing germs, the Schizomycetes or bacillus germs, as I have called them before (perhaps, erroneously, *cf.* below), are always the most numerous in the excrements of animals in which the morbid process is prominently developed in the intestinal canal. The duration of the disease in the individual animals, or the time which elapses from the appearance of the first symptoms till a termination, either in death or convalescence, is reached, seems also to average a longer time this winter—in Henderson County at least—than last year at the other places named. Several circumstances, undoubtedly, have combined to produce this result. Last winter, particularly in the latter part of December (1878) and in the month of January (1879), the temperature of the atmosphere was very low; it snowed considerably; the snow became very deep and covered the ground for a long time; consequently, everything on the surface of the ground remained unchanged and unmoved, and the evaporation of moisture was very limited. The disease-producing germs, or the Schizomycetes, which constitute the cause and infectious principle of swine plague, although not immediately and necessarily destroyed by frost and snow—recent developments have shown that these germs may retain their vitality for a considerable length of time even if imbedded in ice—were prevented from rising into the air, and thus from being carried by winds from one place to another, neither could the same be conveyed from one herd to another in streamlets and currents of water, because everything was frozen and covered with snow; consequently, these germs or Schizomycetes could not propagate; they were kept dormant or in a state of rest, and there can be no doubt that a great many, perhaps most of them, were thus prevented from finding their proper nidus and therefore perished. Consequently, in the latter part of the winter, 1879, but little disease was existing. The plague had almost died out everywhere. Toward spring, however, sporadic cases made their appearance, especially at the borders of timber lands and in swine yards and pastures which contained old straw stacks, or something of a similar nature calculated to give shelter and protection and the means of propagation (warmth and moisture) to the Schizomycetes or disease-producing germs. From such centers, at the close of last winter when snow and frost disappeared, the disease commenced slowly to spread, but in the spring nearly every week or ten days a pouring rain set in and probably washed away most of the germs or Schizomycetes which existed at places accessible to swine, or at which a chance was given to enter the organism of a hog or pig with the food or water for drinking. Be that as it may, one thing is certain, immediately after a heavy or pouring rain a perceptible stop or cessation could be observed in the spreading of the disease, while each time after the lapse of about a week a renewed spreading took place, to be interrupted only by the next heavy or pouring rain. Thus the plague made but little progress until the pouring rains became less frequent or ceased altogether, or till July and August, when a drier season set in, in which heavy dews took the place of heavy rains; but even then, in midsummer, swine plague failed to make as rapid progress as a year ago (1878), because the season very soon became too dry to be favorable to a rapid and extensive propagation and dissemination of the disease-producing elements. Further, during last fall and the larger part of the present winter, the season, with brief interruptions, has been very dry, at any rate in Henderson County; and it seems a dry season is not at all favorable to the propagation of swine plague, unless drainage is very poor and the soil is inclined to be wet. Careful observation has convinced me that continued dry weather on the one hand and pouring rains on the other have a decided tendency to reduce, and a common wet spell, brought about by repeated light rains—a few of about a week's duration were experienced—will invariably promote the spreading of the disease. If it is taken into consideration what has been ascertained in regard to the nature of the Schizomycetes, and the manner in which they are conveyed from herd to herd, and from animal to animal (cf. below), no explanation will be necessary.

Whether the circumstances just related have also diminished the intensity of the infectious principle or the vitality of the Schizomycetes by not affording favorable conditions for development and propagation, or sufficiently frequent changes from within to without, and vice versa, of the animal organism, as seems to be the case, or whether they have only reduced the number of those microscopic parasites by causing a great many to perish, or denying them an opportunity to reach their proper nidus or place of development in the body of a hog, will be very difficult to decide, and is practically immaterial.

At first, it appeared that the disease was milder only in Henderson County, and I thought the sandy soil, the hilly or somewhat broken surface near the Mississippi River, and the, therefore, more perfect drainage might have something to do with it; but this probably is the case only to a very limited extent, because reliable people have assured me that the disease was last year (1878) just as malignant in Henderson County as in any other place. Still, the sandy soil, good drainage, &c., is probably not altogether without influence, especially if the season is inclined to be dry, for, even during the present winter (1879–'80), the disease proved to be more malignant in the eastern parts of the county, in the vicinity of Biggsville, where the soil is darker and heavier and the sur- face less broken than further toward the Mississippi.

One other circumstance may also have contributed somewhat in caus- ing swine plague to be more lenient this year than a year ago. All con- tagious and infectious diseases, in order to affect an animal, seem to require in the latter a certain degree of predisposition; in other words, the disease-producing Schizomycetes, in order to be able to produce mor- bid changes, seem to require certain conditions which do not exist in the same degree in every animal, and which, to all appearances at least, may even be entirely absent in some few animals, or may become par- tially or fully exhausted, or completely destroyed under peculiar circum- stances; for instance, by a previous attack. Further, it is well known that on the first appearance of almost every contagious or infectious disease those animals, as a rule, become affected first and succumb soonest which possess the greatest predisposition or offer the most favorable conditions for the development and the effectiveness of the infectious principle. Swine plague does not seem to make an exception. Wherever it prevailed very extensively a year ago, it may be presumed that the hogs and pigs which possessed a special predisposition, or offered very favorable conditions, and became exposed to the influence of the infectious principle, contracted the disease and have since died, and consequently are out of the way; that most, if not all, of the older hogs at present existing, especially as the disease prevailed last year almost everywhere, are animals with comparatively little predisposition; and that the pigs born since last spring and now living are mostly the offspring of sows which were not much predisposed, or did not offer very favorable conditions for the development of the disease. That such a difference as to predis- position must exist becomes patent by the fact that in nearly every affected herd, no matter how malignant the disease may prove to be, one or a few animals will either remain exempted altogether or will con- contract the disease only in a very mild form, and recover. It receives also some additional confirmation by the fact that wherever swine plague makes its appearance for the first time it usually proves more malignant than at places at which it has been prevailing year after year, provided the quantity and intensity of the infectious principle are about the same. In Henderson County the disease has been an almost regular visitor for twenty-seven years, and in Southern Wisconsin it is a com- paratively new disease. According to a letter received in December (1879) from a reliable person in Bloomington, Grant County, Wiscon- sin, swine plague, notwithstanding a very small beginning—it was intro- duced by one diseased pig from Iowa—in November last caused very severe losses there.

FACTS AND OBSERVATIONS ILLUSTRATING THE MEANS BY WHICH
SWINE PLAGUE IS SPREAD.

1. *Mr. Kennedy's herd, Rozetta, Henderson County, Illinois.*—I made my first visit to Mr. Kennedy's place on October 14, and found a few cases of swine plague. His hogs had been all right till within a few days. The disease had been introduced by three animals recently bought out of an infected herd.

2. *Mr. Forward's herd, near Sagetown.*—I was at Mr. Forward's place on October 20. He has no near neighbors. His farm is a very large one, somewhat isolated, and situated at the head of several ravines. Con-

sequently several small streamlets, so abundant in Henderson County, have their source on the farm, and only one has its source above, and runs through it. The piece of ground used by Mr. Forward as a hog pasture is flanked on three sides by timber, and his herd of swine, thus somewhat protected by the lay of the land against an invasion of swine plague, remained exempted from that disease until last year. Three-quarters of a mile from Mr. Forward, situated at the head of a ravine, which, however, does not extend through the farm, is a rendering-tank, where dead hogs are rendered up into grease or lard-oil. At the tank the carcasses are cut up, pieces are frequently lying about, and those parts which do not contain any grease or which are not worth tanking, such as the lungs, intestines, &c., parts which usually constitute the principal seat of the morbid process, are thrown into the ravine, and are washed away by the water if the season is rather wet, or remain where they are thrown till it rains. Further down this ravine unites with another one, and these two united form a small creek, which empties into the Mississippi River. Every herd of swine that had access to that creek became affected, and nearly every animal died. According to Mr. Forward's statement, his herd of swine, about two hundred head, remained exempted from swine plague till last winter (1878–'79). One morning he found in his hog-lot the head of a dead hog, deposited there, he thinks, by a dog, which picked it up at the rendering-tank. When he found it his hogs were already feeding on it. Exactly six days later some of his hogs exhibited symptoms of swine plague, soon a great many became affected, and finally nearly every hog and pig of his herd died.

3. *Mr. Robert Hodson*, a storekeeper in Oquawka, made the following statement:

I have a farm on the banks of Henderson River, and last year kept quite a herd of hogs. One morning I found lodged at my hog-lot, which joins the river, a dead hog, which had come down stream, and had probably been thrown in some distance above. My hogs discovered it earlier than I, and were feeding on the carcass when I came. Ten days later they commenced to die. My loss amounted to fully $1,500.

4. *Mr. W. H. Lord*, who lives in Warren County, on the county line between Warren and Henderson, stated to me on October 24 that he· had had no disease among his hogs since 1862 except two years ago, when swine plague was communicated to his herd by a drove of hogs, which came from an infected herd, and was permitted, in his absence, to stay over night in his hog-lot. That his swine (his herd is not a large one, and averages only about fifty or sixty head) remained exempt from swine-plague every year except two years ago, notwithstanding the disease prevailed in his neighborhood annually, is accounted for by Mr. Lord as follows: His hog-lot is on high, dry, and bare ground; contains neither straw-stacks, rubbish, half-rotten manure, nor pools of stagnant water, and is kept as clean as practicable. Further, his hogs and pigs are always confined to this yard, and are never allowed to run at large; they receive their water for drinking regularly from a good well, and their food from a corn-crib, situated in the northeast corner of the hog-lot. (I inspected his place afterwards, and found things exactly as stated.)

5. *Messrs. Moir and Peterson* several years ago were engaged in the distillery business, and fed about 2,000 hogs. Their hog-pen, which is still standing, but has not been used for several years, is three hundred feet long, and situated close to the bank of the Mississippi. Swine plague broke out among their hogs and caused a heavy loss. Several times it subsided, or was temporarily stopped by a liberal use of chloride of lime, employed not only as a disinfectant and used externally, but

also fed to the hogs by mixing considerable quantities of it with the slop. As soon, however, as the use of the chloride of lime was discontinued, the disease invariably, in about a week, broke out anew, and was just as malignant as ever. The experiment was repeated several times with the same result. Finally Messrs. Moir and Peterson conceived the idea of dividing the long pen into a dozen separate apartments by putting in partitions, but the feeding-trough, extending through the whole length of the building, from one end to the other, and sloping gently toward the west, was not divided; the slop, as before, was let in the upper, eastern end, and ran down through the whole length of the trough to the lower, western end, where, finally, the refuse was emptied into the Mississippi. After this but very few cases of sickness occurred among the hogs in the upper or eastern divisions, which received the slop clean as it came from the distillery, while in the lower or western divisions, at which the slop arrived after it had passed through the upper and middle parts of the trough, and had been soiled and contaminated by all the hogs in the apartments above, nearly every animal became affected and died. In the lowest divisions not one escaped, while in the upper ones no deaths occurred. It is, however, but justice to state that Messrs. Moir and Peterson, finding much more sickness in the lower than in the upper part of the building, soon commenced to use the lowest division as a kind of hospital, and used it almost exclusively for sick hogs taken out of the upper and middle divisions, which, of course, accounts to some extent for the slight mortality in the upper and middle divisions of the building, and explains why every animal died in the lowest division, but it does not account for the numerous deaths in the second, third, and fourth lowest divisions.

6. *Mr. Sam. Whiteman*, near Rozetta, had swine plague in his herd a year ago last winter, and disposed of every hog and pig he could find on the place. He intended to commence anew, and bought twenty head of healthy shoats. After receiving them one dead pig, belonging to his old herd, was found stiff and frozen in a fence-corner, where it had died. It was immediately buried three feet deep, but in frozen ground, and there the carcass remained frozen till the latter part of winter, when it was found unburied and consumed by the twenty healthy shoats. Ten days later the shoats commenced to die of swine plague.

7. *Captain William Morris*, in Bald Bluff Township, near the county line between Henderson and Warren, gave me the following information: Near his farm, Snake Creek empties into the north branch of Henderson River. About two years ago somebody dumped two loads of dead hogs into Snake Creek, six miles above its junction with the river. The stench soon became almost unbearable, and every hog or pig which had access to the creek or river became affected with swine plague. Mr. Morris at that time had a large herd of hogs, but he kept them shut up in his hog-lot away from the river, and his herd was the only one within six miles on that river which remained exempt.

8. *Mr. Morris's herd of swine.*—I was on his place for the first time on October 31. He had then about four hundred hogs and pigs or shoats, most of them running at large on a farm of 317 acres, and about forty or fifty of his shoats showed such symptoms as are observed during the first stages of swine plague. Only one animal had died (*cf.* below). He had bought, and received on October 18, a drove of hogs and shoats —about thirty head—out of an infected herd. Some of the animals belonging to that herd still exhibited symptoms of disease, but were considered as convalescent, while others appeared to be perfectly healthy, or showed only slight traces of having been sick. When I was there

the whole drove was shut up by itself in a separate hog-lot, but had been driven over the farm, and was fed and taken care of by the same persons who attended to the other hogs. The first symptoms of sickness among Mr. Morris's old herd were noticed a few days ago, probably on October 25.

9. *Mr. Morris's herd again.*—On November 18, Mr. Morris informed me that to test whether a wound would absorb the infectious principle, he had, several days before, contrary to my advice, castrated a few (five) apparently healthy boar pigs, and had kept them separated from the diseased portion of his herd. When I was there (on November 18) three of these pigs were dead, and a fourth one was in a dying condition, notwithstanding the very mild form in which the disease was prevailing, especially in the herd of Mr. Morris.

10. *Mr. John Ragan*, near Biggsville, informed me on November 19 that his pigs commenced to show symptoms of disease just a week after they had been marked by cutting their ears. Swine plague was prevailing in the neighborhood.

11. *Mr. Pendarvis*, an intelligent farmer and dealer in cattle and hogs at Raritan, in the southern part of Henderson County, informed me on November 24 that a few years ago one of his neighbors lost nearly all his hogs. In his hog-lot was an old straw-stack, which served as a sleeping-place for the animals. A few months later this neighbor bought a healthy lot of hogs or shoats, and turned them into the hog-yard which contained the straw-stack. Swine plague very soon broke out among them, and nearly all died. A whole year later this neighbor again bought a healthy lot of hogs and turned them into the same yard which still contained the same old straw-stack, and soon the disease once more made its appearance, notwithstanding the fact that at that time no swine plague was prevailing anywhere in the neighborhood. After this the neighbor inclined to accuse the old straw-stack as the cause of the mischief, removed it promptly, cleaned his swine-yard thoroughly, and kept it free from old straw, &c. He has not had a case of swine plague among his hogs since the straw-stack was disposed of.

12. *Mr. Rickett's* herd, on Henderson River, three miles from Oquawka. I was on Mr. Rickett's farm on November 9. He has his herd of swine divided, and keeps one portion, about thirty head, in an inclosed yard on high, dry, and bare ground, free from straw-stacks and stagnant pools of water, where they receive their water for drinking from a well close to the fence. The other portion of his herd is running at large, and has access to the river. Among the latter swine plague has made its appearance, while the hogs which are kept in the yard are perfectly healthy. Mr. Rickett stated that to his certain knowledge dead hogs have been thrown into the river above and have floated down past his place.

13. *Mr. William B. Graham's* herd, two miles from Biggsville. My first visit to his place was on December 29. Mr. Graham's herd consisted at that date of 127 hogs and shoats, a majority of which had been ringed late in October. The whole herd had the run of a large pasture and of a corn-stalk field, and slept till within two days in a huge straw-stack. The common feeding-place was around a corn-crib in the stalk-field, and the water for drinking was obtained from a small streamlet of running water flowing diagonally from northeast to southwest through the pasture. This small creek or streamlet has its source above, on the farm of one of Mr. Graham's neighbors, who also has his hog-yard or hog-pasture on the same streamlet, but above. In the early part of December, or (more likely) in the latter part of November, swine plague made its appearance in the herd of his neighbor, who immediately sold and

shipped his whole herd, probably to Chicago, as soon as he found his
animals sick and dying, or after he had lost a few. In Mr.
Graham's herd the disease made its appearance, according to his statement, on
December 21 or 22, but probably a few days earlier, because the first
symptoms very likely had been overlooked. Up to December 29 three
animals had died, and were hauled away early in the morning before my
arrival by the "dead-hog man," or tank agent. I found from twenty-
five to thirty animals unmistakably sick, about forty or fifty doubtful,
and about fifty or sixty, to all appearances, perfectly healthy. Among
the sick ones, which were all such as had been ringed—at that time no
sick animal could be found among those that had not been ringed—
about a dozen or more had badly swelled and ulcerating noses, and produced at each breath a snorting or snuffling noise. Although Mr. Graham, having invested in "sure-cure medicines," did not consent at that
time to subject his herd to an experimental treatment, or did not give
them into my charge, I advised him to separate the healthy animals from
those evidently sick, and to remove the former to a non-infected place
out of the influence of the infectious principle. When I visited him
again, on January 10, he had made a separation, but had moved the
healthiest or best portion of his herd to a piece of low ground, full of
hazel brush and low scrubs, situated below and to the southwest of the
old hog-pasture, and traversed by the same small creek. This was undoubtedly the very worst piece of ground to which he could have taken
healthy hogs for protection, because all the water passing through that
piece of ground came from the old hog-pasture, and the animals in consequence had to drink infected water. On January 10 most of the animals taken to that piece of ground, and constituting originally the best
portion of the herd, had died; only a few were still alive.

14. *Mr. Campbell's herd at Monmouth.*—Mr. Campbell informed me on
February 11 that a few years ago he had his hog-lot on the banks of a
creek; swine plague broke out in his herd and nearly every animal
died. He is sure the disease was communicated to his herd by the
carcasses of dead hogs which floated down the creek.

The above facts and observations, which have not been observed by
myself, have been communicated to me by reliable persons, whose veracity cannot be doubted. They corroborate my former conclusions concerning the infectiousness and the spreading of swine plague, as stated
in my previous reports, and demonstrate especially—

1. That swine plague, very probably, is not communicated, at least
not easily, unless the infectious principle (the Schizomycetes) is introduced either into the digestive apparatus with the food or with the
water for drinking, or directly into the blood through wounds, sores,
scratches, or external lesions (*cf.* No. 4, W. H. Lord, and No. 12, Rickett).

2. That the carcass of a hog or pig that has died of swine plague will
communicate the disease to healthy swine, if eaten by the latter before
it is thoroughly putrified (*cf.* No. 2, Forward; No. 3, Hodson; No. 6,
Whiteman).

3. That even frost is not sufficient to destroy the infectious principle,
provided the Schizomycetes, which constitute the same, are not exposed
for some time, for instance, on the surface of the ground, &c., to the
direct influence of the low temperature, but protected against external
influences by some organic substance (*cf.* No. 6, Whiteman, and No. 11,
Pendarvis.)

4. That swine plague is readily and frequently communicated to
healthy hogs by means of the water used for drinking, if it is contami-

nated with the infectious principle either by the carcass or parts of a
carcass of a dead hog, or by the excrements, urine, and nasal discharges,
saliva, &c., of the diseased animals, and that in many places a gross,
and sometimes even criminal, carelessness is prevailing in contaminating and infecting the waters of rivers, creeks, streamlets, &c., by allowing diseased animals to have access to them, and by throwing in the
carcasses of dead hogs, by which a considerable spreading of swine-
plague is effected (*cf.* No. 2, Forward; No. 3, Hodson; No. 7, Morris;
No. 12, Rickett; No. 13, Graham ; No. 14, Campbell).

5. That one or a few diseased swine can, and frequently do, communicate swine plague to a whole herd of healthy animals by infecting the
food or water for drinking by means of their dirty feet and noses, soiled
with their excrements, urine, nasal discharges, saliva, or blood, as the
case may be (*cf.* No. 5, Moir and Peterson; No. 8, Morris; No. 13, Graham).

6. That every wound, scratch, or sore on the surface of the body constitutes a port of entry for the infectious principle of swine plague, if
the latter is prevailing in the immediate neighborhood (*cf.* No. 9, Morris; No. 10, Ragan; No. 13, Graham).

7. That an old straw-stack—any other porous body undoubtedly as
well—may preserve the infectious principle for months, and even for a
whole year (*cf.* No. 11, Pendarvis).

8. That the infectious principle (the Schizomycetes) enters the animal
organism, and communicates the disease more readily and sooner through
external sores and lesions than through the digestive canal or any other
means (*cf.* No. 9, Morris ; No. 10, Ragan; No. 13, Graham).

SWINE PLAGUE NOT LIMITED IN ITS ATTACKS TO SWINE.

That swine plague can be communicated to other animals besides
swine has been demonstrated by the experiments of Dr. Klein, Professor
Law, and myself, and also by several chemical observations; but the
question as to whether swine plague can also be communicated to human
beings is yet undecided, because such experiments, *inoculations*, necessary
to decide that question, can be easily made on animals, but, for obvious
reasons, not on human beings. As to the latter, we have to rely entirely
upon clinical observation and accidental infection. The director (the
late Prof. A. C. Gerlach) and the faculty of the Royal Veterinary School
at Berlin, Prussia, officially gave it as their opinion in a report bearing
date of February 25, 1875, that swine plague—"*hæmorrhagische Follicular-
Diphtherie des Dickdarms*" in the report—can be communicated to human
beings (*cf.* "*Gutachtlicher Bericht ueber verdorbene Leberwuerste,*" communicated in Gerlach's "*Archiv fuer wissenschaftliche und praktische
Thierheilkunde,*" Vol. I, page 182). It may also not be out of place to
relate a case that occurred last summer in Knox County, Illinois. A
well-to-do and highly respectable family, residing near Yates City, lost,
in last July, three children, aged respectively thirteen, five or six, and
two and a half or three years, of a disease diagnosed by the attending
physicians as diphtheria. The two remaining children of the same
family also became affected, but recovered. Five physicians were in
attendance, and made a careful research as to the possible cause or
causes, and could find but one thing which might be construed as such.
The family used ice which had been taken from a creek into which,
above, some hogs (hogs that had died of swine plague) had been thrown
just before the water of the creek became frozen. My informants are a
highly respected physician in Biggsville, Dr. Maxwell, and a near rela-

tion of the afflicted family, Mr. John McKee, who has a drug-store in the same place.

FACTS AND EXPERIMENTS RELATING TO TREATMENT AND PREVENTION.

Considering it as one of the principal objects of my present investigation to ascertain what may possibly be accomplished in regard to treatment and prevention, or rather as to arresting the spreading of swine plague from herd to herd and from animal to animal by such means as are at the command of the farmer, and can be employed by every one who possesses common intelligence and an ordinary degree of watchfulness, I made quite extensive experiments with six different herds of swine, numbering from twenty-odd animals to about four hundred, or, on an average, about one hundred and fifty head each.

1. *Mr. Kennedy's herd.*—My first visit, as already stated, was made on October 14. Mr. Kennedy kept his herd of swine, of twenty-odd head, in a pasture a short distance from his house. He had recently bought a few pigs out of an infected herd, and thereby introduced the plague among his swine. I found three sick animals, among them one that was very sick. These three, on my advice, were immediately taken out of the pasture and put in an open pen by themselves, built expressly for them in the orchard. The other hogs or shoats were also taken out of the pasture and shut up in a yard, which had formerly served as a cattle-yard. The three diseased pigs were treated with hyposulphite of soda, of which they received each, three times a day, a (heaped) teaspoonful in their water for drinking. I further instructed Mr. Kennedy to feed and water each time, morning, noon, and night, first his healthy shoats and then the sick ones, and not to enter or go near the pen or yard of the healthy animals after he had been to the sick pigs. My directions, as far as I could learn, have been faithfully complied with in every particular; at any rate, the medicines have been promptly given according to my directions. My subsequent visits to Mr. Kennedy's herd were on October 17, 18, 21, 24, and 27. One of the sick pigs died on October 23 and another on October 26; only one of the three recovered, which, as I learned afterwards from Mr. Kennedy, was doing well and was again with the herd. Of the latter, only one animal exhibited once slight symptoms as if affected—it coughed some—and although my advice to separate it from the herd was not complied with, no further developments have taken place. The sick animal which recovered had only a comparatively slight attack of the disease, and with the care bestowed upon it—a clean, spacious pen, clean water, and clean food—would probably have recovered even if no medicines whatever had been used. In this herd, therefore, the hyposulphite of soda probably failed to do any good, although it seemed during the first week of the treatment as if a slight improvement was observable, due, very likely, more to a change of quarters, clean water, and clean food—in the pasture the animals had access to a stagnant pool of water—than to the medicine. Still, whenever medicines are given, people, as a rule, are always inclined to ascribe every change for the better to their use.

2. *Mr. Gilchrist's herd.*—Mr. Gilchrist lives on a large farm, twelve or thirteen miles from Oquawka, on the Warren County line. My first visit to his place was on October 14. His herd consisted of over one hundred head of hogs and shoats, of which about forty showed more or less plain symptoms of disease. Several animals had died. I requested a division into at least three different lots, which may be designated as Nos. 1, 2, and 3. Lot No. 1, it was agreed, should only contain such

animals as appeared to be perfectly healthy, and without any symptoms of disease; lot No. 2 was to be composed of such animals as did not appear to be perfectly healthy, but did not show any plain symptoms of swine plague; and lot No. 3 should include all those animals evidently sick. Lot No. 1, Mr. Gilchrist promised should be removed to a piece of ground—a small field without any water--which had been planted to corn, was free from any old straw or other rubbish which might possibly harbor any disease-germs, and was to be plowed the next day. Lot No. 2 he promised to put in another uninfected yard, separate from the regular hog-lot or pasture; and lot No. 3, it was agreed, should remain in the old hog-yard, occupied so far by the whole herd. It was further agreed that lot No. 1 should receive twice a day ten drops of pure carbolic acid in the water for drinking for 150 pounds of live weight, and lots Nos. 2 and 3 each, three times a day, a teaspoonful of hyposulphite of soda for every 150 pounds of live weight. Enough of each medicine was left to last from three to four days. When I made my second visit, on October 18, I found that only a part of the medicine, the carbolic acid, which had been given to the whole herd, had been used, and that no separation had yet been effected, because Mr. Gilchrist, on account of sickness, had been unable to perfect the necessary arrangements. The field or piece of ground intended for lot No. 1, however, had been plowed, and was ready for the reception of the animals. The plan of separation underwent a slight change as to lot No. 3, which it was thought best to divide again by putting the most seriously affected animals, lot No. 4, in a couple of open pens, situated in the barn-yard, and originally built for hog-pens, but unoccupied for a long time, and by leaving in lot No. 3, in the old swine-yard, only such animals as were evidently, though not dangerously, sick. It was farther decided that lots Nos. 1 and 2 should be treated with carbolic acid, and lots Nos. 3 and 4 with hyposulphite of soda. On the whole the herd was not any worse than on my first visit, except that a few more animals were coughing. Only one animal had died, and a *post mortem* examination made of this, but the result was not very satisfactory on account of the high state of putrefaction. Another pig, however, about five months old, was killed by bleeding for the purpose of a *post mortem* examination. Result: Morbid changes (hepatization) in the lungs, and enlargement of the lymphatic glands, as usual; a few ulcerous tumors in the intestines; numerous entozoa (alive) in the choledochus, and some dead worms in the intestines. Went again to Mr. Gilchrist's on October 21 and 24, and found that all arrangements, as agreed upon, had been carried out. Lot No. 1 was on the plowed ground; lot No. 2 in another separate yard; lot No. 3 in the old hog-lot; and lot No. 4, composed of seven very sick pigs, in a couple of small open pens in the barn-yard, separate from any of the yards occupied by the others. The medicines, too, had been given according to directions. On examining the several lots on October 24, I found in lot No. 1 a small pig, a so-called runt, that showed symptoms of disease. It belonged to lot No. 3, but being very small had crawled through the fence. It was immediately removed, and afterwards killed for *post mortem* examination. In lot No. 2 a sick pig was also found, which was likewise removed. In lot No. 3 nearly all the animals showed more or less plain symptoms of disease, but none were very bad or dangerously sick. Lot No. 4, as stated, contained the worst cases, and was originally composed of seven animals, but the number had been reduced to five, for one had died, and another had escaped over the rather low fence. Of these five animals, two, on examination, proved to be very sick.

The *post mortem* examination of the small pig, mentioned above, revealed the usual morbid changes in the lungs and in the heart to a limited degree, and nothing extraordinary except a firm adhesion (or union by firm connective tissue) between the posterior part of the left lobe of the lungs and the costal pleura.

Left more medicines—carbolic acid and hyposulphite of soda—with directions how to use them, with Mr. Gilchrist.

Went again to Mr. Gilchrist's on October 31, and found that the two very sick pigs and one of the others of lot No. 4, and a few of lot No. 3, had died, and that all others were doing well. In order to learn the final result I visited Mr. Gilchrist's herd once more, on December 16, and found that no deaths, and no plain cases of sickness, had occurred in lots Nos. 1 and 2, with the exception of the pig which was removed from lot No. 2 on October 24, and died afterwards. Lot No. 2 had been removed early in November to another yard or place, a piece of plowed ground. In lot No. 3 comparatively very few deaths had occurred.

It must be stated that during the whole experiment none of the hogs or pigs received any water except such as was pumped by a windmill from a deep well. Formerly the animals had access to a little streamlet proceeding from some springs. The medicines (the carbolic acid and the hyposulphite of soda) were given simply on account of their antiseptic and disinfecting properties, because there was reason to suppose that the water used for drinking, notwithstanding it was pumped from a deep well, might become infected, or be contaminated by Schizomycetes or disease-germs floating in the air while exposed. It was pumped by a windmill into a large and open wooden trough, situated in close proximity to the pens of lot No. 4, and adjoining on the other side the old swine-yard, occupied by lot No. 3. From this trough the water was carried in buckets to lots Nos. 1, 2, and 4, and carried by means of a pipe into lot No. 3.

3. *Mr. Henry Rice's herd.*—I visited Mr. Rice's place, about eight miles northeast from Oquawka, for the first time on October 17. His hogs and pigs had the run of the barn-yard and of two large pastures, one of which contained water, and had also access to the stack yard, which contained old straw stacks. I found nine decidedly sick animals, and five dead ones. Of the latter three were very much putrefied; one had died but two hours before, and the other died while I was there. Consequently I had an opportunity of making two *post mortem* examinations. The pig which had been dead two hours—an animal probably six months old and in good condition as to flesh—was examined first. The morbid changes were as follows: Lymphatic glands enlarged, blood coagulated and separated into dark clots and yellowish-colored serum, one-third of the whole tissue of the lungs hepatized, the non-hepatized portions of the lungs partly filled with fluid exudation and containing innumerable small, reddish-brown specks, caused by extravasations of blood (these lungs constituted, in their mottled appearance, and with their numerous embolic hearths and extravasations of blood, a very characteristic specimen, and I regret that I was not prepared to have them photographed); some serum in chest and pericardium; all blood-vessels of the auricles of the heart turgid with dark-colored blood. In the abdominal cavity, one small, incipient ulcerous tumor and numerous worms (*Trichocephalus crenatus*); in cæcum, numerous ulcerous tumors in different stages of development; spleen very large and dark colored, and all other organs destitute of any morbid features.

The other animal, a four months old boar pig, was examined immediately after death. The morbid changes were as follows: Lymphatic glands enlarged; blood of normal color; considerable exudation (serum and coagualtions, the latter principally on the anterior surface of the diaphragm) in the chest; over one ounce of serum in the pericardium; distinctly marked and fully developed hepatization only in the lower portions of both lobes of the lungs, but incipient hepatization or fluid exudation, and numerous small red spots of extravasated blood, each about a large as a pin's head, everywhere through the whole pulmonal tissue, but especially toward the lower border of the lobes; the capillary vessels of the heart, and particularly those of the auricles, turgid with blood. In the abdominal cavity: The whole peritoneum, but especially the serous coat of the intestines, congested, that is, all the smaller capillary vessels turgid with dark-colored blood; the intestines in many places agglutinated (adhering) to the walls of the abdominal cavity; the liver very dark and congested; the gall thick or almost semi-solid; the mucous membrane of the stomach wine-colored, and almost black, and very much swollen toward the pylorus; the mucous membrane of the duodenum black and gangrenous, and that of the jejunum purple and wine-colored; a large number of ulcerous tumors of various size in colon, and a few in cæcum; no entozoa or worms.

Neither of the two pigs presented any external morbid changes, except No. 2, which was slightly bleeding from the nose when it was dragged from the place where it had died to the place where the *post mortem* examination was made.

The following arrangements were made: The whole herd, some sixty odd head, was divided into two lots, No. 1 to contain the apparently healthy animals, or those not evidently sick, and No. 2 to be composed of seven very sick animals. On my arrival I found, as already stated, nine very sick pigs, but one died during my presence, and another was nearly half a mile from the barn-yard in the remotest of the two pastures, and was there left to its fate, as it was expected to die within a short time. The seven very sick pigs or shoats were shut up in a pen especially prepared for them, and lot No. 1, composed of those animals apparently healthy, was allowed to go to pasture (the one nearest the barn-yard) during the day, but was shut up in the barn-yard during the night, from sundown till ten o'clock in the morning, or till the dew had disappeared from the grass. The barn-yard was on high and dry ground, perfectly bare, and destitute of straw stacks, half-rotten manure, or pools of stagnant water. The pasture, too, was destitute of old straw, &c., and contained no water. The animals, therefore, received no water but what was drawn from a well in the barn-yard. Before my arrival the herd had access to running water in the remotest of the two pastures, and also to a stack-yard which contained old straw. All this was stopped. Lot No. 1 was treated with carbolic acid in the water for drinking, the same as lots Nos. 1 and 2 at Mr. Gilchrist's place, because it was thought probable that the animals of lot No. 1 might pick up some disease-producing germs either with their food or with their water for drinking, which was exposed in an open trough after it had been pumped from the well. Lot No. 2, composed, as mentioned, of seven sick pigs shut up in a pen, was treated with hyposulphite of soda, a tea-spoonful three times a day for every 150 pounds of live weight, in the water for drinking. The first dose of that medicine was given in my presence in skimmed milk, and was taken by five of the sick pigs. One of them, after it had taken some, commenced to vomit, but soon went to the trough for more. Two had no appetite and refused to take anything.

The five dead pigs, the two examined included, were cremated in my presence at the place where the autopsies had been made. One sick pig,

4 C D

not under treatment, and apparently convalescent, was examined as to temperature, and was found to be only 101½° F.

Went again to Mr. Rice's place on October 18. The sick pig, left alone in the pasture, was dead, and had been cremated; the seven sick pigs in their pen (lot No. 2) were still alive. A "dead-hog man," that is, a man who travels with a wagon through the country to collect the dead hogs from the farmers for a rendering establishment, had called at noon, and come into the house while the family was at the dinner-table. The stench emanating from his person and his clothes caused the whole family to vomit. No wonder, therefore, that such men are instrumental in spreading the disease wherever they go. Mr. Rice had ordered him to keep away from his premises. There is considerable suspicion in Henderson County, and elsewhere, that these "dead-hog men" find it sometimes in their interest to infect healthy herds of swine, expecially such as are nearly ready for the market, and therefore promise a rich harvest.

Finding that all my arrangements had been faithfully carried out, I left some more medicines (carbolic acid and hyposulphite of soda).

Went again to Mr. Rice's place on October 21. Of the seven sick pigs three had died, one had escaped while the gate was open, and three were still in the pen and alive. Of these latter I found two very sick, and about ready to die, while one was apparently improving, and convalescent. All the animals of lot No. 1 were doing well. My directions had been complied with in every particular.

Made another visit to Mr. Rice's place on October 27, and found that all the diseased pigs of lot No. 2 (the seven shut up in the pen, and treated with hyposulphite of soda) had died. The one that escaped had been caught and again confined, while all others (lot No. 1) which had been treated with carbolic acid in the water for drinking were doing first rate. So it seems that in this case the hyposulphite of soda has done no good as a curative remedy; even the one pig which seemed to be convalescent or improving on October 21 had died. Afterward, on December 18, I made one more visit, and learned that no loss whatever had occurred in lot No. 1, while in lot No. 2 every animal had died.

4. *The herd of Capt. Wm. Morris*, which, as has been mentioned, consisted of about four hundred animals. When I made my first visit on October 31, about forty or fifty animals were coughing, and exhibited more or less symptoms of swine plague. Only one animal had died, and as it had been dead but a few hours a *post mortem* examination was made. Morbid changes: Externally nothing abnormal, except a little redness between the fore legs. Internally: Enlargement of the lymphatic glands; extensive hepatization and numerous small extravasations of blood in the lungs; considerable exudation in the not yet hepatized portions of the pulmonal tissue, which caused the lungs to fill the whole thoracic cavity. The substance of the heart appeared to be in a state of congestion, and not only the auricles, but also the ventricles of that organ presented their capillaries turgid with blood. In the abdominal cavity the only morbid changes that could be found consisted in incipient ulceration in the mucous membrane of the stomach. No entozoa or worms, neither in the bronchiæ nor in the intestines.

Mr. Morris promised, at my request, to divide his herd into several lots, and to separate the healthy animals from the diseased ones. I left two pounds of carbolic acid to be given to the healthy animals in the water for drinking, and a quantity of hyposulphite of soda for the diseased ones.

Went again to Mr. Morris's place on November 3. My directions had been complied with. The whole herd had been divided into six different lots, the diseased animals by themselves and the healthy animals by themselves, in different yards or inclosures. The medicines, the carbolic acid and the hyposulphite of soda, had been used and no deaths had occurred. It even seemed as if the appetites of the diseased animals had somewhat improved, and as if the coughing had become a little less. Still how much of the apparent improvement should be accredited to the medicines, how much to the clear and cold atmosphere—the thermometer indicated several degrees below the freezing point; how much to the continued dry weather—it had not rained for several weeks; and how much to the strict separation and the division of the large herd into smaller lots, is very difficult to determine. At any rate the disease prevailed in a very mild form in Mr. Morris's herd.

My next visit was on November 7, when I found the strict separation broken up on account of the scarcity of water—it had not rained for nearly a month. During the night the pigs were still kept in their respective pens and yards, but during the day four of the different lots had access to a common trough to get water. The arrangement was as follows: One well, separated by a fence, supplied two troughs with water. One of the troughs was placed in a swine-yard, occupied by about one hundred healthy hogs and pigs, and the other one was outside and furnished water for the four lots mentioned. The hogs and shoats recently bought by Mr. Morris, as before mentioned, composed the sixth lot, and were kept by themselves in the barn-yard, more than forty rods distant from the other swine-yards, and received their water for drinking from a well near the barn. I found nearly one hundred hogs and pigs, or about 25 per cent. of the whole herd, more or less coughing, a few thumping, a few limping, and some very much emaciated, but only one animal dead. The carbolic acid had been used freely in the water for drinking, and Mr. Morris is inclined to ascribe the unusual mildness of the disease (caused undoubtedly by a combination of circumstances) to its effect. Left more carbolic acid and some hyposulphite of soda.

. On November 8 it commenced to rain, and continued for a week.

Went again to Mr. Morris's place on November 18. Found about 50 per cent. of the whole herd affected and coughing, but the disease was of such a mild type that only three animals, which had been castrated, had died, as has been mentioned before, or rather had been killed by Mr. Morris when he found them past recovery. One of these three had been dead only a day and had not yet been buried. It was in a first-rate state of preservation, and therefore a good subject for *post-mortem* examination. Morbid changes: Externally, a little redness of the skin on the lower surface of the body and between the fore-legs. Internally, lymphatic glands swelled; adhesion, though not extensive, between left lobe of lungs and diaphragm; both lobes externally very mottled in appearance, and diseased in about an equal degree, but the hepatized portions and lobules alternating with almost healthy portions, and in the latter numerous extravasations of blood presenting themselves as small red spots.*

---

* The illustration, Plate II, presents the external surface of the left lobe, photographed from nature, but a little reduced in size, and somewhat distorted in shape on account of its weight while suspended before the camera. The central portion of the plate is a good representation, and shows the mottled appearance of the lobe, but the left and upper portion of the plate are poor, and were evidently not in focus. The microphotographs, Plate III, figs. 1 to 6, were taken from slides of transversal sections of pulmonal tissue of the right lobe of the same lungs. Fig. I, Plate III, taken from a very thin section, shows nearly all the pulmonal vesicles perfectly closed; fig. 2 shows diseased pleura and small blood extravasations; fig. 3 is from a very thin sec-

Those parts and lobules not yet hepatized, or perfectly impassable to air, contained considerable exudation still in a fluid condition, and some of the smaller bronchiæ, especially those in the posterior portions of the lungs, harbored a very large number of very fine thread-shaped worms (*Strongyli paradoxus*). As other morbid changes, may be mentioned several ounces of serum in the chest, and more than one ounce in the pericardium. There was nothing very abnormal in the abdominal cavity. The dead pig was an animal about seven months old, and in a fair condition as to flesh.

November 22.—The disease had developed a more malignant character. Over two hundred animals were sick; from forty to fifty were very bad, and several deaths had occurred. Ordered once more a strict separation, which had become possible, since abundant rains had removed the scarcity of water; gave some carbolic acid, and some salycilic acid, and some hyposulphite of soda, to be tried on different lots, with instructions how to use it.

The continued rainy weather, the exceedingly bad conditions of the roads, and the great distance of Mr. Morris's place from Oquawka, (about fifteen miles), prevented me from making another visit during the month of November. I was, however, informed by letter, dated December 1, that the separation had been carried out, the medicines used according to directions, and that only a few animals had died, and all those slightly affected were recovering. I made afterwards, on December 18, another visit, and found the separation broken up again as soon as an improvement became apparent. Still, comparatively few animals, the exact number I was unable to learn, had died. Found one dead animal when I was there. The *post-mortem* examination revealed the usual morbid changes in the lungs, heart, pleura, pericardium, and lymphatic system, but no morbid changes of any consequence in the abdominal cavity, except enlarged glands and somewhat degenerated pancreas.

5. *The herd of Mr. Ely Beaty*, about four miles from Oquawka. This herd consisted of about one hundred head of hogs and pigs, and was visited for the first time on December 20. Found two pigs dead and five very sick. The herd was divided into two portions, and had been for some time. About thirty of the older and larger animals were kept in a high and dry pasture close to the house, and were said to be healthy; consequently I had nothing to do with them. The other portion of the herd, composed of about seventy, mostly small and young animals, among which the cases of sickness and death had occurred, occupied the barn-yard, and were allowed to roam on a large uncultivated tract of very broken land partly timbered. This (infected) part of the herd was immediately divided; all animals apparently healthy were removed to the orchard, and the diseased ones were allowed to stay where I found them. Two other pigs were kept in an open pen adjoining the barn, and separated from the barn-yard only by a fence. These were not disturbed, received no medicines, and did not become affected. The apparently healthy pigs removed to the orchard were treated with car-

---

tion, and shows diseased and thickened pleura. The normal pulmonal tissue is recognizable, notwithstanding the pulmonal vesicles are more or less filled with exudation; fig. 4 shows extravasation of blood and hepatization, but the structure of the pulmonal tissue can still be recognized: fig. 5 shows in its upper portion a blood vessel completely filled with blood; and fig. 6, which was taken from a thicker section, not stained, presents partially hepatized pulmonal tissue, in which the shape and form of the pulmonal cells are yet visible. Figs. 1 to 5 were taken from sections stained with Kleinenburg's solution.

bolic acid in the water for drinking, the same as lots Nos. 1 and 2 of Mr. Gilchrist's, and lot No. 1 of Mr. Rice. No medicines were given to the five sick pigs, because they were small, emaciated, of little value, and were not expected to recover. On December 22 three of the sick pigs had disappeared and could not be found; they had probably wandered off to a nook in one of many ravines, and there died.

December 27.—No new cases of disease, except in that portion of the herd composed of the older animals which were kept in the pasture. An old stag, which had been castrated only a few weeks before, was found dead. According to the information received, it had shown symptoms of inflammation of the brain for about twenty-four hours before its death. It died on December 26. Putrefaction had set in, but a *post-mortem* examination was made for the purpose of learning whether the animal was affected with swine plague or had died of another disease. External morbid changes: the skin on the lower surface of the body purplish-black; the castration wounds not fully healed; and the spermatic cords inflamed, enlarged, and almost black. Morbid changes in the chest: distinctly limited hepatization in the lower anterior or small lobes of the lungs; numerous unusually large extravasations of blood in the pulmonal tissue; the heart very much enlarged, about twice its normal size, and full of dark-colored blood, only partially coagulated; several ounces of serum in the pericardium. In the abdominal cavity nothing could be ascertained with certainty, as putrefaction was very much advanced. The skull was not opened, because no instruments were at hand. There can be no doubt the animal was affected with swine plague, but it is doubtful whether death was caused by that disease or by something else—for instance, by an overdose of salt or brine; at any rate, the symptoms observed during life, if correctly reported, would justify such a conclusion.

Visited Mr. Beaty's herd once more in January, and found that the five pigs which were sick on December 20 had all died, and that none of the others, with the exception of the old stag, had exhibited any symptoms of disease. After the 20th of December all received their water for drinking from a well close to the house. Before that date, the larger portion of the herd (the one kept in the barn-yard) had access to several small streams of running water.

6. *Mr. William B. Graham's herd,* two miles from Riggsville.—This herd has already been mentioned. I made my second visit to Mr. Graham's place on January 10, principally for the purpose of obtaining fresh material for experimental purposes. About fifty animals of his herd had died, notwithstanding the use of "sure-cure medicines," and most of the hogs and pigs still alive were more or less affected. Mr. Graham was not at home, but as several dead pigs were lying about unburied, I was permitted to make a *post-mortem* examination and to take all the material wanted. I chose a barrow about six months old, which had died, probably, not over an hour before. The carcass, which was but little emaciated, presented the following morbid changes: The skin on lower surface of abdomen, between the fore-legs, &c., was of a reddish-purple color; lymphatic glands enlarged; about five or six ounces of yellowish or straw-colored serum was found in the chest; about three-fourths of the whole pulmonal tissue of the left lobe and more than one-third of the pulmonal tissue of the right lobe of the lungs was morbidly changed, that is, presented hepatization in various stages of development, and, therefore, a mottled or marbled appearance; at several places there was adhesion between pulmonal and costal pleuras, and between pulmonal pleura and diaphragm; a few *Strongyli paradoxi*

in the finer bronchiæ of the posterior portions of the lungs, but none anywhere else; about half of the whole surface of the pulmonal pleura, even where not adhering to the wall of the chest, coated with exudation and rough; heart but little changed; some serum in pericardium. In the abdominal cavity, stomach entirely destitute of any food, but containing several entozoa and a quantity of yellowish mucus; spleen enlarged to about three times its normal size; pancreas somewhat changed, presenting on its external surface small yellowish specks, resembling detritus; bile dark colored and thick; liver, sub-venal glands, and kidneys without any visible morbid changes; large intestines almost without any food or feces, but containing a considerable quantity of intensely yellowish-colored mucus; no ulcerous tumors anywhere, but mucous membrane of large intestines somewhat swelled or thickened. I took pieces of the anterior portions of both lobes of the lungs and one of the glands of the mediastinum for further examination and for experimental purposes (cf. below).

Went again to Mr. Graham's, at his solicitation, on January 12, when he asked me to take charge of his herd and promised to comply with my directions in every particular. Of the one hundred and thirty hogs and shoats originally constituting his herd about seventy had died, forty-nine were yet eating, though by no means perfectly healthy or free from any infection, and the others were more or less sick, some of them dangerously. The forty-nine animals mentioned were removed to a large pen or inclosure made for them in the orchard, on high and dry ground, where they received clean food (corn) and no water except such as was pumped from a well. The inclosure was free from old straw, rubbish, &c., and the ground was bare. In my former experiments I treated the diseased animals with hyposulphite of soda, and met with very poor success, and used carbolic acid as a preventive medicine. In this present case I concluded to put the well-known antiseptic and disinfecting properties of the hyposulphite of soda to a test, and to give that drug as a preventive and in somewhat larger doses than formerly—three times a day a tea-spoonful to every hundred pounds of live weight—in the water for drinking to the forty-nine animals shut up in the orchard. The other shoats which were evidently sick I left to the treatment of Mr. Graham, with some "sure-cure medicine" of some Iowa men, styling themselves, I believe, "The National Hog Cholera Company." It may be remarked here, to avoid repetition, that Mr. Graham's hogs (those separated and kept in the inclosure in the orchard) consumed from January 12 to February 8 twenty-five pounds of hyposulphite of soda furnished by me and several pounds (four or five) bought by Mr. Graham; or each animal, on an average, consumed about three-fifths of a pound.

January 17.—Mr. Graham had been obliged to remove from the inclosure in the orchard (out of the forty-nine) one decidedly sick animal. Among the forty-eight thus left several were coughing and looking rather gaunt, but most of them were doing very well.

January 19.—Nearly all of the forty-eight hogs and shoats are doing well, and only a few show symptoms of disease, while most of the others (those found to be sick on January 12) are dead. A heavy rain converted the ground of the inclosure into mud and made it very uncomfortable for the animals. Mr. Graham, therefore, with my consent, enlarged the inclosure and took in a larger piece of the orchard. The treatment otherwise remained the same.

January 24.—Went again to Mr. Graham's. Of the original forty-nine hogs and pigs, forty-four were yet in the inclosure; four sick ones had been taken out and removed to the barn-yard, where a few other sick ones were kept. In all forty-six of the original forty-nine were alive; five had been removed, and of these two were still living. Of the others only a few were yet alive. Eleven or twelve dead hogs and pigs were piled up in a heap close to the barn and two others were lying in a fence-corner. I would have insisted on their burial or cremation, but the whole premises were already about as much infected as possible, and I thought that by allowing things to go on in the same way it might make the test of the hyposulphite of soda more severe and more valuable.

My next visit was on January 27. Found a few animals among the forty-four in the orchard still coughing. The dead pigs were still lying unburied.

January 30.—Found everything unchanged. The forty-four pigs in the orchard were doing well.

My last visit was on February 4. Found the forty-four hogs and pigs in the orchard all doing very well; those that had been coughing or showing other slight symptoms of disease were recovering, and all others had gained in flesh. When called up to be fed in my presence every one responded with alacrity and appeared greedy. Of the original forty-nine, forty-five were alive and doing well; five had been separated, and of these four had died. Of all others two or three were alive, except two boar pigs, which had not been with the herd, but had been kept in a separate pen and had not become affected. The dead shoats were still lying around in the same places, and one pig which had died two days before was even lying in the barn. I therefore urged Mr. Graham to burn every dead animal at once and not wait any longer for the "dead-hog man" to come around. I learned afterwards, by letter, dated February 16, that one more of the sick pigs had died, while no new case of disease had occurred.

Several other herds of swine, either diseased or reported to be diseased, were visited, partly for the purpose of selecting suitable herds for experimental purposes, but principally to obtain all the information possible by my own observation and from what the owners might have to communicate of their experience. Said herds belong to the following persons—most of them large farmers—and have been visited on the following dates:

Mr. M. Mills, about seven miles from Oquawka, October 14.
Mr. Duke, about eight or nine miles from Oquawka, October 21.
Mr. Crouch, about seven or eight miles from Oquawka, October 27.
Mr. Haley, about seven or eight miles from Oquawka, October 27.
Mr. Carpenter, eight miles from Oquawka, November 3.
Mr. McGaw, near Biggsville, November 21.
Mr. Weigand, near Biggsville, November 21.
Mr. Smith, near Oquawka, November 15.
Mr. George Curry, Olena, November 23.
Mr. Cortleyou, Raritan, November 24.
Mr. W. Stanley, twenty miles from Oquawka, November 24.
Mr. Jamieson, near Biggsville, January 10.
Mr. Gibler, Oquawka, January 11.
Mr. Radmacher, Oquawka, January 28.
Mr. Welsh, seven miles from Oquawka, February 5.
Mr. Schulze, near Biggsville, February 6.

At Mr. Cortleyou's place I found one dead hog, and a *post mortem* examination was made, which, however, revealed nothing new, except that the lungs adhered almost everywhere to the walls of the chest. The

plumonal and costal pleuræ were, at several places, so firmly united as
to make it impossible to open the chest, and to remove the ribs without
tearing the tissue of the lungs, or separating them by means of the knife
from the walls of the chest.

At Mr. Radmacher's three *post mortem* examinations were made at his
solicitation, but the disease did not prove to be swine plague; no morbid changes were found except an immense number of small acephalous
cysts or hydatids, situated on the serous membrane of the small intestines.

At most of the places named I obtained some valuable information as
to the means and vehicles by which swine plague is spread, corroborat-
ing in every respect what I found before. As of special interest the experience of Mr. John Haley deserves a brief mention. Mr. Haley is a
farmer of superior intelligence, a good observer, and well acquainted
with the symptoms of swine plague in all its various phases. His herd
of swine consisted (in October) of sixty animals. Last July one of his
pigs exhibited plain symptoms of the plague. In order to prevent, if
possible, further mischief, he killed the affected animal at once, and
buried it four feet deep. He acted in time, because no other animal
had yet become infected, and his herd was saved. If swine plague were
everywhere dealt with in like manner as soon as it makes its appearance the losses would be very few.

EXPERIMENTS WITH HEALTHY PIGS.

The observations in regard to the spreading of swine plague from
place to place, from herd to herd, and from animal to animal, and the
results obtained in the treatment and the measures of prevention applied to the six herds experimented with, made it desirable to determine
with certainty three important points. First, although my observations
left no doubt in my mind as to the disease being frequently, and probably in a majority of cases, communicated by means of contaminated or
infected water for drinking, absolute proof was yet wanting, which, it
seemed to me, was obtainable only by a direct experiment, in which
positively healthy pigs, while protected against any other possible
source of infection, are compelled to drink water contaminated or infected either with parts of a carcass of a hog that has died of swine
plague, or with the excrements, excretions, or secretions of an animal
affected with that disease. Second, the results of my experiments in
regard to measures of prevention proved to be exceedingly favorable,
and notwithstanding the uncommon leniency of the disease in four or
five herds experimented with—in the sixth one, that of Mr. Graham,
the plague proved to be malignant enough—it cannot be denied that
the measures of prevention employed were attended with very good
and satisfactory results, and effective; but the question remained to be
answered whether the good results were due mostly, or exclusively, to
the strict separation, which undoubtedly was of the greatest importance,
as it prevented, to a certain extent at least, a further influx of the infectious principle, or whether the same were chiefly the effects of the
disinfecting and antiseptic properties of the carbolic acid and the
hyposulphite of soda, administered in the water for drinking, notwithstanding that three drops of carbolic acid added to three ounces of
water infected with one drop of pulmonal exudation (in an experiment
made June 9, 1879), did not seem to be able to destroy the Schizomycetes,
or to prevent altogether their development. This question, too, could
be answered only by subjecting an undoubtedly healthy pig to the in-

fluence of the infectious principle in such a way as would surely bring on the disease, and by treating it at the same time with one of those medicines. Third, I considered it as desirable to subject the vitality, or rather effectiveness of the Schizomycetes, or of the infectious principle, to a practical test after they had been preserved for a long time (nearly a year) in a suitable vehicle outside of the animal organism. In order to decide these points, I procured three perfectly healthy and vigorous pigs, free from any wounds or sores, each about four or five months old, and weighing about ninety pounds. (The three together weighed 265 pounds.) In the midst of a large empty lot, on high and dry ground, and inclosed by a good substantial board fence, which permitted no animal larger than a chicken to crawl through, I built them a pen of new lumber especially designed for my purpose. The pen measured 12 by 8 feet, had a good board floor and a board roof, and was divided by partitions, 4 feet high, into three separate apartments, each 8 by 4 feet. The troughs, one for each apartment, were so made as to facilitate cleaning and removing of the same if desired. I received those pigs on January 9, kept them a few days for observation as to their health, and commenced my experiments on January 11. It was at first my intention to procure for two of these pigs some water from an infected creek or streamlet, but winter weather, frost, bad roads, and a distance of seven miles or over between Oquawka and the nearest streamlet known to be infected, caused me to abandon that project, and induced me to substitute well-water infected on purpose with infectious material. I procured the latter—pulmonal exudation, which, undoubtedly, is neither as infectious, nor does it contain such an immense number of Schizomycetes as the excrements and other excretions of a diseased pig—from Mr. Graham's herd. The water was drawn fresh (three times a day) from a good well, poured into the troughs, and then contaminated or infected for each pig, Nos. 1 and 2, with a few drops of the pulmonal exudation. Pig No. 1, which occupied the west end of the pen, and pig No. 2, which was kept in the middle apartment, received such infected water for drinking during four successive days, or from the evening of January 11, till noon January 15. Pig No. 1 received no medicine whatever, and pig No. 2 received each time, together with the pulmonal exudation, six or seven drops of the concentrated carbolic acid (95 per cent. of Mallinckrodt's cryst. carbolic acid, and 5 per cent. of water) in the water for drinking. Pig No. 3, occupying the eastern part of the pen, was subjected to another experiment, and received clean water, clean food, and no medicines of any kind. Last winter (in January, 1879), while investigating swine plague in Lee County, I had occasion to filtrate infectious pulmonal exudation, somewhat diluted with water, for the purpose of freeing the same from the Schizomycetes or disease-producing germs. I made repeated efforts, and used filtering paper, but did not succeed. The filtrate, however, which was perfectly clear, was preserved in a vial with a tight-fitting glass stopper. Repeated microscopic examinations (April 13, July 26, November 15,.and January 11) showed that the Schizomycetes had not disappeared, and had undergone but very slight changes (cf. drawing, Fig. XV). It was therefore to be supposed that the filtrate, which was yet clear and transparent, and without any smell whatever, might still possess its infectious properties. Consequently I concluded to use that old filtrate for the purpose of testing the tenacity of life of the infectious principle, and injected on January 12, 20 minims of the same by means of a graduated hypodermic syringe into the cellular tissue, just behind the fore-leg of pig No. 3.

January 12 to January 25.—All three pigs appeared to be perfectly healthy, and had good appetite; at least none of them exhibited any signs or symptoms of disease.

January 26.—Pigs Nos. 1 and 2 appear to be healthy, and have good appetite. Pig No. 3, the one inoculated with the old filtrate, shows plain signs of indisposition, such as drooping of the head, considerable coughing, and partial loss of appetite.

January 27.—Pig No. 1 shows unmistakable symptoms of disease, such as coughing, drooping of the head, loss of appetite, &c., and is less lively in its movements than formerly. Pig No. 3 coughs repeatedly, and has very little appetite, but is apparently thirsty. Both pigs (Nos. 1 and 3) have lost the curl in the tail. Pig No. 2 is perfectly healthy, and has a vigorous appetite. This pig has received no carbolic acid during the last four or five days, but now receives it again regularly three times a day, each time six to seven or eight drops in its water for drinking.

January 28.—Pig No. 1 is worse, coughs a good deal, droops its head, eats scarcely anything, and is very thirsty, but is still savage when touched. It may be remarked here that all three pigs, being of "scrub" stock, with a slight mixture of Poland, China, and Berkshire, and having been raised in a large herd, accustomed to large fields or pastures, were naturally very savage, which made a close examination exceedingly difficult. For the same reason I found it impossible to ascertain the temperature without using force, and causing thereby an abnormal rise. Pig No. 3 coughs repeatedly, breathes faster than usual, eats very little, is thirsty, and its coat of hair looks rough. Pig No. 2 is perfectly healthy; its coat of hair is sleek and smooth.

January 29.—Pig No. 1 is worse, and evidently suffering. Its respiration is very much accelerated; the coughing is worse and more frequent; the thirst is increased; appetite is wanting, and the coat of hair is looking rough and unclean. Pig No. 3 eats a little more, but still coughs very much. Pig No. 2 has a vigorous appetite, and is perfectly healthy.

January 30.—Pig No. 1 coughs a great deal, is evidently in distress, and does not touch its food. Its coat of hair is rough, and emaciation is visible. Pig No. 3 has more appetite, eats about half its usual ration, and appears to be livelier, but is still coughing. Pig No. 2 is perfectly healthy.

January 31.—Pig No. 1 about the same as yesterday; at any rate not worse. Pig No. 3 is improving. Pig No. 2 is all right in every respect.

February 1.—Pig No. 1 eats some, but is coughing a great deal, and is thirsty. Pig No. 3 eats more, but not a full meal, coughs less, breathes freer, and its coat of hair looks less rough. Pig No. 2 eats vigorously, and is healthy.

February 2.—Pig No. 1 eats some, about half a meal, and don't seem to be as thirsty as formerly. Pig No. 3 eats nearly a full meal, and coughs very little. Pig No. 2 is all right.

February 3.—Pig No. 1 is improving, eats about half a meal, but is still coughing. Pig No. 3 has regained its appetite, and coughs very little. Pig No. 2 is as healthy as ever.

February 4.—Pig No. 1 is improving, eats more than half a meal, and coughs but little. Pig No. 3 has a good appetite, and seems to have fully recovered from its slight attack. Pig No. 2 is perfectly healthy, and the use of carbolic acid is dispensed with. It has received in all a little over half an ounce. Have never heard it cough.

February 5.—Pig No. 1 is improving, eats nearly a full meal, coughs very little, and its coat of hair is smooth again. Pig No. 3 has fully recovered, and pig No. 2 is healthy.

February 6.—Pig No. 1 eats well again, and shows no sign of disease, except now and then a little coughing. Pigs Nos. 2 and 3 are healthy and ready for another experiment. When the three pigs were received, pig No. 1 was slightly heavier than pig No. 3, and weighed over ten pounds more than pig No. 2. At present (February 6), pig No. 2 is the heaviest, and weighs at least twenty pounds more than No. 1, and ten pounds more than No. 3.

When pig No. 1 commenced to show signs of recovery, pig No. 3 had only a very slight attack, and was not expected to die. I thought of killing it for the purpose of ascertaining the extent and the nature of the morbid changes existing, but as I intended to make more experiments, and as it is not easy to obtain suitable material when wanted. I concluded to allow the animal to recover, and to save it for further use, but on February 8, I was taken sick myself, and was thus compelled to abandon my plans for the time being. From February 6 to February 12, the day on which I left Oquawka, all three pigs were doing well and improving.

MICROSCOPIC EXAMINATION.

Special attention has been paid to the examination of the water of such running streamlets, small creeks, and stagnant pools as were accessible to diseased hogs. Most of the water examined under the microscope was taken from small runs, ravines, and pools of the hog-pastures of Messrs. Kennedy, Gilchrist, Rice, and Morris, but samples from other places have also been examined. The objectives used were a one-ninth imm., or No. 8 Hartnack, a one-sixteenth imm., or No. 10 Hartnack, and a one-tenth oil or balsam-index imm. of Tolles, of recent make and superior quality. The results, in all cases, have been essentially the same. Besides some algæ and other minute growths, I always found some globular and rod-shaped Schizomycetes, to all appearances identical to those which occur invariably in the blood, lymph, exudations, other morbid products, excrements, urine, &c., of the hogs and pigs diseased with swine plague (*cf.* drawing, Fig. XVIII). It must be stated, however, that every sample of water was taken from such a place in the streamlets or pools where the water was known to have been more or less defiled by diseased hogs.

The blood, exudations, &c., of nearly every animal of which a *post mortem* examination was made, were also subjected to a microscopical examination. Special care was taken in every instance to collect the blood directly out of a blood-vessel in perfectly clean, so-called homœopathic vials, which were closed immediately with new corks. The results have been essentially the same as those obtained last winter (*cf.* next chapter).

The fact of finding globular and rod-shaped Schizomycetes, apparently identical to those occurring in the diseased hogs, in all the samples of water taken in perfectly clean vials from such small creeks, ravines, and pools as had been used as drinking and wallowing places of diseased hogs, convinced me still more that a communication of swine plague from one animal to another, and from one herd to another, is frequently, and may be in a majority of cases, effected by means of the water for drinking, if defiled by diseased hogs, or contaminated with the morbid products of the disease or the carcasses of dead hogs. That such is the case received additional proofs by the results of my experiments and

the good services of the carbolic acid and the hyposulphite of soda
mixed with the water for drinking, but particularly by the result of the
experiment with experimental pigs Nos. 1 and 2.   If it were not so, the
carbolic acid especially, it must be supposed, could not have had much
effect.

THE SCHIZOMYCETES.

In my former reports I adopted the name of "*Bacillus-suis*" for those
globular and rod-shaped parasites of the Schizomycetes family which
are invariably found in the blood, the morbid products, the excretions,
the secretions, &c., of swine diseased with swine plague, and which, for
reasons stated in my previous reports, I am obliged to look upon as constituting the infectious principle and the real cause of that disease, for
the following reasons : I called those Schizomycetes, when presenting
themselves in their rod-shaped or higher developed form, "*Bacilli*" (little sticks), partly on account of their stick-like shape, and partly on
account of their, in many respects, close relation and similarity—by
constituting the cause of an almost equally destructive disease—to the
well known "*Bacillus anthracis*," and I gave them the name "*suis*" from
sus (swine) as constituting the cause and infectious principle of a disease peculiar to that animal, though communicable, under favorable
circumstances, to others.   I was then, however, not sufficiently acquainted with the exact classifications of the various disease-producing
Schizomycetes, as made by different European authors, such as Cohn,
Billroth, Klebs, and others, and I would not have attempted to name
them at all had I found them named or sufficiently described by any one
else.   Since then I have worked considerably with the microscope, and
have become somewhat familiar with the classifications adopted by the
Europeans, which, it must be regretted, differ widely, and, owing to
almost insurmountable difficulties, are yet unsatisfactory.   All, however, seem to agree that those Schizomycetes classed by them under the
name of "*Bacillus*" do not form clusters or colonies (Rasen, *zooglœa*-
masses, *gliacoccos*, or *coccoglia*), and do not undergo metamorphoses from
globular to rod-shaped Schizomycetes, two things decidedly characteristic of the microscopic parasites of the Schizomycetes family as found
in swine plague ; consequently the name adopted, *Bacillus*, was not well
chosen, and is not suitable.   Cohn, considered as one of the best authorities, discriminates *Sphœrobacteria*, *Microbacteria*, *Desmobacteria*,
and *Spirobacteria*, and divides the former in *Chromogenic*, *Zymogenic*,
and *Pathogenic Sphœrobacteria*.   According to his classification the Schizomycetes of swine plague, in their globular form, would come under
the head of *Pathogenic Sphœrobacteria*, and in their rod-shaped or stick-
like form under the head of *Microbacteria*, under which he arranges
only two species, *Bacterium termo*, and *Bacterium lineola*, neither of
which is identical to the rod-shaped Schizomycetes of swine plague.
Bacterium termo is much more lively in its movements than the latter,
which, also, invariably disappears as soon as *Bacterium termo* or putrefaction *bacterium* makes its appearance in large numbers, and *Bacterium
lineola* is considerably larger.   *Bacillus anthracis* is classed by Cohn
among the *Desmobacteria*, of which he gives it as a characteristic that
the same never form any clusters or *zooglœa*-masses.   *Spirobacteria* (see
sub. IV in drawing in my first report) have been found a few times,
but I am convinced their occurrence was accidental.

Billroth makes a different classification, and objects to a separation of
the globular from the rod-shaped Schizomycites, as both forms belong
to several species, and constitute only different stages of development.

According to his classification the Schizomycetes of swine plague would come under the head of *Coccobacteria*, and might be called *Coccobacteria suis*.

Two of my drawings (one sub. VII in my first report, and another one in my second) present club-shaped bacteria or Schizomycetes. These, according to Billroth, are *Helobacteria*, and constitute a higher development of *Coccobacterium*. The bright and light-refracting granules or globules which constitute the knob, situated in some of the *Helobacteria* at one end, and in others further toward the middle (*cf.* drawing in my second report), are capable, according to Billroth, of enduring high degrees of heat and cold, and may even completely dry up without being destroyed, or losing their germinating power. If moistened and swelled again in a sufficiently wet or watery substance, they will produce a great many very fine and pale spores (*micrococcos*), which are usually enveloped in a pale viscous substance (glia) ; such a group of spores, colony, or cluster (*gliacoccos* or *coccoglia*), remains for some time together, or at rest. These spores, or *micrococcos*, can multiply by division, and afterwards either remain together as irregular-shaped clusters, or *coccoglia-masses*, or in form of chains (*streptococcos*); or, if the glia formation is destroyed, multiply further as single *micrococcos*. In most cases, however, these spores, or *micrococcos*, which have come forth from the *Helobacterium*, soon stretch or grow lengthwise, and become rod-shaped bodies (bacteria). In fluids these bacteria, after some time of rest, commence to move, work themselves out of the glia or viscous substance, and swim about in the fluid (swarming vegetation-spores, or swarming bacteria). During this swarming period and later, after they have found somewhere a place of rest, these bacteria begin to stretch or to grow longer—sometimes very long ; a traverse furcation appears in the middle; the ends commence to swing to and fro, and finally a separation takes place, and the long bacterium is divided into two. Usually the ends divide again and again, till finally the joints sometimes become so short as to be not much longer than thick. In other cases only lines of demarcation are formed, and an actual separation does not take place, at least not at once. In such a case the joints are first square, but soon become round, and then the *bacterium* (*strepto bacterium*) resembles a *coccos* chain. If the *glia* which envelops the crecos-colonies, from which the bacteria come forth, is very thick and tough, and if the bacteria part from each other by separating continually, new glia and no chains are formed, then confluent and resting (motionless), *Glia* and *Petalobacteria* are produced, and their elements, especially on the surface of a fluid, can vegetate for a long time without obtaining motion.

According to Billroth the final changes of the bacteria and bacteria chains, which have been without motion from the beginning, or have come to rest after a period of swarming, may differ as follows : 1. The plasma may pass out of the envelope in form of a sterile viscid substance, and the empty envelope remains. 2. The plasma may become crenated in different directions (moniliform), while the envelope remains; the crenation leads to separation, and pale globules (*micrococcos*) are formed, which multiply more and more within the envelope, and growing, roundish-shaped, palmelloid-ramified, and cylindrical cells or sacs, full of micrococcos (*ascococcos*) are produced. These cells or sacs finally break or dissolve, and the micrococcos become free. What becomes of them Billroth leaves undecided. 3. The plasma of a bacterium contracts to one or more shining or light-refracting globules, with dark outlines, and thus the *Helobacteria*, commenced with, are formed.

As nearly all these various changes and formations have been repeatedly observed, it may be that these club-shaped formations, or *Helobac-teria*, of Billroth, which I was inclined to look upon as foreign to the disease, are only a higher development, or another form of the swine-plague Schizomycetes.

Klebs has the same objections to Cohn's classification as Billroth, but does not agree with the latter. He divides the Schizomycetos (Naegeli), which he calls Schistomycetes, into two groups, *Microsporines* and *Mo-nadines*. According to his classification the Schizomycetes of swine plague would come under the head of the former, of which it is a main characteristic that the same do not develop any offensive gases, like the members of the second group, the *Monadines*.

It may be asked, which of the various forms presented in the draw-ings represent the true swine-plague producing Schizomycetes? I will try to answer as well as I can. In the first place, the drawings, especi-ally those accompanying my first report, are not as accurate as I de-sired them to be, because I am no draughtsman, and in the beginning of my investigation had to work with objectives—my best lens was a No. 8-im. Hartnack, much inferior to those I have now; a one-tenth Tolles and No. 10 Hartnack; therefore I had difficulty to get sharp and accurate outlines, and still greater difficulty in reproducing them on paper. Secondly, a great many Schizomycetes, being so exceedingly small, appear to be very similar to each other; especially *Bacterium ter-mo*, so often met with, is very similar in appearance to the rod-shaped Schizomycetes of swine plague, and unless a very superior lens of high magnifying power is used, a Tolles $\frac{1}{16}$, for instance, can sometimes hard-ly be distinguished from the latter, except by its livelier movements, which, of course, cannot be shown in a drawing. Two or more species of Schizomycetes may be very similar in appearance, or so nearly alike as to make a discrimination impossible, and still they may possess en-tirely different properties, and produce very different effects. The fol-lowing may serve as an illustration: The pigment-bacteria, or *chromo-genic sphaerabacteria* of Cohn, are all alike and cannot be distinguished, and still the various species produce different colors, but each species invariably the same. One, for instance, produces green, another red, another blue, another orange, and so on. The Schizomycetes of swine plague and those of putrefaction, though similar in form, are entirely different in their effect, and even antagonistic to each other in so far as the former, as has already been stated, commence to disappear as soon as the latter make their appearance. To show the visible difference be-tween the swine plague Schizomycetes and *Bacterium termo* I refer to drawing, Fig. XIX *a* and *b*, which represents the latter after Siedamgro-tzky and Cohn.

Not all the small particles, however, which are sometimes seen swim-ming in the blood or blood serum, and may even show a slight motion, are micrococci, or globular Schizomycetes; some of them are products of detritus, or minute fat globules, &c. But those minute globular bodies, which strongly reflect the light, are sometimes even more nu-merous than the blood-corpuscles, and are not destroyed by adding a weak solution of caustic potash, must be considered as micrococci, or globular Schizomycetes. Further, where clusters or colonies (*gliacoccos or zoo-gloea*-masses) are existing, and where rod-shaped Schizomycetes or bac-teria are accompanying the spherical forms, then no doubt can remain. There are yet several other means by which their true nature can be determined, but to enumerate them would lead too far, and will not be necessary. One other point, however, may yet be mentioned. If the small globular bodies, seen under the microscope, are detritus or minute fat-globules, they will, at best, remain as they are, but will not propagate or increase in numbers like the micrococci, which multiply or propagate in many other fluids besides blood, if put into them, and even in fluids entirely different from that in which they have been found. In other words, the micrococci can be cultivated, and detritus and fat-globules cannot.

Under some circumstances the swine plague Schizomycetes are easily destroyed, for instance, if exposed on a bare, dry surface to the rays of light, or a free access of air; but in a suitable vehicle or nutritious substance, or protected by a porous body which keeps them moist and warm, their tenacity of life seems to be very great (cf. experiment with experimental pig No. 3, and the case communicated by Mr. Pendarvis). It is also a well-known fact that in the spring, particularly if a hard or dry winter has been destructive to the infections principle, swine plague almost invariably makes its first appearance again in swine-yards which contain old straw-stacks, &c., at the edge of the timber, or in some other sheltered nook or corner where the Schizomycetes have found protection. As to a proper vehicle or nutritious substance or fluid, the Schizomycetes of swine plague do not seem to be as particular as many other kinds. Water, especially if it contains a slight admixture of organic substances, is about as suitable as anything else.

The swine-plague Schizomycetes, even in a very sick animal, or in the carcass of one that has just died, are never as numerous in the (fresh) blood as in the morbid products, the urine, the excrements, the lymph, the pulmonal exudation, &c., and are usually found in the fresh blood taken from a vein or artery and examined immediately, only in their globular form. As I stated in my first report, and recent observations have confirmed it, the globular Schizomycetes undergo changes, develop, and, at a certain stage of their development or growth, form clusters or colonies (zoogloea-masses or coccoglia); these clusters, it seems, are formed principally in the smaller, narrower vessels, get stuck in the finer or finest capillaries, and obstruct them; then these zoogloea masses divide, or parts are torn off by the pressure of the blood-current, are carried a little farther, and cause new obstructions, or form emboli, a proc ess which, it appears to me, causes most if not all the morbid changes. At any rate, in such embolic hearths, not only in the lungs, but in other tissues just as well, whole nests of partially developed (globular, double, and rod-shaped) Schizomycetes can be found. It is therefore but natural that the circulating blood contains comparatively (not positively) few Schizomycetes, because the latter are carried into, get stuck, congregate, and accumulate gradually in the capillaries already obstructed by the viscous clusters. Besides, in the emboli in the affected tissues, the Schizomycetes can always be found in great abundance in the lymphatic glands, in the kidneys, in the extravasated blood, and in the ulcerous tumors; in the kidneys, probably, because through those organs a great many Schizomycetes are eliminated. They are also thrown off or discharged through the intestines, the lungs, and the skin, and, perhaps, through the salivary glands and the mucous membrane, but that I do not know.

Most of the German investigators claim for the disease-producing Schizomycetes the same principle as the putrefaction and ferment-bacteria, that is, chemically acting or poisonous properties. Whether the swine-plague Schizomycetes act also as a poisonous substance, and not simply in a mechanical way, I do not know, because I have no proof of their chemical action, but that the many millions or billions existing in one diseased animal are able, notwithstanding their minuteness, to produce important changes by robbing the animal organism of nutritive material, and by depriving it of large quantities of oxygen, directly as well as indirectly, is very evident.

One other question may be asked, Where do the swine-plague Schizomycetes come from? This question cannot be satisfactorily answered. Only one thing is certain, they immigrate or enter from the outside, and are not developed, as has been claimed by several authors, under the influence of the disease from germs pre-existing in the normal blood. *Wherever swine-plague Schizomycetes do not find an entrance, there no swine plague will appear.*

RESULTS AND CONCLUSIONS.

These are in perfect harmony with those obtained in the fall of 1878, and in the winter and spring of 1879, and may be briefly stated as follows:

1. The most effective means of prevention that can be applied by the individual owners of swine consists, first, in promptly destroying and burying sufficiently deep and out of the way the first animal or animals that show symptoms of swine plague, if the disease is just making its appearance, and in disinfecting the premises, or, as that is difficult, in removing the herd at once to a non-infected place, or out of the reach of the infectious principle. If possible the herd should be taken to a piece of high and dry ground, free from any straw and rubbish—if recently plowed, still better—and should there receive clean food and no water except such as is freshly drawn from a well. If this is complied with, and if all communication whatever with any diseased hogs or pigs is cut off in every respect, which is absolutely necessary, and still danger should be anticipated, for instance, if one or more animals should have become infected before the herd was removed, or a possibility of either food or water for drinking being or becoming tainted with the infectious principle should exist, the danger may be averted, or at least be very much diminished by administering three times a day in the water for drinking either some carbolic acid (about ten drops each time for every 150 pounds of live-weight), or some hyposulphite of soda (a tea spoonful for every 100 pounds of live-weight), till all danger has disappeared. Second, where swine plague has been allowed to make some progress in the herd, or until the presence of the disease is not discovered until several animals have been taken sick or have died, others have become infected, the best that can be done is to separate at once the healthy animals from the diseased and suspected ones; to place the healthy animals by themselves and the doubtful ones by themselves; to separate, disinfect, and treat the animals in the way just stated. Special care must be taken to prevent any communication, direct or indirect, between the three different parts of the herd. If one person has to do the feeding, &c., he must make it a strict rule to attend always first to the healthy animals, then to those considered as doubtful, and last to the sick ones, and must never reverse that rule, or go among the healthy hogs or pigs after he has been in the yard or pen occupied by the others. If possible each portion of the herd should have its own attendant, who should not come in contact with any of the others. The separation must be a strict one in every respect; even dogs and other animals may carry the infectious principle from the diseased animals or from the yard occupied by them to the healthy hogs and pigs. Buckets, pails, &c., which are used in feeding the sick hogs should not be used for the healthy ones, because the infectious principle may be conveyed by them from one place to another. Last but not least, it is very essential that the

yard or hog-lot occupied by the healthy portion of the herd be higher than that occupied by the others. If it is lower, and especially if it is so situated that water and other liquids from the other hog-lots can flow into it, or over it, the separation is worse than useless, for then the healthy portion of the herd will surely become infected unless the ground is exceedingly dry. Third, whenever swine plague is prevailing in the neighborhood, any operation, such as ringing, marking by wounding, or cutting ears or tail, and castration and spaying particularly, must not be performed, but should be delayed until the disease has disappeared, or does not exist anywhere within a radius of two miles. If such operation should become absolutely necessary, the wounds must be dressed at least once a day with an effective disinfectant, for instance, with a solution of carbolic acid or thymol, till a healing has been effected.

2. Swine plague is very often communicated from herd to herd and from place to place by a careless, and, in some cases, even criminal contamination of running streamlets, creeks, and rivers with the excrements and other excretions of diseased hogs and pigs, and with the carcasses and parts of the carcasses of the dead animals. This source of the spreading of the disease can be stopped only by declaring such contamination of streamlets a nuisance and making the offense punishable by law. Allowing swine affected with the plague to have access to such streamlets should be considered as constituting good evidence of such a contamination, as also the throwing of dead hogs, or parts of a carcass, into such streamlets, creeks, or rivers.

3. The rendering tanks established in almost every locality in which swine plague is or has been prevailing, contribute very much, directly and indirectly, to the spreading of the disease. They contribute directly by disseminating the infectious principle wherever the tank-agents, who collect the dead hogs from the farmers, travel with their wagons; and by contaminating and infecting, in many instances at least, the waters of streamlets, creeks, and rivers with such parts of the dead hogs as are not worth rendering, but which constitute the principal seat of the morbid process. Indirectly they contribute by inducing the farmers to leave their dead animals lying around unburied, thus remaining a source of infection until the " dead-hog man" comes and takes them away. If transportation of swine that have died of the disease is prohibited by law, the numerous rendering tanks will soon disappear, and another source of infection will thus be closed.

4. The disease is spread not only by the transportation of dead hogs, but also by that of diseased ones. That such is the case becomes apparent by the fact that swine plague in its spreading not only follows the course of streamlets, creeks, and rivers, but also travels along the lines of railroads and public highways. All traffic in, and transportation of, diseased hogs and pigs, and of animals that have died of swine plague, should, therefore, be stopped; and sending diseased swine to market—a very common practice at present—should be made a criminal offense. Further, a law which would compel every owner of swine to take care of them, to confine them to his own premises, and not allow them to run at large on public highways, &c., would, if executed and complied with, do a great deal of good, and prevent a great many infections. It has happened very often that a stray hog or pig has carried the disease into a healthy herd; and, vice versa, it has happened also—perhaps just as often—that a hog or pig has become infected while among other swine and, coming home again, has introduced swine plague into the herd to which it belonged.

5. As to sweeping and effective measures of prevention, I would only

5 C D

repeat what I said in my last report. No authenticated case of a spontaneous development of swine plague has yet come to my knowledge, and the disease, I am more convinced than ever, can be stamped out, but only by adopting the most stringent measures, such as I advocated in my last report. If the stamping-out process is begun in the winter, after the fat hogs have been sold or butchered, and before the spring pigs have been born, the difficulties will not be insurmountable, and the hardships or inconveniences necessarily to be imposed upon the owners of diseased swine will not be as great as might be supposed. The best method, as stated before, would be to kill and bury or cremate, immediately, every hog or pig that shows symptoms of swine plague. Where this cannot be done, the diseased or infected herds must be isolated from all healthy animals for a period extending at least to two months after the last sick animal has died or recovered. All dead hogs or pigs must be immediately buried or burned; persons attending to sick hogs should be prohibited from going among healthy swine; infected strawstacks, &c., should be burned.

6. As to medicines for the purpose of prevention, carbolic acid given twice or three times a day in the water for drinking has proved to be of value, and has done good service, and so has hyposulphite of soda. These medicines are valuable, and have proved to be effective, especially in cases in which there is reason to suppose that the water for drinking or the food may become infected, but it is doubtful whether the disinfecting and antiseptic properties of those medicines will be sufficient to destroy the infectious principle or its effect, if its influx is a great one, and continued by keeping healthy and diseased swine together in the same lot, yard, or pen. A strict separation is necessary, at any rate of the greatest importance, because the effect of the infectious principle, like that of an accumulative poison, seems to increase and to become more intense or violent after each new influx; in other words, the disease, as a rule, will be the more malignant, and the time of incubation or period of colonization will be the shorter the greater the amount of the infectious principle introduced. Other disinfectants, such as salicylic acid (rather doubtful) and thymol have probably a similar effect as carbolic acid, but are rather expensive, and therefore can be made use of only on a small scale.

7. Salt and ashes, sulphate of iron or copperas, sulphur, assafœtida, black antimony, lime, coal, carbonate of soda, soap, oil of turpentine, and quite a number of other similar substances, singly, and in various combinations, have been used very extensively by a large number of farmers in different parts of the State, and at different seasons of the year, but notwithstanding diligent inquiry, I have failed to learn of a solitary case in which any of those substances, or any combination of them, has produced favorable results, or in which their use has been followed by a decrease in the mortality that might not be ascribed more reasonably to other causes. Sulphate of iron especially is of no value, neither as a preventive nor as a remedy. Mr. Bassett, an intelligent farmer in Champaign County, tried it thoroughly, and has used it extensively, and lost 96 per cent. of a very nice herd of shoats. Others have met with similar results.

8. As to a treatment of the diseased animals, there can be no doubt that a good hygienic treatment—a strict separation of the diseased animals from each other, so as to prevent any further influx of the infectious principle is advisable. Swine diseased with the plague evince very often a vitiated appetite for the excrements and the urine of their companions, and as these excretions contain immense numbers of Schizomycetes,

SWINE PLAGUE.

Lung of Experimental heifer, showing hepatization

A Hoen & Co Lithocaustic Baltimore

External surface of left lobe of lung of Mr. Morris' pig

A Hoen & Co Lithocaustic. Baltimore.

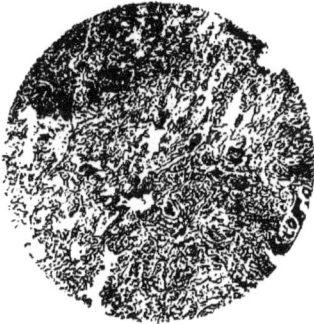

N°1.

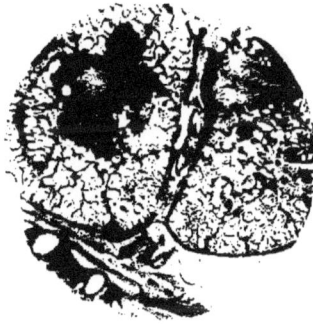

N°2.

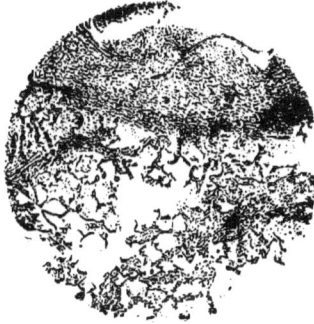

N°3.

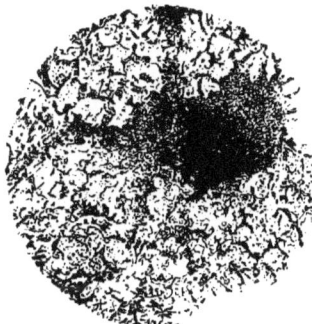

N°4.

N°5.

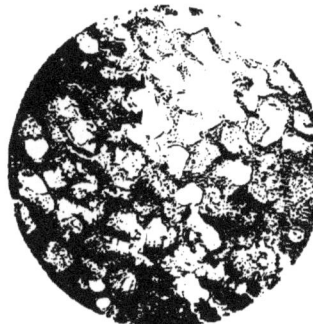

N°6.

Nos. 1, 2, 3, 4, and 5 are thin transversal sections of lung stained in Kleinenburg's solution. No.6, same, but not stained.

s from the bronchia
ig, killed by bleed-
ne IV, exam.at 6 P.
10.
es.

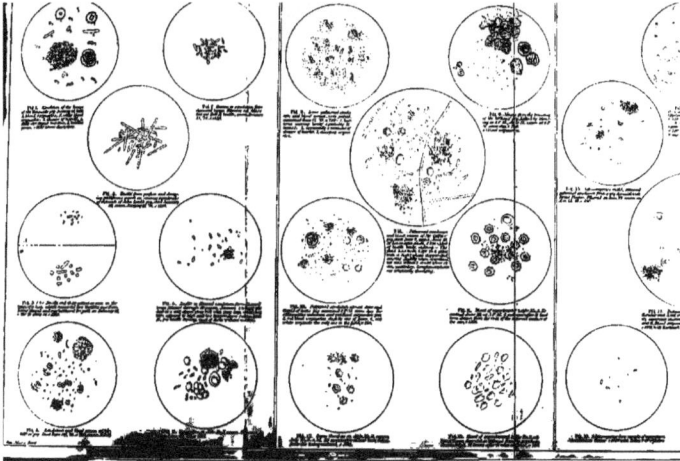

spherical and rod-shaped, and are therefore highly infectious, more and more infection or disease-producing elements will be introduced into the animal organism if that vitiated appetite is satisfied. Clean quarters and clean troughs (it is very important to clean the troughs after each meal), clean and fresh well-water to drink, clean food to eat, reasonable and adequate protection against the inclemency of the weather (against heat as well as against cold, rain, snow, &c.), and pure air to breathe will go a good way and may save many an animal.

As to a medical treatment, it would be necessary, if anything at all is to be accomplished, to subject every individual animal to a special treatment, dictated by circumstances; but as this is impossible, especially in a large herd of swine, not much can be expected from the use of medicines unless a "specific" is discovered that is simple of application. If such a specific remedy is existing, one would suppose, from the nature of the disease, it must be among the antiseptics or disinfectants. I have tried several of them, and so far have met with very poor results, as most of the animals thus treated have died, probably because the morbid changes had become irreparable. Hyposulphite of soda especially, which afterwards, in Mr. Graham's herd, proved to be very effective, even under aggravated circumstances—at least, its use was attended with very satisfactory results—and which is comparatively cheaper, and easy of application, was used extensively in three different herds, but failed to produce any visible good results, and so with all other medicines that were tried.

All the medicines, secret and otherwise, used so far—and their number is legion—have not done a particle of good, or, if they have, I have been unable to hear of it. Usually those farmers who have used the most medicine, or the greatest variety of medicines, have lost the largest number of hogs, possibly because, relying upon the medicines, they neglected all other sanitary measures. Good results, in somewhat mitigating the morbid process and improving thereby the chances of recovery, have been produced by feeding boiled, cooked, or steamed food, and also by feeding animal food. The wholesome effect of the former seems to be due to the fact that in cooked, boiled, or steamed food, if fed as soon as cold enough, no disease-producing Schizomycetes are apt to be existing, and if it is fed exclusively none are introduced through the digestive canal into the animal organism. Animal food has had in some cases a good effect, probably because it is rapidly digested, and a rapid digestion, it seems, is not favorable to an introduction of the disease-producing Schizomycetes into the animal organism by means of the digestive canal. Still, feeding animal food constitutes by no means a sure protection, because hogs fed in slaughter-houses, and hogs fed with the offal from a hotel-table (for instance, those belonging to the Doane House, in Champaign, in the fall of 1878) became affected and died of swine plague.

Very respectfully,

H. J. DETMERS, V. S.

Chicago, Ill., *February* 28, 1880.

REPORT OF DR. JAMES LAW.

SECOND SUPPLEMENTAL REPORT ON SWINE FEVER.

Hon. WILLIAM G. LE DUC,
Commissioner of Agriculture :

SIR: At the time when I made my first supplemental report, several of the experiments referred to in that paper were incomplete, while others had just been started, so that it becomes necessary to furnish a second addendum to give the final results of my observations. As the simplest mode of dealing with these supplementary facts I shall refer · to them *seriatim,* beginning with those which are merely complementary of the last report.

INFECTION BY COHABITATION.

In my last report an instance of this kind was furnished, and a deduction made that the disease was most virulent when at its height, inasmuch as that the exposed pig seemed to resist the contagion from an animal in process of convalescence, but speedily (in twelve days) fell a victim when placed along with a pig in which the malady was actively advancing. In the present report (No. 1) is given the necropsy of the pig infected by such exposure. The characteristics of the disease were sufficiently well marked, for though the bowels showed little more than a catarrhal inflammation, with an excessive secretion of glairy mucus, dirty, greenish-black pigmentation and two small circular blood extravasations, yet the other organs presented distinct swine-plague lesions. Thus, there were the characteristic blotches on the skin, the petechial discolorations on the heart and in the kidneys, the deep purple patches on the liver, and above all the pigmentation or deep red congestion of the groups of lymphatic glands and all parts of the body. The significant congestion of the lungs was also present. The only remarkable feature of the case was the excessive bloody engorgement and enlargement of the spleen, which is a constant feature of malignant anthrax, but is usually found in other affections (malarial fevers, septicæmia), in which there are profound changes in the blood. While, therefore, this lesion is an unusual one, yet it is one to be expected in this disease whenever the destruction of the blood globules or material changes in the albuminoids of that fluid reach a certain point of extension. In an animal that has been exposed to the infection of the swine-plague and which presents all the other characteristic lesions, this one superadded manifestation must be accepted as only implying a more than ordinarily profound modification of the blood elements.

POST-MORTEM EXAMINATION OF INFECTED LAMB.

In my last report I gave the record of the inoculated lamb up to the end of January. (See supplementary report: pages 101 and 112.) I now add No. 2, the remainder of the record and the necropsy of the same. The intestinal irritation and catarrh as manifested by the tenderness of the anus and the mucous discharges with the feces, together with the elevated temperature and enlarged lymphatic glands, presented much in common with the affection in the pig. The marked eruption in the ears might be accepted as representing the skin lesions of the pig.

After death the more characteristic lesions were the purple mottling of the liver, kidneys, and heart, the grayish consolidation of portions of the lungs, and the deep pigmentation of the lymphatic glands in general. The nodular caseous masses scattered so profusely along the coats of the bowels in this case, and which are far from uncommon in sheep, appear to consist of diseased and overdistended mucous crypts, and cannot be held as in any way connected with the contagion of the swine fever.

The absence of acute lesions, like red congestion of the lymphatic glands, in this lamb may be partly accounted for by the mildness with which the disease manifested itself, and by the fact that nearly four weeks had passed since the last inoculation, and three weeks since the last manifestation of abnormally high temperature. That the lamb suffered from the poison may be safely assumed from the fact that the pig No. 4 sickened with the specific fever after inoculation from it.

POST-MORTEM EXAMINATION OF THE INFECTED MERINO NO. 3.

Like the lamb, this was left with an imperfect record in our last report. In this the life record is completed and the necropsy given, with results very similar to those furnished by the lamb. Here again the main changes consisted in purple mottling of the liver and heart, and the deep pigmentation of the lymphatic glands. The yellowish-brown coloration of the kidneys implied antecedent changes probably of the nature of inflammation or extravasation. The caseous rounded masses found in the bowels of the lamb were remarkable here by their absence in further corroboration of the remark that these are independent lesions resulting from pre-existing disease, and in no way connected with that now occupying our attention.

PIG SUCCESSFULLY INOCULATED FROM SHEEP AND LAMB NO. 4.

(See supplementary report, page 101.)

This experiment, referred to in the text of the last report, is now furnished *in extenso* with the necropsy of the infected pig. It will be seen that the pig was inoculated twice, at an interval of fifteen days, with the mucus from the anus of the infected sheep, and one with scabs from the ear of the lamb. Enlarged lymphatic glands were observable before the last inoculation, and six days after it there was a febrile temperature, and the more violent manifestations of the complaint.

After death the following characteristic lesions were observed: The intestines had patches of congestion, the follicles were enlarged and the rectum ulcerated; purple discolorations were present on the liver, kidneys, and heart; the lymphatic glands were enlarged and congested, of a deep red, in some cases almost black.

This evidence as to the nature of the disease was clear enough, but to substantiate it the following experiment was undertaken:

SUCCESSFUL INOCULATION FROM THE PIG INFECTED BY THE SHEEP NO. 5.

The pig designated as No. 5 in this report was inoculated with scabs from the ear and eyelids of pig No. 4, and though this caused little change of temperature, it was followed by all the other prominent symptoms of the disease. On *post-mortem* examination on the twelfth day after the inoculation, the characters of the plague were found well marked. The red and black blotches on the skin were extensive, the ears blue, the intestines extensively congested with enlarged follicles in the cæcum and colon and blood extravasations and ulcers in the rectum. Purple discolorations and petechiæ were numerous on the liver, kidneys, and heart, and finally the lymphatic glands in general were in part congested of a deep red, and in part pigmented as the result of a previous congestion.

Here, then, in the second generation of the poison from the sheep we have the symptoms as well marked and the course of the disease as rapid and severe as in its first remove from the ovine subject.

PIG SUCCESSFULLY INOCULATED FROM THE INFECTED LAMB.

In experiment No. 6 of the present report is given the record of a pig inoculated from the swelling in the axilla (near the seat of inoculation) of the lamb. The inoculation produced fever, with the general malaise, moping, peevish grunt, inappetence, and cutaneous blotches of the swine-plague. When killed, on the eleventh day, the skin presented a number of red and purple blotches, and was covered with the black unctuous exudation so frequent in this disease. The bowels contained patches of congestion, the cæcum and colon enlarged, and the follicles and the rectum ulcerated. The liver, kidneys, lungs, and heart had purple blotches. Finally the lymphatic glands in the abdomen were enlarged and congested of a deep red or black, and those in the chest and guttural region darkly pigmented. This is as unequivocal a case as those already recorded, and fully confirms our position that the virus of this disease may be transmitted through the sheep and conveyed back to the pig with active effect.

INOCULATION FROM RAT AND LAMB NO. 7.

In my last report, page 101, I reported suspicious lesions in a rat inoculated with swine-plague virus and more characteristic symptoms in a pig inoculated from this rat. In No. 7 of the present report will be found the full record of the pig in question. By a reference to this record it will be seen that without much elevation of temperature this pig showed a purple cutaneous eruption on the fifth day and enlarged glands on the twelfth, when it was inoculated with bloody mucus from the anus of the infected lamb. After this the symptoms became much more severe, and when killed, twenty-two days after, the pig showed unequivocal symptoms of the affection. Whatever may be concluded as to the result of the infection from the rat in this case, it is at least a further corroboration of the position that the inoculated sheep is infecting. To further test the susceptibility of the rodent, the following experiments were undertaken:

INOCULATIONS FROM PIG, RAT, AND LAMB.

The subject of this experiment was a female Suffolk pig presented by Cornell University, having been the smallest of the litter. It was about three months old, small for its age, and very fat and sluggish. It was first inoculated with albumen which had been charged with a drop of blood containing bacteria, from pig No. 13 (see report page 90), and had been cultivated in three succeeding portions of albumen drawn on each occasion from a fresh, newly-broken egg, through a tube that had been previously heated to redness to destroy all organic life. For fifteen days nothing more was shown than a few purple spots and patches on the rump, tail, and hocks.

The subject was again inoculated with the congested intestine of the rat which had died two days after the inoculation. The intestine had been frozen over night. For thirteen days more the same equivocal symptoms continued.

A third inoculation was now practiced, this time with bloody mucus from the anus of the lamb, diluted with a weak solution of common table-salt.

Twenty-two days after the third inoculation the pig was sacrificed, and beyond some pigmentation of the lymphatic glands presented no distinct lesions that could be held characteristic of the specific fever. In short, the animal had suffered so slightly, if at all, that it might well be set down as a case of insusceptibility. This is only what was to be expected, as in the case of all plagues and contagia a certain number of animals will successfully resist exposure and escape, though the infection is most virulent and concentrated. The number of my subjects was too small to allow of any satisfactory general estimate; but, so far as it goes, it shows one insusceptible animal in twenty-five, or at the rate of five per cent. It may, however, be questioned whether the pigmentation of the lymphatic glands did not imply a previous mild attack of the disease, and whether the apparent immunity in the later inoculations was not due to the protective influence of the previous illness.

SUCCESSFUL INOCULATION OF A RAT—NO. 9.

This subject was inoculated February 5, 1879, with virulent matter that had been preserved for seventy-eight days, closely packed in dry wheat-bran. The rat was preserved for thirteen days, and finally killed February 18, and dissected immediately after death. The guttural lymphatic glands were deeply congested, so as to be mottled with red. The inguinal glands had a brownish-red hue, the sublumbar lymphatic glands were enlarged and pigmented. The small intestines presented patches of congestion and redness. The right lung was in greater part red and consolidated. The liver was deeply mottled with purple, and the kidneys of a very dark red externally. Finally, the spleen was greatly enlarged and gorged with blood.

Taken all in all, these symptoms are so closely in keeping with those of swine plague that there seemed no reasonable grounds for doubting that it was really this disease. The one drawback to this conclusion is the condition of the spleen; but the enlarged and blood-gorged condition of this organ is not unknown in the pig itself, as shown in No. 1 of the present report. A second reason for not attaching undue importance to the engorged spleen, nor accepting it as indicative of malignant anthrax, is that the pig inoculated from this rat developed all the symptoms of the hog-fever, while the spleen was rather shrunken and puckered than enlarged. Had it been inoculated with the virulent products of malignant anthrax, engorgement and distension of the spleen had been inevitable.

SUCCESSFUL INOCULATION FROM THE RAT.

On February 19 a healthy pig was inoculated with the congested lymphatic glands and lungs of the above-mentioned rat, the morbid products having been inserted in a pouch under the skin.

On the sixth day there was much malaise, with redness of the skin and the appearance of the black unctuous exudation on the ears and legs. These went on increasing, and black spots and patches, ineffaceable by pressure, appeared on the inside of the thighs and hocks.

The subject was destroyed on the twentieth day and showed the usual symptoms of the disease. The stomach and bowels were congested, with glandular swellings in the large intestines, the lymphatic glands corresponding to the congested bowels were of a deep red, almost black, and elsewhere the lymphatics were enlarged and pigmented. Purple blotches appeared on the liver, heart, lungs, and air passages, while the spleen was small, rather bloodless, puckered, and shrunken.

The symptoms of the disease were, in short, as unequivocal as when inoculation was made from the sick pig direct, and, taken along with the less conclusive evidence furnished by case No. 7, may be held to prove that the rat is capable of contracting this disease and of conveying it back to the pig.

PROBABLE CONVEYANCE OF THE DISEASE BY RATS.

In my report for 1878 I expressed an apprehension of this disease being conveyed by rats, which fear is only too fully justified by the more recent developments. The danger of the conveyance of diseases by these vermin not only from pen to pen, but from farm to farm, can never be lost sight of, as rats do not by any means confine their depredations to a circumscribed locality, and are quite ready to emigrate and found a new colony if their present habitat is unproductive or closely beset by their natural enemies. Rats, therefore, that pass from one piggery to another may convey the specific poison on their surface or in their systems, and may not only leave the germs in the troughs while sharing the feed with the pig, but even inoculate it direct while gnawing the horns of its feet.

The importance of exterminating rats from the vicinity of piggeries cannot be too strongly insisted on. Rats are probably the main source of trichinia spiralis in pigs, as the infected rat, with its muscles wasting as the result of the lodgment in their fibers of myriads of the encysted trichinia, becomes correspondingly weak and inactive, and is easily caught and devoured by the omniverous animal. If, then, we take the observation of Dr. Bellfield and Mr. Attwood as a basis, and accept as a fact that 8 per cent. of the hogs killed in Chicago are trichinous; add to this that the discovery of these worms in American hams and bacon has led to the closure of several European markets against these products; and, finally, that this specific fever of swine may be contracted and conveyed by rats, and we have cause enough for the closer supervision of the breeding of swine, and for a systematic destruction of rats wherever either trichiniasis or the swine plague has manifested itself. The swine breeders themselves should be warned against this source of disease and loss, but the sanitary authorities should follow up every case of trichiniasis and hog cholera, and see that the rats are not allowed to become active in its propagation.

INFECTION BY INOCULATION OF CULTIVATED VIRUS.

My last experiment was made with a material which might have been supposed to have been thoroughly disinfected. A little pleuritic fluid swarming with actively moving bacteria was added to some milk and boiled for five minutes. When cold, a drop or two of ammonia was added, the neck of the glass vessel plugged with cotton wool, and the whole placed in an incubator at 98° F. for two days. A second and third portion of boiled milk and ammonia were successively inoculated from the first, and second, and a little of the milk with the third generation of the cultivated poison was injected under the skin of a healthy pig.

The subject suffered from illness with red and purple spots on the skin and a greasy black exudation; was killed on the twenty-first day and immediately dissected. The stomach was extensively discolored, of a dark brownish red, becoming a bright red at the margins. The small intestines were congested and showed punctiform pectichiæ, especially on the duodenum and on the ilieum near the ileo-cæcal valve. The large intestine had enlarged follicles and patches of congestion, and the lymphatic glands of the bowels were discolored a deep red. Elsewhere the lymphatic glands were either reddened or pigmented. The lungs and liver showed little change, but there were purple discolorations on the kidneys and heart.

This case was evidently one of the specific hog fever, and, unless some source of fallacy entered, seems to imply that the germs of the disease may on certain conditions resist for a time the heat of boiling water. A single experiment is, however, too narrow a basis for the support of a theory, and I shall therefore content myself with merely recording the result, and leave the matter to be made the subject of a more crucial test at some time in the future.

Perhaps the most remarkable feature of these experimental inoculations is the fact that the pigs inoculated from the infected sheep and rats appeared to take the disease in a mild form, and in all such cases it seemed probable that, had the animals not been sacrificed by the knife, a recovery might have ensued. In the case of the pigs inoculated from the sheep, one (No. 4) was twice inoculated, thirty-eight and twenty-three days before it was killed, and although the disease showed itself in an unequivocal form, yet it was not severe and did not promise a fatal result. A second (No. 7), inoculated from the lamb twenty-one days before it was sacrificed, proved more severe, but did not reach a fatal result; a third (No. 8) was inoculated twenty-two days before its death, and showed such slight symptoms and *post mortem* lesions that it might have been questioned whether it really had the disease; a fourth (No. 5), inoculated twelve days before its death, had very slight symptoms and lesions of the disease; while in a fifth (No. 6), that lived for the same length of time after inoculation, the symptoms were more severe, but there was no certain indication of a fatal result.

Of the pigs inoculated from the rat, two (Nos. 1 and 7), inoculated from the first rat, had such slight symptoms that they were afterward inoculated from the lamb, and the third (No. 10) inoculated from the second rat, twenty days before it was killed, showed moderate but distinctly marked symptoms, and was manifestly improving when it was sacrificed.

Inoculations from the infected rabbit were more redoubtable. One pig (No. 8, Add. I), inoculated twice from the rabbit and killed on the fourteenth day after the last inoculation, was suffering severely and might

have died. Another (No. 9, Add. I) was so ill on the twentieth day
after the inoculation that he could not have survived many hours longer.

These facts point to the most important conclusion that the poison
of the swine plague, when passed through the system of the sheep or.
rat, becomes lessened in virulence, and usually conveys the disease
back to the pig in a non-fatal form. Should this be sustained by
further experiment, and should this, like some other bacteridian diseases, so affect the system that a second attack is rendered much milder
or entirely prevented, it will open the way for a system of vicarious inoculation that will save our swine breeders from the yearly losses of tens
of millions that now threaten the very existence of this industry. It is
noticeable that the pig (No. 5) inoculated from the pig infected from the
lamb, and therefore by poison the second remove from the ovine subject, though showing symptoms of the disease, did not suffer severely
in the twelve days it was allowed to survive, so that the mitigation of
the poison may remain for some generations after it has once passed
through the sheep.

That a further inquiry in this direction promises valuable results
may be further deduced from recent developments in anthrax and
chicken cholera.

In February, 1878, Burdon Sanderson and William Duguid, at Brown
Institution, London, inoculated guinea-pigs with the poison of the anthrax, and conveyed the disease from guinea-pig to guinea-pig for several generations of the poison. From different guinea-pigs it was
inoculated back upon two yearling heifers and a six months' calf, and
in all produced active disease, but in no case with a fatal result. Recovery in all cases might be said to have occurred by the fifth day.
The liquids from the guinea-pigs thus inoculated on the cattle were subjected to the counter test of inoculation on other guinea-pigs, and in all
cases with fatal results. To test the effect on the system the calf was
reinoculated fifty-two days after the first, and the heifers nine days after
the first, with the effect of producing a milder attack on the heifers, and
a severe but not fatal illness in the calf.

Again, twenty-five days later, they were inoculated with anthrax
poison, cultivated in grain infusions, which makes a most virulent and
fatal preparation; but, though two sickened, in none did a fatal result
ensue.

Dr. Greenfield, who was in charge of the Brown Institution in 1879,
continued these experiments. A steer was inoculated four times in
succession with anthrax poison from the guinea-pig with steadily-
decreasing results, and then a fifth time with blood from an anthrax
sheep; but he survived all, and did well.

A six-months' calf was inoculated with anthrax fluids that had passed
through the guinea-pig and been afterward cultivated in an albuminous
fluid for four generations of the poison, and nine days later, with blood
direct from the spleen of an anthrax guinea-pig; but a recovery ensued
in both cases. An old emaciated and pregnant cow inoculated with
the blood of an anthrax guinea-pig died on the fourth day. Age, debility, and pregnancy were charged with the result. Finally, a sheep
was inoculated with the anthrax poison that had been passed through
the guinea-pig, and thereafter cultivated to the fourth generation in an
albuminous fluid; but the result was not fatal. It should be added,
that the anthrax liquids used on all these animals were tested by con-
temporary inoculations on guinea-pigs and mice, and invariably with
fatal effect.

As a sequel to these, it should be noticed that Pasteur claims to have mitigated the poison of *chicken cholera* and the bacteridian disease by cultivation in different fluids; and to have conveyed it back to the fowl, not only without producing a fatal result, but with the effect of rendering the system of the fowl unimpressible by the same poison for the future.

The close analogy between these two diseases and the swine fever in their mode of causation by bacteria suggests very strongly a common pathology for all; and as the mode of reproduction and development of the different bacteria which respectively cause the three plagues is probably the same or closely allied, the promise is held out that the specific swine fever may be anticipated and prevented, as the above experiments imply that the other two affections can be.

Nor are the above-named observers alone in their tentative results. Wernich, Bauman, and Neucki find it highly probable that bacteria are destroyed by certain products of the putrefaction to which they themselves have given rise; so that the continued existence or propagation of a specific bacterium in an individual system is rendered difficult or impossible by the previous generation of that microphyte in the same animal body.

But at this point still another question arises. In view of the mild effects produced by inoculating the cultivated virus (Nos. 8 and 10, present report), may the poison of this disease not be mitigated by cultivating it in particular solutions, so that when inoculated on the pig it will come short of destroying life, and yet prove a protection against the ordinary fatal form of the poison? Klein's cultivations were made in the aqueous humor of the rabbit, and though he has not stated how violent were the inoculated cases, yet it would not be surprising if they proved fatal, as did our own cases of inoculation from the rabbit. My inoculations with the swine-plague virus preserved in bran produced severe symptoms and a fatal result, in keeping with the virulence of anthrax virus which had been preserved in a similar medium. In my other cases, inoculated with virulent egg-albumen (No. 8) and ammoniated milk (No. 10), the resulting disease was moderate, and did not threaten fatal result.

While, therefore, it cannot be confidently affirmed that we can at will induce a mild form of this affection which shall protect the system against a severe one, we have in the above facts a sufficient warrant and inducement to carry this experimental investigation to a certain and reliable conclusion. It remains for the experimental pathologist to determine the exact conditions under which such immunity can be acquired, if at all, and how long the protection to the system is vouchsafed. From present appearances it seems oversanguine to expect of our legislators any sufficiently vigorous and persevering system of extinguishing our imported and indigenous animal plagues, so that it becomes the more desirable that we should bend our energies to ascertain what measures will rob the more prevalent ones of their terrible mortality, and if the plague germs must be produced and preserved in our midst, what will assure us that only the mitigated form of the poison shall be laid up, and not the deadly one, as heretofore.

JAMES LAW.

ITHACA, N. Y., *June* 10, 1880.

Poland China pig, infected by cohabitation. (*See experiment No. 6, Addendum I.*)

EXPERIMENT No. 1.

Date.	Hour.	Temperature of body.	Remarks.
1879.		° F.	
Jan. 31	10 a. m..	103	
Feb. 1	...do	102	
2	...do ....,	101. 5	
3	...do	102	
4	...do	101. 75	
5	...do	100. 5	
8	9 a. m..	102. 25	
9	...do	101. 25	
10	...do	102	
12	...do	102	
14	... do	101	Looking badly.
15	...do	100	
16	...do	99. 75	

Post mortem examination, February 17.—*Skin:* Deep red blotches beneath the belly, inside the fore and hind legs, under the jaws, and at the entrance of one nostril.

Digestive organs: Mouth natural. Submaxillary and gutteral glands congested and pigmented.

Stomach: Contains food meal, with a little hay and an excess of yellow viscid mucus.

Small intestines: Contains much glairy, yellow mucus, with some food in the lower part of the ileum. *Ileo-cæcal* valve is pigmented of a deep dirty green beneath the mucous membrane. *Rectum* has two small circular blood extravasations.

Liver: Mottled with purple spots. Gall-bladder full of dark green viscid bile.

Spleen: Enormously enlarged (eleven inches long by two inches wide at its broadest part). Gorged with blood.

Inguinal, circumflex ileac, pelvic, sublumbar, mesenteric, and *omental lymphatic glands* pigmented.

Kidneys: Left, natural on the surface; cortical substance brownish yellow; medullary substance with numerous purple ecchymosis. Right has patches of congestion on the surface of its outer border, and otherwise bears the same lesions as the left.

Lungs: Have congested lobules partially collapsed.

Heart: Endocardium mottled of different shades, from purple to yellowish brown. Above the tricuspid valve is a considerable straw-colored exudation.

Subdorsal lymphatic glands: Deeply congested and discolored. *Prepectoral and pre-scapular glands* in a similar condition, and surrounded by a semi-liquid straw-colored exudation. A similar exudation is found around the guttural glands in the pericardium and around the base of the heart.

EXPERIMENT NO. 2.

Long-wooled lamb. (*See No. 11, Addendum I, page 112.*)

Date.	Hour.	Temperature of body.	Remarks.
1879.		° F.	
Jan. 31	10 a. m..	104	Scours, passing much mucus; iliac glands enlarged.
Feb. 1	...do	104	
2	...do	104	
3	...do	103	
4	...do	104. 8	
5	...do	104	A hypodermic needle brought a greenish cheesy-looking matter from the center of the axillary swelling.
6	...do	102	Increased axillary swelling.
7	...do	103. 75	
8	...do	103. 75	
10	...do	103	
12	...do	103. 25	
14	...do	102. 75	
15	...do	103. 75	
16	...do	104. 25	

Killed by bleeding, February 18, 1879.—*Post-mortem* examination immediately after death.

Digestive organs : Tongue sound; *stomach* sound.

Small intestines and, to a greater extent, the cæcum and colon studded with hard, spherical nodules containing a caseous material, and some of them communicating with the cavity of the intestines by a narrow orifice. *Guttural œsophagean* and *mesenteric lymphatic glands* gray from pigmentation.

Liver: Mottled with purple and yellowish spots. Bile of a bright green. *Hepatic lymphatic glands* deeply pigmented. *Spleen* natural.

Kidneys: Cortical substance slightly purple on the surface. Medullary substance of a pale yellowish white, surrounded by a purple zone. Peritoneum contained three hydatids.

Lungs have a number of lobulets of a dark-red congested appearance, but still firm and tough. There are also a number of hard nodules of a dirty grayish color on the surface of the organ. No parasites. *Bronchial lymphatic glands* pigmented.

Heart: Purple spots on the endocardium of the right and left ventricle, especially the latter.

EXPERIMENT NO. 3.

Merino sheep. (*Continued from page 112, No. 10, Addendum I.*)

Date.	Hour.	Temperature of body.	Remarks.
1870.		° F.	
Jan. 31	10 a. m..	103. 8	
Feb. 1	...do	102. 5	
2	...do	103. 75	
3	...do	102. 75	
4	...do	100. 5	
5	...do	103. 5	
6	...do	102	
8	...do	103	
9	...do	102. 5	
10	...do	102. 5	
12	...do	102. 75	
14	...do	103	
15	...do	102. 5	
16	...do	102. 75	

Merino sheep: Killed by bleeding, February 18, 1879.—*Post-mortem* examination immediately after death.

Digestive organs presented nothing abnormal. *Mesenteric lymphatic glands* deeply pigmented.

Inguinal glands deeply pigmented, especially to the medullary portion.

Liver has purplish and reddish mottling on the surface.

Gall-bladder: Partially full of a bright-green bile.

Spleen: Normal.
Kidneys: Yellowish brown in cortical portion.
Right lung: In great part congested, of a bright red color.
Right heart: Has endocardium marked with purple spots.
Left heart: Mottled extensively with spots of a dark purple.

EXPERIMENT No. 4.

White male pig.

Date.	Hour.	Temperature of body.	Remarks.
1879.		° F	
Jan. 7	9 a. m..	102	
8	...do	100. 5	
9	...do	101	
10	...do	100. 75	
11	...do	102. 75	Inoculated with mucus from anus of sheep.
12	...do	104	
13	...do	102	
14	...do	103	
15	...do	104. 75	
16	...do	103. 3	
17	...do	102. 75	
18	...do	102. 25	
19	...do	103. 5	
20	...do	102	
21	...do ...	102. 5	
22	...do	102	
22	5 p. m..	102	Inoculated with scab from ear of lamb.
23	9 a. m..	101. 5	
24	...do	100. 5	Inguinal glands enlarged.
25	...do	101. 75	Inoculated with anal mucus from sheep and lamb.
26	...do	99. 75	
26	4. 30 p. m	102	
27	9 a. m...	101. 75	
28	...do	102	
29	...do	103	
30	...do	102	
31	...do	103. 75	Enlarged inguinal glands; purple spots on belly.
Feb. 1	...do	104	
2	...do	103	Off feed; livid spots on teats.
3	...do	102. 5	Livid spots; enlarged glands; unctuous secretions from skin.
4	...do	102	Off feed; pink papules at hair-roots; black skin exudation, concreting in scabs.
5	...do	102. 75	
6	...do	104. 5	
8	...do ...	103	
9	...do	102. 25	
10	...do	102. 75	
12	...do	102. 75	
14	...do	103. 25	
15	...do	103	
16	...do	103	

White male pig: Killed by bleeding, February 17.—*Post-mortem* examination immediately after death.

Digestive apparatus: Mouth healthy. *Guttural lymphatic glands* pigmented.
Stomach: Full of food; mucous membrane slightly congested.
Small intestine: Slightly congested at isolated points.
Large intestine: Has patches of congestion and enlarged follicles, the latter especially in the colon. Rectum bears an ulcer, the scab of which is marked on the outer coat of the bowel by a liquid exudation and a congested lymphatic gland.
Duodinal lymphatic glands very black; *Mesenteric lymphatic glands* pigmented, of varying shades of gray. *Rectal lymphatic glands* blood-red.
Liver: Mottled with purple spots. Gall-bladder full of dark-green liquid bile.
Spleen: Normal in size, but with yellowish-white shrunken portions along the edges.
Kidneys: Yellowish-brown on the cortical part, but with petechial spots on the surface and internally on the medullary portion.
Lungs: Normal. *Endocardium* mottled with purple spots. Subdorsal and internal pectoral glands deeply congested.

EXPERIMENT No. 5.

White male pig.

Date.	Hour.	Temperature of body.	Remarks.
1879.		° F.	
Jan. 30	10 a. m.	102	
31	...do	103. 6	
Feb. 1	...do	102. 5	
2	...do	102. 5	
3	...do	104	
5	...do	103. 5	Inoculated with scab from ear and eyelids of sick pig No. 4 (infected from sheep and lamb) and placed in same pen with it.
6	... do	104	
8	... do	103. 5	
9	...do	102. 25	
10	...do	103	
12	...do	103	
14	...do	103	
15	...do	103	
16	...do	102. 75	

White male pig: Killed by bleeding, February 17.—*Post-mortem* examination at once.

Skin: Dark-red blotch inside the left fore leg extending from near the carpus to the sternum. Bright-red spots over the anterior part of the sternum and inside the hocks, on the prepuce and lower part of the scrotum. Ears slightly blue.

Digestive organs: Tongue, tonsils, and larynx sound. Guttural and submaxillary lymphatic glands pigmented of a grayish color.

Stomach: Full of food, great curvature has its mucous membrane congested.

Duodenum: Congested of a deep red. Jejunum and ileum somewhat less so. Similar patches of congestion on the ileo-cæcal valve.

Anterior mesenteric glands the seat of dark-gray pigmentation.

Colon: Congested at intervals with many enlarged follicles. Rectum presents red discoloration and ulcers, one of the latter containing a blood-clot. Rectal and colic lymphatic glands pigmented, some red with congestion.

Intestinal parasites: Small intestines contain thirty-eight ascarides, one measuring thirteen and a half inches in length.

Liver: Has purple spots and patches, especially on the right lobe. Gall bladder is full of greenish bile.

Spleen: About natural. One spot of brownish-red congestion.

Kidneys: Have purple spots on their surface extending about one line into the cortical substance.

Heart: Left ventricle has large petechiæ on its internal surface, also on the edge of the mitral valve. Right healthy.

Lungs: Posterior border of the Linder lobe of the right lung is bluish and contains lung-worms.

Mediastinal lymphatic glands: Pigmented and congested. Some perfectly black.

EXPERIMENT No. 6.

Female pig.

Date.	Hour.	Temperature of body.	Remarks.
1879.		° F.	
Jan. 30	10 a. m.	103. 25	
31	.. do ...	104	
Feb. 1	.. do ...	104	
2	...do	104	
3	...do	104	
4	...do	103. 75	
5	...do	103. 5	Inoculated with dry, greenish, cheesy matter from axillary swelling of lamb.
6	...do ...	105	
8	9 a. m.	105	
9	...do	104. 75	
10	...do	104. 5	
12	...do	104. 25	
14	...do	102. 75	
15	...do	103	
16	...do ...	103. 75	

Killed by bleeding February 18.—*Post-mortem* examination immediately after death.
Skin: Purple blotch on the left flank over a globular caseous mass.   Under the black, unctuous cutaneous exudation red flaques appear upon the ear, also a slightly bluish tinge upon the nose.
Digestive organs: Mouth healthy.   *Guttural lymphatic glands* pigmented and enlarged.
Stomach moderately filled; contents very acid; considerable reddish and brownish discoloration of the mucous membrane along the great curvature.
Duodenum: Congested in its mucous folds, with thickening of the mucous membrane.
Jejunum and Ilium: Have patches of congestion and contain eight ascarides.   *Cæcum* and still more the *colon* have enlarged follicles.   Rectum shows congestion and one small ulcer.   Cæcum contains thirteen whipworms.
Mesenteric lymphatic glands: Congested of a deep red or black and greatly enlarged. Colic lymphatic glands perfectly black.   Sublumbar and inguinal glands darkly pigmented.
Liver: Has purple patches near the free border; its cut surface is yellowish brown.
Gall bladder: Full of orange-brown bile.   Spleen almost normal.
Kidneys: Right has purple spots on the surface, medullary substance and papillæ. Cortical substance less pale than usual.
Lungs: Have purple spots on their surface.   Subdorsal and bronchial glands pigmented.
Heart: Right side has a large loose clot and purple mottled endocardium.   *Left* ventricle holds a loose clot and many of its carunæ columnæ are black throughout, as if they were but clots of blood.

EXPERIMENT No. 7.

Female pig.

Date.	Hour.	Temperature of body.	Remarks.
1879.		° F.	
Jan. 7	9 a. m...	103	
8	...do	103. 5	
9	...do	103. 25	
10	...do	104	
11	...do	103. 75	
12	...do	102	
13	...do	102	
14	...do	101. 75	Inoculated in flank with congested small intestine of rat which died two days after inoculation from sick pig. (See page 101, Addendum I.)
15	...do	101	
16	...do	102. 5	
17	...do	102. 5	
18	...do	100. 5	
19	...do	103. 25	Purple spots on teats.
20	...do	102	Purple spots on teats and belly in size from that of a pin's head and upwards.
21	...do	101. 5	
22	...do	102. 23	Enlarged inguinal glands.
23	...do	102	
24	.. do	102. 23	Pink spots like pins' heads mostly around roots of bristles.
25	...do	101	
26	...do	101. 5	Inguinal glands materially enlarged.
27	...do	101. 5	Purple on teats.
28	...do	103	Injected 1 dram saline solution with bloody mucus from rectum of lamb.
29	...do	102. 8	Red spots like pins' heads along the belly.
30	...do	102	
31	...do	102. 5	
Feb. 1	10 a. m..	101. 75	
2	...do	102.75	
. 3	...do	102	Purple spots beneath breast-bone; inguinal glands enlarged.
4	...do	102. 9	
5	...do	102. 5	Purple flaques around the seats of inoculation; pink papules around the bristles inside the thighs, and on the belly; purple spots on one ear.
6	...do	101	
8	...do	102. 25	
9	...do	101. 75	
10	...do	102	
12	...do	103	
14	...do	102	
15	...do	102. 75	
16	...do	103. 5	

Female pig killed by bleeding February 18.—*Post-mortem* examination just after death.
Skin: Ineffaceable red spots one-third of a line in diameter on the belly, teats, in-

ner sides of the thighs, forearms, and ears. Bristles are very erect and rough. Skin covered with an unctuous secretion.
Digestive organs: Mouth, normal. Right guttural lymphatic glands enlarged and pigmented. Left, normal.
Stomach: Contains little food; sour; mucous membrane on the great curvature discolored, red and dirty brown. Small intestines with patches of congestion, especially along the folds. Large intestine has enlarged follicles and patches of congestion.
Liver: Bears purple patches. Gall bladder full of orange-brown bile.
Spleen: Nearly normal. Slightly shrunken and puckered at its thick extremity.
Mesenteric, gastric, hepatic, and mesocolic glands: Darkly pigmented, and some discolored of a deep red.
Kidneys: Nearly normal.
Lungs: With a few patches mottled of a deep red.
Heart: Left side, nearly normal; right side, with purple spots of ecchymosis.
Subdorsal and bronchial lymphatic glands: Pigmented and partially reddened.
Prepectoral and prescapular glands deeply pigmented.

EXPERIMENT No. 8.

Suffolk pig.

Date.	Hour.	Temperature of body.	Remarks.
1879.		° F.	
Jan. 1	9 a. m...	101.5	
2	...do	102	Injected hypodermically ¾ dram of inoculated albumen, 4th generation, in inoculation apparatus from blood of pig (experiment 13) which contained moving bacteria. In emptying and recharging the apparatus the liquids were drawn from a newly-broken .88 through a tube previously heated to redness.
3	...do	101.5	
4	...do	102	Purple spots on rump and tail; papules and flaques; purple patches on the hocks.
5	...do	102.5	
6	...do	102.75	
7	...do	100	
8	...do	101	
9	...do	103	
10	...do	102	Skin has many hard and brownish black scabs covering a red, slightly depressed surface.
11	...do	102.5	
12	...do	102.5	
13	...do	101	
14	...do	102	
15	...do	101.75	Inoculated with congested intestine of rat which had been frozen over night.
16	...do	103.25	
17	...do	101	
18	...do	102.5	
19	...do	101	
20	...do	101.75	
21	...do	102	
22	...do	101	Has not eaten its food.
23	...do	102	
24	...do	101	Pink spots on skin; black crusts; dung fetid.
25	...do	101	
26	...do	102	Purple spots on rump and thighs.
27	...do	102	
28	...do	101	Injected 1 dram saline solution with rectal bloody mucus from lamb;
29	...do	103	Tail has red spots; is soaked with urine and fæces.
30	...do	102	
31	...do	103	
Feb. 1	...do	100.5	
2	...do	100	
3	...do	100.5	
4	...do	102	
5	...do	102.5	Purple spots on ears.
6	...do	102.5	
8	...do	102.75	
9	...do	102	
10	...do	102.5	
12	...do	102	
14	...do	102	
15	...do	102.75	
16	...do	103.25	

Suffolk pig killed by bleeding February 18.
Tongue, especially in its posterior portion, furred of a brown color.
Stomach and intestines: Bore little evidence of change.
Lymphatic glands: Pigmented.
Liver: Discolored purple patches, and, towards the margin, yellowish staining. Bile, moderate in quantity, orange brown.

6 C D

Spleen : Small—a little puckered at the edges.
Kidneys : Very pale; firm and resistant, as if they had undergone fibrous degeneration.

EXPERIMENT No. 9.

Rat killed February 18, 1879.—*Post-mortem* examination immediately after death.
Guttural glands : Mottled with red and dark lines. Inguinal glands of a brownish red.
Right lung : Firm and gorged with blood. Left lung nearly natural. Liver deeply mottled with purple.
Spleen : Excessively large and gorged with blood.
Kidneys : Cortical substance of a very dark red ; medullary substance, pale.
Sublumbar lymphatic glands : Enlarged and pigmented.

EXPERIMENT No. 10.

White male pig.

Date.	Hour.	Temperature of body.	Remarks.
1879.		° F.	
Feb. 3	9 a. m ...	103.75	
4	...do	103.75	
5	...do	103.5	
6	...do	103.5	
7	...do	102	
8	...do	104	
9	...do	104.25	
10	...do	104	
11	...do ...	103.75	
12	...do ..	103.25	
13	...do	103.25	
14	...do	103	
15	...do ...	102.75	
16	...do	102.5	
17	...do	102.25	
19	...do	....	Inoculated to-day with the congested and reddened lymphatic glands and congested lungs of a rat (No. 9) which showed lesions corresponding to those of the swine fever. The infecting matters were inserted in a pouch formed under the true skin.
20	...do	102.5	
21	...do	102.75	
22	...do	102.5	
23	...do	102.75	
24	...do	102.75	
25	...do	103	Is very uneasy. Molasses-like exudation on ears and legs.
26	...do	103.25	
27	...do	103.50	
28	...do	103.75	Exudation increased and extended over nearly the whole body.
Mar. 1	...do	103.25	
2	...do	102.75	
4	...do	102.5	
6	...do	102.25	
7	...do	102.25	Exudation drying up.
8	...do	102	
9	...do	102.25	Shows much uneasiness.
10	...do	102	
11	...do	102.25	

Killed by bleeding March 11.—*Post-mortem* examination immediately after death.
Skin : Inside both thighs extending down to the hocks are discolored spots and patches, not effaceable by pressure. The molasses-like exudation on the skin is nearly dry on the body, but still soft and unctuous on the legs.
Digestive organs : Tongue healthy. Guttural lymphatic glands enlarged and pigmented.
Stomach : Has its mucous membrane mottled of a dark-purplish brown on its great curvature.
Duodenum : Slightly congested in its upper portion. Remainder of the small intestines present patches of slight inflammation. Ilio-cæcal valve normal.
Large intestines : Present small globular elevations like enlarged solitary glands. These are especially abundant in the colon.
The duodenal lymphatic glands : Of a deep red, almost black. Mesenteric lymphatic glands enlarged and deeply pigmented. Sublumbar lymphatic glands and the inguinal are similarly enlarged and pigmented.
Spleen : Normal, except that it is very firm and puckered along its border.
Liver : Firm. Patches of purple discoloration are seen, especially at the borders. Gall bladder full; bile of a bright-yellowish green.
Kidneys : Nearly normal. Cortical substance a little pale.
Urinary bladder : Full. Density of urine 1026.

Heart : Empty. Endocardium of left ventricle with numerous dark petechial spots.
Those are less numerous on the right ventricle, but of a deep purple color.
Lungs : Present petechial spots on the pleuræ and bronchi.
Parasites : Five ascarides in small intestines ; one hairheaded worm in cæcum.

EXPERIMENT No. 11.

White female pig.

Date.	Hour.	Temperature of body.	Remarks.
1879.		° F.	
Feb. 3	9 a. m ...	103. 75	
4	...do	103. 75	
5	...do	103	
6	...do	102. 5	
7	...do	103. 5	
8	...do	103. 75	
9	...do	102. 75	
10	.. do	103	Has been in *rut* for several days.
11	...do	102. 75	
12	...do	103	
13	...do	102. 5	
14	...do	102. 75	
15	...do	102. 5	
16	...do	102. 75	
17	...do	102. 5	
18	...do	....	
19	...do	....	Inoculated hypodermically with a solution of milk and pleuritic effusion of sick pig (both boiled) with ammonia, cultivated in isolation apparatus to the third generation._
20	...do	102. 25	
21	...do	102. 25	
22	...do	102. 5	
23	...do	102. 75	
24	...do	102. 5	
25	...do	102. 75	
26	...do	103	
27	...do	103. 25	A little exudation on the ears.
28	...do	103. 5	
Mar. 1	.. do	103. 5	Is very uneasy. Peevish grunt.
2	...do	103. 25	
3	...do	....	
4	...do	103. 25	
5	...do	....	
6	...do	103	
7	...do	103. 5	
8	...do	103. 5	
9	...do	103. 25	
10	...do	103. 25	
11	...do	103. 25	
12	...do	....	Killed by bleeding.

Post-mortem examination immediately after death.
Skin : A few purple discolorations on the inner side of the hocks. The molasses-like exudation has dried up into a black incrustation.
Digestive organs : Mouth and connections normal.
Guttural lymphatic glands : Slightly pigmented.
Stomach : Has several extensive dark-reddish patches on the mucous membrane covering the great curvature, shading off with bright red at the margins.
Duodenum : Congested along the margins of the folds of mucous membrane with patches of bright-red punctiform petechiæ.
Jejunum and ileum : Congested along the folds of mucous membrane, especially in the middle part of its course. Near the ilio-cæcal valve are bright-red punctiform petechiæ.
Duodenal lymphatic glands : Of a dark-red hue, almost black. *Anterior mesenteric* glands are deeply pigmented, and in many cases of a deep red.
Large intestine : Has follicles enlarged. These are especially numerous in the colon. The rectum bears patches of congestion and the lymphatic glands adjacent are of a deep red.
Spleen : Small and firm, ridged or puckered at its free border. Not gorged with blood.
Liver : Firm, nearly normal. Gall bladder filled with a bright, yellowish-green bile.
Kidneys : Nearly healthy. Medullary substance a little more highly colored than natural.
Lungs : Normal. Contains two lung worms.
Prepectoral lymphatic glands : Slightly pigmented.
Right inguinal glands : Of a deep red. Left the seat of grayish pigmentation.
Heart : Left ventricle deeply discolored internally by ineffaceable deep purple and crimson stains. Right ventricle normal. Right auricle contains a large clot.
Intestinal parasites : One ascaris in jejunum ; four tricocephali in cæcum.

CHARACTER OF SWINE PLAGUE IN THE SOUTHWEST.

By Dr. C. C. THORNTON, *Chew's Landing, Miss.*

At this time (June, 1879) charbon and hog cholera, so called, rage fearfully in this locality. I have myself lost three mules and over one hundred dollars' worth of hogs. I am satisfied the name of hog cholera is a misnomer, and I think I can very readily substantiate this from facts. With an instrument magnifying from 1 to 800 diameters, I have worked assiduously with the sole end in view of identifying these diseases, and I propose to give you the results of my labors, with the hope that benefit may be derived therefrom. With from 1 to 300 diameters, I find a dark and nucleated cell or cells, with dark outlines and a dark center. The cell appears circular at first, but in its different stages of aggregation or fermentation it assumes an egg shape from the piling of one upon another; or a better comparison, perhaps, would be that of a cauliflower or head of curled lettuce. The tissues of these cells appear ecchymosed. A red globule seldom appears after death, but yellow serum and dark grumous blood are generally found. The blood before death is pale and watery. A drop shows one half yellow water or serum, and the other half a disorganized clot of a purplish cast. This was particularly noticeable in a cow that died of charbon and hog cholera combined. I suggest the name of putrid measles for this disease, as I think it describes better the character of the malady, and will be more satisfactory to the profession. In some districts it should be designated as malignant or black measles, which is not considered so fatal. I am satisfied that it is a measles of the most virulent type, as mortification sets in almost with the beginning of the disease. The animal appears drawn up and drooping behind, is desirous of lying down, and seeks brush or a secluded place for concealment; shows a constant disposition to eat dirt; is hoarse and has a slight cough; manifests great thirst but no disposition for food; seems to have no blindness or brain trouble, but a great aversion to being disturbed. It is easily aroused, but is disposed to seek quiet and sleep. Notwithstanding this, intense pain is apparent, and the result is almost certain death.

As to the lesions. The cough, which is not constant or permanent, causes you to lose sight of the condition of the lungs, which are congested. The eyelids now become inflamed and the eyes watery, and the disease assumes somewhat the appearance of a catarrhal trouble. There appears to be no particular throat affection. After death an examination reveals a miliary eruption of the lungs, presenting the appearance of miliary tubercles, and frequently congestion and mortification of the same. Purple spots cover the abdomen and ears, and frequently extend to the lungs and liver and other internal organs. In the charbon case one half of the spleen was covered with fresh eruption, while the other half was but slightly affected. In another case the liver had advanced to a complete state of decomposition. These vesicles contained a yellow water similar to that found in other organs. The heart showed endocarditis, and the kidney was really the only organ not diseased, although its investing membranes contained yellow water. In the charbon case the liver was congested, with a yellow deposit in the sack. This bile was frothy, and appeared in a full state of fermentation. The bowels had similar eruptions in different stages of ulceration. The entire mucous membrane of the stomach was denuded, and exuded an unhealthy-looking secretion much resembling pus. The same cells I have tried to describe above were observed here. No worms showing animation, under a power of from 1 to 300 diameters, were discernible. I am satisfied this morbific product is generated by a process of fermentation in the blood, and wherever this germ is deposited there an eruption or ulceration appears. The stomach being the seat of great excitement may be the primitive seat of the disease. In rooting around where diseased animals may have been, these germs may be taken into the stomach, and thus produce the malady. The disease was on my place among some tenant hogs, all of which, with the exception of one boar, died. Two of my sows that were lousy, and in a bad state generally, bedded with the boar and took the disease. With the constant use of chlor. ammonia I kept one alive for two months, and I feel assured that this had a beneficial effect in the lung and throat trouble that generally accompanies this disease. Some had sore and swollen noses, and four sucking pigs had a pustular eruption much resembling chickenpox. If this eruption could be kept out, I believe it would prove beneficial to the animal. I have used coal-oil and turpentine with marked effect, a teaspoonful of turpentine or a tablespoonful of coal-oil in a little gruel. I have also used caustic potash and concentrated lye. Sulphur seems to increase and augment lung trouble, unless given in very pleasant weather. I believe the sulphites (bisulphite and hydrosulphite of soda) would prove of benefit, if properly administered.

In about all of the cases examined I found that mortification had taken place in some one or more organs before death had occurred. There were pleuritic lesions in one case, and adhesion of the lung in the case that lingered so long. In this case the lung and liver gave way and became purple and softened with vesicles or bullæ of yellow water. I have preserved sections of the different organs, which, if they could be forwarded safely, I would send to the department.

I am satisfied that my observations so far overturn the received opinions of this disease. The hog with the charbon complication presented the poisons above. Whether the cauliflower or curled lettuce cells were different poison, constant with one or the other disease, I am unable to say, as I could not give the case the attention I desired. I am satisfied that charbon is produced by a blood poison—that it is a morbid product introduced by the bite of horse-flies, which are now present in myriads as they were in 1868, when the disease was so prevalent and fatal. The punctures have the appearance of leech bites, though the tendency to decomposition is much more rapid, and produces gangrene in a very short time. Bruised or purple spots also appear on the muscular tissues and soon extend to the subjacent tissues, first on the same side and then on the opposite side. Horses and mules bitten badly by these flies show circumscribed swellings of small size. Some of these raised spots are as large or larger than marbles. A lady bitten on the arm by one of these pests was seriously affected. The arm became greatly swollen, turned purple, and was very painful from the shoulder to the hand. A gentleman bitten on the ankle was laid up for two weeks. The leg was greatly swollen and very painful. Now, if these single bites produce these symptoms in the human subject, how much more likely to produce similar troubles and poisonous effects in stock running out unprotected and at their mercy. In the mornings and evenings we have swarms of a small yellow fly which are very troublesome. Whether these are young horse-flies or not I am unable to say. Then we have the large horse-fly with green or striped head, and a smaller one that stings badly.

There is still another one which we call the cattle-fly—an extra large black fellow that looks much like a hearse or death carrier.

Since writing the above I have had another shoat die of the so-called cholera. I found purple spots on tongue and inside of lips, but no throat trouble. A portion of the right lung was hepatized, the liver congested, and the bowels ulcerated, inflamed, and mortified before death. The feces were impacted, and the large intestines inflamed with adhesions of outer surface. The sheath was contracted and the opening in the prepuce retracted, which caused a retention of urine. The bladder was greatly distended. The spleen and kidneys seemed free of disease, but there was much bile in the gall-sack. The lungs, bowels, and mesentery were greatly congested, and the abdomen had a gangrenous point over the ulcerated and mortified bowels. This hog was sick but a very short time. It was in perfect health, and took the disease three or four days after being exposed to animals affected with the malady. Two other healthy shoats were infected at the same time, and showed symptoms within from three to five days. The symptoms generally were the same as heretofore described. The character of the lesions does not seem constant in the lungs of those examined so far, except in a disposition to mortification. The extent of the hepatization would certainly have engendered mortification at an early period, as in the case of the bowels, and the wonder to me is how life can possibly last while such extensive and destructive lesions and decomposed organs, essential to the vital functions, exist. It is almost incredible that the animal should still live with the lungs, liver, and other vital organs in a pulpy and decomposed condition—with the bowels perforated, internal organs in many instances softened, and the intestines mortified.

*　　*　　*　　*　　*　　*　　*

I have had further opportunities of confirming my ideas of the nature of the two terrible scourges of the farmer, putrid measles and charbon. and I am now more convinced than ever that putrid measles better describes the disease than any other designation. That the disease is propagated by contagion has been clearly proven by confining hogs in an infected pen or with animals suffering with the disease. In two cases thus tried the time of incubation was found to be from two to six days. The first symptoms are those of catarrh with slight cough. This cough is not very constant or very severe. In some the eruption is clearly defined. in others it is shown on the mucous membrane of the under lip and on the abdomen. Some show slight ecchymosed spots, while others exhibit a deposit of a melanotic character, which seems to be confined to no surface, organ, or tissue. The lungs suffer most. If the animal resists the first stage of destructive infiltration of the lung tissues it generally terminates in a miliary eruption or deposit of the lungs of a cheesy, tubercular character, which produces further decay, as miliary tubercles generally do. This is the reason they fail to fatten, if they recover, and are afterwards useless as breeders. A few recover where the eruption is kept on the external surface.

The disease attacks hogs of every age; none seem to escape except pigs of a healthy mother. In dissecting a sow which had died of this disease I found the pigs in the uterus unmistakably affected with the malady. The pigs were but two months advanced and the skin free from hair, so that the eruption was plainly shown. The eruption showed cysts of a cheesy deposit even in the skin. A magnifying power of from four to eight hundred diameters failed to discover an animal germ as the cause of the trouble, but a deposit of a melanotic character, showing pigmentary cells and blood globules in a disorganized and decomposing state, with occasional pus and pyemic cells in a disorganized condition, were found. This melanotic deposit was also found in the spleen, as shown by an enlargement of its papillæ and a change in color. * In one case one half of the spleen was shown to be in a state of cicatrization, while the other half presented the ulcerate character of the disease. This papular eruption, when confined to the surface, ends frequently in a cheesy deposit, which eventually ulcerates and empties itself with but little disturbance. If driven in by exposure to inclement weather, like measles the trouble is carried to the lungs, giving them a mottled appearance, and portions of the lungs become congested with a dark grumous-looking blood. In two cases out of six examined a further state of decomposition and gangrene of the lungs was found. One exhibited considerable enteritis of an eruptive and ulcerative nature, impaction of feces, and ulceration and mortification of bowel and wall of abdomen, or peritoneal covering. All seem to have retention of the urine, though the kidney generally appears less diseased than any other organ. The appendages of the heart and liver are generally congested. In one or two cases bullæ were presented over the surface of the liver. The character of the inflammation in all the internal organs seemed of an eruptive nature, with papulæ in state of ulceration. When externally apparent these ulcerations look like a white center with an inflamed areola, which, under the microscope, have the appearance of melanotic deposits, clustered pigment cells, and blood globules in a state of decomposition. Much of the deposit had a greenish, dirty-brown appearance, and some ragged and even blackened with a greenish cast, like flesh in a state of decay.

As constant symptoms in the six cases examined there was found more or less congestion of the lungs in all, retention of urine and a disposition to gangrenous inflammation. The liver was more or less affected in all; biliary deposit in one, without any liquid bile; excess of dark bile in three; gangrene in lungs of four, of bowels in two, and of liver in four. Slight congestion of kidneys in only two, occasioned, I think, by intense inflammation of the bowels and mesentery; enteritis in four; miliary tubercles in two—the cases that suffered longest. I think in those cases that linger the longest the eruption takes a miliary or melanotic character, or both, as the deposit which, to the naked eye, appears white, under the microscope has a dark, dingy, dirty-brown appearance. These seem to be massed together with occasional black and greenish-black spots of decomposed blood and fibrin.

I have made other microscopical and *post-mortem* examinations of mules and hogs which had died of charbon, a disease which has raged here fearfully for the last month, and also on other portions of Yazoo and Sunflower Rivers. I find the disease no other than gangrenous or malignant erysipelas, of the lowest possible grade. This is a most direful scourge to the mule and horse race. As the disease progresses I find pus corpuscles and pyemic matter and cells. Not only this, but I find the fibrin and blood globules in the last stages of degeneration and decay. The gangrene has a greenish-brown and black appearance; the swellings are ecchymosed at times throughout; at others a patch of this is surrounded by an indurated swelling; the tissues are infiltrated with a yellow serum; the blood contains an excess of white corpuscles; and toward the end frequent pus cells, pyemic matter, and decomposed blood globules may be found. The red blood globule is greatly diminished in quantity to the other globules of the blood. At first the disease seems to be one of the subcutaneous tissues, but rapidly spreads to the internal organs, like malignant erysipelas. There is this difference, however: it seldom attacks the brain except when a dissolution is evident, then the symptoms indicate a general apopleptic condition of nearly every vital organ. The kidneys suffer but little, the lungs and heart most. The lungs soon become pulpy from the infiltration of the lung tissues, and the appendages of the heart are distended with blood of the blackest character; the arteries are filled with black blood, and the veins in many instances are empty of blood, but in a very short time after death are distended with gas. The blood, under a power of from 200 to 400 diameters, shows decomposition in the globules when taken from the living animal. At this stage gas is thrown off, which I suppose to be carb. ammonia, as chlorine water causes a white deposit over the surface of the blood when a drop is added to it. I made a *postm-ortem* examination of a mule which died of the disease. After taking out the lungs, and while washing other organs, they became glazed over and presented all the appearances of erysipelas, confirming my theory for years as to the true nature of the disease. A more marked case I have never seen in a human being, and I attended hundreds in hospital practice during the war. I certainly cannot be mistaken as to outward appearances, although I might be as to structural lesions, chemical changes, or microscopical appearances not constant or common with the disease. In a more minute study of these appearances I have discovered the deposit arranging itself like the crystals of a chemical compound, disengaging gas and assuming a nucleated form, with sometimes a crystalline appearance, or with crystals mixed through it. These were in every stage of decomposition and decay. Gas appears in the veins soon after death, showing the blood to be in a state of decomposition. It sometimes passes freely from the cut vein in bubbles, while its appearance indicates a turbid or poisoned condition of the fluid.

I find that the excrement of the horse-fly will thus decompose blood, as the mucus of the mouth, stomach, and intestines of the fly contains crystals of a similar appearance when observed under the microscope. It presents a palpable alkaline appearance to test-paper. What this alkaline is yet remains to be determined. I inclose you several specimens of these birds of torment to the horse and mule. The medium-sized brown fly I regard as the most irritating in its bites. The excrement from it is frequently of a greenish dirty mixture, and causes rapid changes when mixed with the blood of a healthy animal. Perhaps decomposition has not advanced so as to present the characteristic appearances of this matter from mouth to anus through the internal structure of the insect.

Preparations of iron seem to be demanded in this state or condition of the blood as being the most readily and easily assimilated. I find the muriatic tincture in sweetened water or molasses and water superior to everything else. The system demands iron, and I give it freely in small and repeated doses. One-half ounce of the tincture in a pint of sweetened water every two hours, with but little water to drink until the swelling shall have subsided. If the lungs are depressed and the urine suppressed I advise from one-half to one drachm tartar emetic, one-half to one ounce nitrate potash, and one ounce of cream tartar in a pint of bland fluid. I oppose blood-letting as injudicious and destructive, except in a local way, and also deep blistering or firing. Both are equally fatal in their consequences. So are hot applications and actual canteries. A superficial and rapid blister and free incisions I think advisable in all cases, and the quickest possible means of producing these the sooner will the condition of the animal be improved. The preparation of iron should be continued, and doses administered every two or three hours, as the necessities of the case may seem to demand. It may be found necessary to continue the iron preparations for some days, and even weeks, in order to thoroughly eradicate the poison and clear the system of its deleterious effects. The disease is one of the subcutaneous tissues and lymphatic glands, and the animal suffering with it should have entire rest and but little water.

REVIEW OF SPECIAL REPORT NO. 12.

By Dr. John M. McGehee, of Milton, Fla.

The reports of the commissioners appointed to investigate diseases incident to swine seem to be so analogous that their labors appear almost concomitant, at least so far as their means of investigation would seem to have been equal. Some of them, however, have reported a singular exemption of the lungs from disease, as shown in their examinations.

The history of the disease given by Dr. Detmers seems to be so very thorough in all respects, and as it is so generally supported by the other commissioners, I may as well confine my criticisms to his report; and as I have had no experience with the disease, unless it may be considered an experience to have always kept it away from about me, I have no criticisms to make concerning any part of his report except as relates to his treatment; and my views must be predicated upon his statements and my own knowledge of pathology and therapeutical agents generally.

His investigations point clearly to a specific infection, which he defines to be an animalcule which lodges only in the mucous membranes of the alimentary organs or on exposed capillary tissues, yet the question will arise, Are the lungs exempt from all liability of contracting the disease?

The infection once lodged the whole vascular system is affected in proportion to their evident reproduction. They produce the most widespread inflammation and attending consequences, the excretory organs singularly escaping, the liver standing heroically at its work until all of its material is exhausted. The absorbents, overloaded with work, are clogged and engorged, the invading *bacilli* destroying the adipose tissue and the albuminous elements, of which latter they rob even the bones. This description is rather figurative than strictly scientific.

Such are the conclusions which these reports clearly teach us. The indications, then, are to destroy the animalism doing the mischief as soon as possible. Of course my knowledge of the remedies necessary must be comparative. But with the sincerest respect for their position, and the highest regard for their skill in this investigation, and gratitude for their most valuable discoveries, which all must feel who properly understand them, I must say that not one of the commissioners have used the remedies that I would have used, or in the manner in which I would have used them. And I invite the most rigid criticism of my suggestion of remedies.

In this disease it appears that the lower portion of the bowels are the most seriously affected—perhaps the first lodgment is here. This portion of the bowel is less active than the upper portion, and is less affected by the gastric juices. These facts suggest that the infection is favored by a condition of things as nearly in a state of rest as possible, and otherwise negative. If, then, it be that the virus lodges first in the lower bowels, ordinary doses of an acrid poisonous nature given in such proportions as prudence may recommend would be so diluted by the juices of the stomach and upper bowels as to have but very little poisonous or acrid effect on the infectant lodged in the lower bowels. These remarks apply to the disease, perhaps, before any visible symptoms appear, and Dr. Detmers has given us only the visible evidences of the incubation, which only can be discovered. The sensibility of the hog being weak, before any constitutional disturbance would be noticed the multiplication of the *bacilli* would amount to millions. Then our remedial agents must be directed to reaching the infection after it has reached the fluids and produced evident constitutional disturbances. Assuming, then, that the infectious principle is an animal organism, how to destroy its vitality is the question for consideration.

Dr. Detmers proposes only one agent which seems to promise any remedial effect at all—and that is carbolic acid—and so far as the manner he proposes any probable benefit to be derived from it he might as well have proposed fire; yet, with some exceptions, his deductions from his pathological statements are good. One very grave error seems to be that he has a theory, and leaps too far from cause to remote effect to establish it—a common error with almost all men of whatever trade or profession. He speaks of the proposition of "quacks" to cure the disease, and the morbid changes which must be repaired, as though these changes were invariable and always present in the commencement of the disease. If such were the fact there could be no recovery in any case, and yet many cases do recover, although generally thereafter valueless. And yet there are marked cases to the contrary, as, for example, those who buy them to fatten, supposing to possess some immunity from a second attack. He may, however, suppose the quacks to propose to cure under all circumstances. Such a construction would be placing their unlearned pretensions in a very ridiculous light—perhaps in a very unfair position. This would be unnecessary, as in the absence of learned veterinarians we should encourage, as we often do receive much service from the multitude.

Dr. Detmers speaks of a general hygienic plan of treatment as far more promising of good effects than any agents of *materia medica*. If the physiology of the hog is analogous to that of the human race, and his pathology as given is correct, I think his inferences are decidedly wrong. I understand his pathology to be the introduction of a foreign element into the circulation which produces the most intense inflammation, and all the morbid changes of which he speaks are entirely consequential. Then the treatment must begin through the medium of circulation ; the curative agent on which he most relies is carbolic acid. It is needless to discuss the action of all acids on the chyle ; the profession understands that, and the common people require only the leading facts. "Who," the doctor asks, "has the audacity to assert that he is able to destroy these *bacilli* and their germs without disturbing the animal economy to such an extent as to cause the immediate death of the animal ?" Then what can carbolic acid be administered for ? All the commissioners speak of the great destruction of the animal tissue by the disease. Acids are the most powerful agents to interfere with the functions of replenishing these tissues.

Without criticising farther the treatment of the commissioners, and expressing here my opinion that their agents are simply temporizing and dilatory, I shall proceed to give my views of the proper treatment of the disease as indicated by the most thorough and satisfactory diagnosis of the commissioners themselves, and in doing this I shall speak of those therapeutical agents as I *know* them to be from my own personal experience.

I have for several years past endeavored to abstract my mind from all things relating to medicine, and I shall, therefore, have to rely upon ineffaceable recollections of my experience, which will hardly entitle my opinions to the respect due to members of the profession.

Whether we propose to destroy the *bacilli* already in the fluids, or whether we propose to treat the disease as an intense inflammatory diathesis, involving the whole animal tissue, the most potent remedies known to the profession for either are alike indicated; that is, the introduction into the system of mercury in some shape or form. Recognizing the practical wisdom of placing this subject under the charge of learned veterinarians and other scientific men, and supposing that this investigation will be continued until some satisfactory solution is reached, my propositions will be very general. In the early stages of the disease the bowels are observed to be constipated. It is well known that all the animal organs are in the most effective condition for health when they are in a soluble state. Therefore, when a purgative is needed I would advise that mercury be its principal component. This first indication being accomplished, I would advise alterative doses of calomel to be carried to such an extent as to decidedly impress the whole system with its specific influence. To accomplish this the more certainly and speedily I would advise a bath of warm water containing a certain portion of mercury (bichloride) and rubbing the inguinal and axillary regions with mercurial ointment.

In the early stages of the disease it might become necessary to give an emetic and even to bleed, in order that the system may be more readily and speedily brought under the influence of the mercury.

I have heard it stated with positive reliance that strychnine is a valuable curative agent in the affection. It may be, though contrary to Dr. Detmers' hypothesis, that the poison does enter the circulation and destroys the germs before destructive morbid changes have taken place; mercury I know will do it.

The white blood cells seem to be the first consumed by the disease. The great loss of adipose and fibrous tissue it would seem would theoretically have suggested to Dr. Detmers a food of this kind in his dietic treatment. I would even theoretically suggest a trial of raw eggs as an experiment in some cases.

Whether influenced by the prejudice of a preconceived theory or not, Dr. Dunlap insists that the disease is typhus in some form or other. There may be some cases of typhus in which the alterative action of mercurials is not advisable as a remedial agent. I certainly have never seen one. I have never seen it given in such cases as an alterative without discovering some benefit from it, and I have had my views placed under the severest test in the same house, in the same family, and at the same time, and in seven cases the expectant treatment lost four and the mercurial none. And yet I have ever since been held responsible for these deaths, because I prescribed it with certainty, and did not set aside the orders of those who had a better right to direct. Whether the views of Dr. Dunlap, or the diagnosis of Drs. Detmers, Law, and others be correct, I see no cause for any very material change of treatment.

In all my experience in the treatment of diseases, I can state with certainty that in all cases of active inflammation I have never had any cause to regret anything so much as a temporizing policy. It is possible that our therapeutics may not apply in the cure of diseases incident to swine, as we are justified in inferring from comparative anatomy. The pathology and diagnosis of the commissioners seem to be very thorough and clear. The treatment and prevention are the next points to be considered. These can only be decided by experimentation. In reviewing the history of the preventives as reported by the commissioners, though very contradictory, it is nevertheless instructive, and in some respects satisfactory. It is very apparent that the disease is extremely variable in its intensity, and I think that nothing is clearer than that this intensity is to a very great extent owing to the diet. The inference we would naturally draw from the facts stated is that those hogs whose food was of such a character that it tended to constipation were the most severely and fatally affected. Perhaps it is only necessary to refer to two cases, which seem to me illustrative to some extent. In the case referred to by Dr. Voyles (p. 119), of the toll-gate keeper who had a few hogs, and they all escaped the infection. Now the common surroundings of persons so situated generally are that they have a little garden patch well tended; that they generally keep a house of entertainment and refreshment. The result is a slop-tub and a considerable amount of greasy slops. Under such circumstances hogs become truly domestic animals. I have long since observed that the thrift of hogs which get slops in quite small amounts was due more to the hygienic effect of the slops than to its nutrition, perhaps by keeping the bowels open, and thus facilitating the passage through of the germs before they have time to colonize, or it may be that they are destroyed by the slops before they find a lodgment in the bowels.

. The next point is the statement of Dr. Dyer in the case of Mr. Green's hogs, which were all fed on grass, and in which case the death-rate was very great. All farmers know the effect of grass as a food in constipating the bowels of hogs. So far as my information serves me, those animals that live in the woods entirely on such food as they can obtain generally suffer most with the plague. I will here state certain facts which may prove instructive. In those years in which mast is very abundant (notably post-oak, which is very astringent) persons who are familiar with the facts tell me that when hogs get very fat on them in the fall they are sure to die in the spring. This, however, they say is caused by worms. Hogs are always best able to resist disease when their bowels are kept in good condition and they are free from worms.

Dr. Payne says Mr. Ish had a tan-yard, and he lost no hogs. The decaying flesh off the hides, when eaten by the hogs, is suggested as a probable agent in destroying the *bacilli*. (See Dr. Detmers's report, p. 30.) Mr. Carter, sr., gave his hogs coal-oil and lost none, while Mr. Carter, jr., gave his hogs the same and lost all. The inference in this case is that Mr. Carter, sr., with the experience of age, gave his hogs oil in such manner and in such proportions as to keep its influence at all times present in the alimentary canal, where it acted as a disinfectant and prevented the lodgment of the *bacilli*. Mr. Carter, jr., seeing the good effect of the old gentleman's remedy, gave it after the infection was lodged and in such imprudent proportions as to repel the animals from all food, doing no good, and thus destroying all nutritive support. Dr. Detmers (p. 181) cites cases of where the living fed on the carcasses of those that died of the disease with good results. We should reasonably expect such results. This does not, as some may suppose, contradict his statement of the introduction of the disease by feeding the intestine of an infected hog. Though the *bacilli* may not be killed by

putrefaction, the change which the flesh undergoes in the stomach may destroy them, or they may have some degree of affiliation and follow the abundant supply of albumen to its discharge, which is accelerated by the food of flesh. At all events, the reproduction of the *bacilli* is so rapid that when once lodged no new supply can ever overtake them, and the space through which they have passed has no food left for a new invading force.

On page 48 Dr. Detmers states how easily the *bacilli* are destroyed by simple agents. If, then, he is not mistaken in his views as to the cause of the disease and its destructibility, then it follows that some opposing agent kept at all times in the bowels previous to the lodgment of the infection will certainly keep it off. I have already stated the circumstances which might prevent the successful use of these disinfectants.

Dr. Detmers is disposed to regard with great confidence the skill of his predecessors, and I think he too readily surrenders his science to the force of certain diseases, of which he mentions pleuro-pneumonia and glanders. In the last-named case let him try a thorough mercurial treatment in conjunction with powerful vegetable alteratives given in large quantities. For this purpose I will send some fresh Florida sarsaparilla root, and see what will be the result.

We are now learning to cure blind staggers here, a malady which we once thought incurable. We keep this fact at all times in view, that no temporizing policy in any severe affliction will result in much good.

The reports all agree as to the great difficulty of tracing the infection to a source in certain named cases. This does not go far with me to prove that the disease is at any time sporadic. Men who have a theory will always find an argument in support of it. When acting as quarantine physician I had the same difficulty in tracing the origin of certain cases of yellow fever, yet I never failed eventually to trace its connection with a previous infection.

Some persons have been confused by the great virulence of swine plague in some cases and its mildness in others, and have been half inclined to doubt its identity. I have noticed the same facts in yellow-fever epidemics. I have known it to almost wipe out large families in a very short time, while in other seasons or cases it was so mild as to be hardly recognized. While I had my theory as to the cause, I knew it would be idle to state it without being able to substantiate my statements.

I have had no experience in the treatment of swine plague, but I have observed in all cases where the vital energy is weakened it is in those cases where the fatality is greatest. The same holds good in yellow fever in the human family. Those who succumb quickly and recover quickly from other diseases are most fatally affected by this fever, and they are generally persons of sanguine lymphatic temperament. This, I suppose, is the reason that some breeds of hogs are more fatally affected than others of tougher organization. Without reviewing the subject any farther, I will repeat what I have already intimated; that is, if the pathology given of the disease is correct, and the physiology of the hog is analogous to that of the human race, my suggestions of agents will *absolutely* prove both preventive and curative.

I know of nothing that promises greater results for the good of the whole country than the eminently practical manner in which this investigation has been ordered and conducted. The respect for veterinary surgery will be very much elevated by the skill which those of its profession and other scientists have shown in this investigation of swine plague. I know physiology, pathology, and medicine to be sciences, and that therapeutics will fail to be curative only so far as it is not understood, or so far as it is impossible to replace defective organism.

I have intimated that the cause of preventives not being uniformly successful is that they have not been used with sufficient regularity and intelligence. Human testimony is very weak, and physicians know that even mothers cannot always be relied upon to look carefully after the diet of their children.

Much has been said of malarial effects upon swine. I think that question should forever be discarded as applied to swine. I own some land at the mouth of the Escambia River. On each side of the mouth of the river cities have been laid out. The lots were sold and settled upon; but the settlers have all died or abandoned them, and the proposed cities are so thoroughly dead, that there is scarcely a known owner for any of the land, save a few claimants who hold by right of possession. Among these are the Messrs. Murphy, who raise fine hogs at less expense than I have ever seen them raised elsewhere. I never heard of one being lost by disease, and yet in the summer season no one can be induced to sleep within two miles of this locality. The animals have clean water if they want it, but they seem to prefer living in the mud. In order to sleep dry, they build their beds of rushes higher than the tide-water reaches.

SWINE PLAGUE IN FLORIDA.

Hon. WM. G. LE DUC,
    Commissioner of Agriculture:

SIR: Since my last communication to you on the subject of swine plague, the disease has prevailed in this town to some extent, and is still affecting a few herds. I have had an opportunity to make many *post mortem* examinations, which have been very instructive to me. Only four of these cases I propose to refer to, as I regard them as characteristic of the disease in its destructive stages.

1st. Pig of Mr. William Allen, first noticed to be weak and staggering for two or three days, and found dead about the third day. *Post mortem* showed the whole mucous membrane of the intestinal canal highly inflamed and in some places congested; the pyloric extremity of the stomach in the same condition. The cardiac portion was quite healthy. The stomach contained about a pint of cracked corn, and was offensively pungently acid. The spleen was small and dark. The brain and surrounding integuments were entirely bloodless; the brain was a white pulpy mass, without any appearance of a blood vessel. All the other organs were perfectly normal, except that pallor of some of the glands so often noticed by Dr. Law and others. The bowels were distended with flatus, and contained no parasites or ulcers. The lungs were entirely healthy.

2d. Pig of Mr. William Allen; sick a few days and found dead. *Post mortem* showed the mucous membranes of the stomach and intestines perfectly healthy; no parasites. All the serous membranes were intensely inflamed and the whole substance of the lungs equally so. The large bronchia contained many worms floating in frothy mucus (I regret that I did not examine the brain in this case.) All the other organs healthy; bowels contained but little food, but were distended with flatus. This case being so different from the first, I was induced to examine critically to see how far the inflammation extended. I removed the serous from the fibrous coat of the intestines to the length of about six feet, and found that the inflammation had extended to the muscular coat only in a few small spots, producing a pale blush. The intensity of the inflammation of the serous membrane tended to a solution of continuity between it and the fibrous coat, which facilitated the dissection. Both of the pigs had been fed liberally on corn.

3d. Pig of Mr. S. B. Howel; had been sick eight or ten days and found dead. After death investigation showed the intestines packed with sand and mud, particularly the cæcum and colon; parasites of all sizes were found in the bowels, but none in the bronchial tubes. Only small portions of the lungs were healthy. All the blood vessels of the viscera very much congested; the bladder was enormously large yet flaccid, showing that the muscular energy was too far exhausted to overcome the pressure of the mud-packed bowel upon the urethra. The testicles were very much shrunken; the brain was normal.

4th. Pig of Mr. William Fleming; sick ten or twelve days; found dead; extremely emaciated and great functional derangement of the respiratory and nutritive organs; no parasites found in lungs or bowels. In this case as in the second, where the gall cyst lay in contact with the intestines, the latter was colored like the bile in the gall bladder. Extensive adhesions and formations of false membranes had occurred, showing that there had been active serous inflammation of the pleuræ. This pig died from extreme exhaustion of all the repletive functions.

Having seen these cases only after death I cannot treat of the symptoms, but only of the lesions as they appeared after death; and as my remarks are intended to be suggestive and practical, I shall be as brief as possible in my inferences. In the case of the first pig, where there was food in the stomach, great acidity of a pungent and volatile character was observed. This symptom was often observed by the veterinary commissioners. I have found it only in the early stages of the disease. This volatile acid seems to be the element which enters the bowels and distends them, and whether it is the effect of the *bacilli* or not it makes but little difference. Assuming this acid to be an anomalous or acetic acid we may readily understand what great constitutional derangement must occur upon its elimination through all the fibrous tissues, dissolving the fibrine and albumen of the animal, or at least disorganizing it.

The intensity of the inflammation and its location in different organs is a very marked feature of the disease, which, when seated in any of the viscera, operates as a most powerful revulsive, depriving the nervous system of its repletive source and superinducing death from atony in a few days. This seems to be the result in cases where the inflammation is very great, and notably in those animals that have been fed freely on corn. The last named seem to be the only cases which die in the early or active inflammatory stage of the disease. I am satisfied that it is the acid formation in the stomach and bowels which causes the great desire to eat dirt, which certainly has a tendency to neutralize the acid.

In the case of the third pig the lesions were not so extensive as to produce death when compared with the condition of the fourth pig. I am therefore led to believe that death in this case was the immediate result of exhaustion from the repletive organs being packed with mud.

The fourth pig died from the impaired organism of all the functions of the animal economy.

Dr. Detmers attributes the sero-sanguineous effusion to embolism caused by the "agglutinated bacilli" in the capillaries. We frequently find this character of effusion in the human subject in case of very severe congestive fever. I regard the theory of *error loci* of Boerhaave a more satisfactory solution.

The treatment as indicated by the examination of the first pig in the beginning should have been antiphlogistic, an emetic, followed by the mecurial, as suggested in my last communication, with the continued administration of an alkali sufficient to neutralize the acid formed in the stomach. The diet should have been albuminous and fibrinous. This course, if pursued on the first appearance of the disease, would most likely have relieved the animal. At a later period the treatment should have been similar, with the addition of active local irritation over the region of the dorsal vertebrae. The same treatment would apply to the second case and perhaps to the third and fourth cases in their early stages, but at the stage which the diseased organs were observed by me a different treatment was evidently indicated several days before the death of the fourth pig. The third pig might possibly have been saved by giving it, a few hours after it had eaten the mud, a stimulating injection followed by a mild and stimulating aperient, and applying a counter-irritant over the whole spinal column, succeeded by a nutritive antacid diet and mercurial alterative. At no stage of the disease should a carboniferous diet, such as Indian corn, be used, as it very much assists the formation of the obnoxious acid.

When the disease has approached the condition of that noticed in the fourth pig, the lacteals and absorbents are so far obliterated, or indurated and generally inactive, that no good could result from a cure.

While making the above investigations I observed that Mr. R. R. Smith, of this place, had many fine hogs around his mill. The disease appeared among them about the time that they began to lie under the old houses in cold weather, as they generally do every year. He lost about twenty-five of them. I did not propose to treat any of them, as he had many remedies of his own in which he had great confidence. I, however, urged him to diligently try his own prescriptions, which were mostly strychnine and tar. His pigs continued to die and he became discouraged. I suggested the addition to his remedies of strong ashes. At this time many of his hogs showed decided symptoms of the infection, and five fine sows lost their pigs. All of the sows recovered, however, and are now a fine lot of thrifty, fat hogs.

I think that the acid feature of the disease a leading symptom to be opposed. Great discrimination should be used in compounding whatever preventives may be used, and regulating the proportions so that the remedy should be at all times in the stomach and bowels of the swine in such form as to destroy the animalculæ and not be deleterious to the animal's health and thrift. A general error will be found in making the compound too offensive to be regularly eaten with the food. I think I have discovered this error in Mr. Smith's treatment, and if the disease does not recur in his herd it will be owing to this error of his compound.

Assuming the statement of Drs. Detmers, Law, and others, that the propagation of the disease is from the vital spores, as mentioned by them, to be a fact, and dealing only with facts as known to the profession, what are the conclusions which we must arrive at? Only this: to destroy the vitality of the germ, first, by a potential cautery before the germ is lodged. To do this potash is known to be most destructive to all animal tissue. The insensible cuticle of the hands can never become accustomed to its weakest solutions until protected by a lifeless thickened covering. Will the delicate *bacilli*, then, be likely to resist its influences? Supposing the germ has entered the circulation, the indications are, then, to follow it and poison it with remedies less deleterious to the animal than to insect life. I know of no remedy more likely to do this than mercury, as I have before stated. I have found chloroform singularly fatal to the lower order of insect life. Frequent administration of this might be experimented with. Salt should be used. Inflammations, congestions, &c., should be treated according to indications.

As all of this may be regarded as the theory of a visionist, let us see how it has worked with me in former years. During the war, when the Confederates evacuated this section, I returned to Montgomery, Ala. An old friend, Mr. William Maning, invited me and insisted that I should make his house my home during the war. The swine plague was fatally prevalent at that time, and was daily approaching nearer. Being naturally fond of experimenting, I directed much of my attention to the meat question. Salt and ashes was a popular remedy on the plantation for colic in horses, and copperas was recommended by others as a remedy in the swine disease. I blended them together, and gave them with the food two or three times a week with certainty; and during the three years I continued this treatment but one hog was known to die with any disease on the plantation. I then attributed the success to the copperas only; I find that it was due only to the salt and ashes.

I often met Representative Lowe, of North Alabama (the nephew of my friend), while he, too, was a refugee, and convalescing from serious wounds. He doubtless remembers the splendid success of his uncle in meat-raising and fine swine generally, while every farmer around us lost nearly all of his hogs. Yet I know that one man may use this remedy and not lose a pig, and another will say he has used it the same way and will lose all. Reverse the parties and the results will be reversed. All seems to be in the manner of application; whenever the remedy is properly used I have yet to see it fail. Let it be remembered that this remedy is not recommended as a curative, but only as a preventive. When the disease is seated, it will do but little good. In the case of Mr. Smith, referred to above, I think that it was the nux vomica which did most to destroy the poison, while the ashes only supported the health by destroying the acid formation.

There is too great a disposition to look on the curative art with a reverence more becoming a believer in the art of necromancy.

It has always been the habit of us of the South to discourage all appearances of an innovation upon old usages; yet it gives me pleasure to know that by your departure from this line you have been enabled to inaugurate a system by which the swine disease has been fully understood; and whatever may be its present attainments, its ultimate cure is a foregone conclusion.

Respectfully,

JOHN M. McGEHEE.

MILTON, FLA., *April*, 1880.

INVESTIGATION OF SOUTHERN CATTLE FEVER.

REPORT OF D. E. SALMON, D. V. M.

Hon. WM. G. LE DUC,
Commissioner of Agriculture:

SIR: I have the honor to submit the following report, which is intended to give, in a connected manner, all that is known in regard to, and to include the results of my own investigations of,

THE SOUTHERN CATTLE FEVER.

Synonyms.—Spanish fever, Texas fever, splenic fever, periodic fever, gastric fever, acclimation fever, American cattle plague, red water, black water, distemper, murrain, dry murrain, yellow murrain, bloody murrain.

Definition.—An exceedingly fatal epizootic, specific, communicable fever of cattle, at present confined to regions south of the 37th parallel of north latitude, except as communicated to cattle north of this line by cattle brought from south of it; it is characterized by peculiarities of extension, of transmission, of symptoms, and of pathological lesions, which will be discussed, as thoroughly as the present state of our knowledge will admit, under the respective headings.

HISTORY, DISTRIBUTION, AND MORE APPARENT PHENOMENA.

The disease in the permanently infected district.—Remarkable as it may appear, our first accounts of this disease, as at present recorded, did not come from the region to which it is generally supposed to be indigenous, but from those sections beyond this to which it had been conveyed by the movements of cattle. It is hoped that an effort will yet be made by some one having access to historical documents to trace either the introduction of the disease or its effects during the early settlement of the Southern States; for, surely, if the disease then existed in these States, destroying imported cattle to the same extent as at present, there must be some record of its ravages. Aside from their general historical value, the knowledge of such records is important as the only means of determining the districts to which this malady was originally indigenous—knowledge of great usefulness in deciding the measures to be adopted in an attempt to arrest its destructive work.

Early accounts.—The earliest account which has so far been noticed is in a lecture before the Philadelphia Society for Promoting Agriculture, by Dr. James Mease, November 3, 1814, quoted by Mr. Dodge in a report to the Department of Agriculture.* It is there stated that the

---

*Statistical and Historical Report of Splenic Fever, in Report of Department of Agriculture on Diseases of Cattle in the United States, Washington, 1871, p. 176.

cattle from a certain district in South Carolina so certainly disease all others with which they mix in their progress to the North that they are prohibited by the people of Virginia from passing through the State; that these cattle infected others while they themselves were in perfect health; and that cattle from Europe or the interior taken to the vicinity of the sea were attacked with a disease that generally proved fatal. In a paper read before the same society September 20, 1825, * he says, "The circumstance of cattle from a certain district in South Carolina affecting others with this disease *has long been known*," but the precise locality, or its extent, he was unable to ascertain, notwithstanding inquiries on the subject. The country of the long-leafed pine had been said to be the native place of the infection, and the cattle alluded to were said to emit a peculiar smell.

Mr. Dodge also mentions the fact that old residents of the Piedmont region, between the tide-water areas and the Blue Ridge, are familiar with this form of disease; and the cattle drovers who have brought stock from the country of the long-leafed pine to greater elevations and higher latitudes testify with remarkable unity to the constancy of its appearance and the uniformity of its prominent characteristics. Mr. J. Wilkinson, of Athens, Ga., stated to the Department of Agriculture, in April, 1867, that cattle seldom contract the disease unless removed from where they were raised; that if taken from the mountain country to the low country, they soon take the fever, and die without communicating any disease to the native cattle; that cattle taken from the low into the elevated country will continually improve, while they will communicate a fever that will kill the cattle they come in contact with; that, after remaining in the elevated or colder country a certain time, they lose the power of communicating disease. As an example, he stated that he had been in the habit of driving cattle from Florida to Virginia, and that, while these cattle did well, after he passed the line of 34° they began to spread the fever all along the line of travel, till he struck the Virginia line, a distance of 250 miles, when there was no more trouble. He considers from these facts that it must be due to a change of climate.

Though little was known at the North of this fever prior to 1866, and though many there were even skeptical of the existence of any such malady, its presence and the dangers from it were recognized by the States of Virginia and North Carolina many years previously. Thus, it seems from Dr. Mease's lecture, just referred to, that the people of Virginia would not allow the passage of cattle from a certain district in South Carolina as long ago as 1814, and I have been unable to determine how much earlier.

Successive laws showing extension of infected district.—At the session of the general assembly of North Carolina in 1836 and 1837 a law was enacted to prevent the driving of cattle into the State from either South Carolina or Georgia between the first day of April and the first day of November; also, to prevent cattle from being driven from those parts of North Carolina where the soil is sandy and the natural production or growth of timber is the long-leafed pine into or through any of the highland parts of the State where the soil or growth of timber is of a different kind, between the same dates.

At the session of 1873 and 1874 a new law was passed prohibiting the driving of cattle from South Carolina or Georgia into any of the counties west of the Blue Ridge between the first day of April and the *last*

* Loc. cit., p. 177.

day of November—two very important changes, since the vast region
between the country of the long-leafed pine and the crest of the Blue
Ridge is no longer specified, and the time during which the danger is
recognized is extended from the first day of November to the last day
of November.

Again, at the session of 1878 and 1879, it was found necessary to
make alterations, and the prohibition is now for driving cattle into any
of the counties west of the Blue Ridge that had come not only from or
through Georgia or South Carolina, *but from any of the counties east of
the Blue Ridge,* and the prohibition to continue *during all seasons of the
year.*

Here, then, is practical evidence and recognition of the extension of
the infected district across the great central belt of the State, from the
country of the long-leafed pine, in the eastern part, to the Blue Ridge,
in the western part; but even after the passage of these laws the advance of the disease does not seem to have occurred to any one, or, at
least, to have been clearly realized, until the publication of Mr. Lenoir's
letter, on page 253 of Special Report No. 12 of the Department of Agriculture, in 1879.   And yet, in the forty-two years between the passage
of these acts, the disease must have progressed westward fully two hundred miles, *and infected an area, in this State alone, of not less than twenty
thousand square miles.*

The germs show an increased power of resistance to cold.—There is
another fact here of equal importance which seems so far to have entirely escaped attention. As the disease has advanced, its essential cause,
of whatever this may consist, has shown a greater resistance to cold, or,
in other words, has shown itself capable of surviving a greater degree
of cold.   Thus, in 1836 it was not contracted after the first of November,
even in the middle belt of the State; in 1873 it was found best to extend
this time to the last day of November, even for the mountain section;
and, finally, in 1878, it was found to be no longer safe to drive across
the Blue Ridge even in winter.

Southern fever in Virginia.—Maj. R. L. Ragland, of Hyco, Halifax
County, Virginia, has kindly written to me at length in regard to the
introduction and continuance of this disease in that section. All au-
thentic accounts, he says, agree that it was brought from the eastern
part of North Carolina; it was introduced there sixty or seventy years
ago and prevails annually if the weather is hot and dry.   It seldom appears before July, and is most fatal in August; the last cases heard of
in 1879 occurred in October, and it rarely prevails after frost.   There
are farms on which it has never occurred, while on adjoining ones it
kills more or less every year; again, it is more prevalent and violent on
some farms than on others.   He considers the description of the characters and progress of the disease as given by Mr. Lenoir to be correct,
and adds that in fatal cases the manifolds seem constricted and the con
tents almost dry, and the animals are constipated.   It exists in adjoining counties; he has reports of its extending towards the north and west,
and thinks the late law passed by the legislature of Virginia is further
evidence of this extension.

This law was approved April 2, 1879, and undoubtedly has for its object to prevent the introduction of cattle from any infected district,
though, like the North Carolina laws, it is so imperfect as to be practically useless in checking the introduction and extension of the malady.
Cattle, with certain exceptions, brought from south of the line separating Virginia from North Carolina and Tennessee, and a prolongation
thereof west to the Rocky Mountains, if brought into the State be-

tween the 10th day of March and the 10th day of October are to bo inspected by inspectors appointed by the governor; these may require an affidavit that the cattle are not brought from the proscribed district; if condemned, the cattle must be removed or killed.

Peculiarities of the fever.—As I have learned by quite extensive inquiries and personal investigation during the past few months, the disease does not affect native cattle in North Carolina to any considerable extent, except within a limited district along the northern boundary line of the State and a similar narrow district along the Blue Ridge Mountains—mostly east of them, but in a number of instances in the gaps or passes, and beyond them. At Morganton, Burke County, forty miles east of the Blue Ridge, the disease was introduced about 1870, and for a number of years was very destructive, but the native cattle finally became inured to it, and now the losses are slight.

Somewhat earlier, if I am correctly informed, it invaded Rutherford and Polk Counties, immediately east of the Blue Ridge, but farther south than Morganton, and, as usual when first affecting a district, the losses were very severe; but here, too, the native cattle have mostly become inured to it. As Mr. Lenoir has told of its presence in Wilkes County, we have here a line, along the Blue Ridge, drawn nearly across the State.

In South Carolina I find that its history has been similar, and it has there, also, advanced till checked by the Blue Ridge.

When cattle beyond the infected district are affected as a consequence of the transportation of the disease by other cattle, the fever is very intense, and is seen in its most acute form; in from three to four days after the appearance of the first symptoms the vast majority of the affected animals die. As the disease slowly advances over new territory adjoining the affected district this is equally true; but, like other epizoötics, its attacks are generally milder after it has been a few years in a locality; its course is longer, and a larger proportion of the affected recover; finally, it nearly disappears, native cattle being seldom affected, except to so slight a degree as to be scarcely, if at all, noticed. The people now conclude that they have exterminated it, as they do under similar circumstances with swine plague, by dosing with peach-leaf tea, saltpeter, red clay, and other nostrums, or even by boring the horns and sliting the tail.

From this circumstance of its becoming milder and almost disappearing in from ten to fifteen years after its introduction, one sees the reason for Mr. Lenoir's comparison of it to ringworm, slowly advancing, with an angry external border line, and apparently dying away in the district over which it has passed.

The disease remains permanently in newly infected districts.—Now, in what does this dying away consist? Does the disease completely die out, and the section over which it has passed become once more free from it, as before its invasion; or does its subsidence depend upon the native animal becoming gradually inured to it, and comparatively insusceptible to its effects? Unfortunately, the latter condition appears to be the result. Not only do cattle brought here from beyond the infected district contract the disease with all of its original virulence, but animals from this newly infected district, though apparently healthy themselves, if taken to uninfected localities convey the scourge to all cattle with which they pasture, or which pasture on lands traveled by them since the preceding winter.

The facts on which this assertion depend are easily obtained, and are only too numerous. It is almost universally admitted that cattle from west of the Blue Ridge seldom survive the first summer if taken into the counties east of it; a well-informed gentleman at Charlotte told me that they generally sickened in ten or fifteen days, if brought there in summer.

In regard to the converse of this, I had occasion, in October, to investigate an outbreak at Mr. Tom's, in Buncombe County, but a few miles west of the Blue Ridge, caused by cattle brought from a neighboring county just east of these mountains; it is also claimed that cattle from Polk County have, more than once, conveyed the disease into the adjoining county of Henderson. Besides, the passage of a law prohibiting the driving of cattle from any counties east of the Blue Ridge into any counties west of it is in itself evidence that this is the case.

Even in the eastern part of North Carolina this fever is spoken of as the South Carolina distemper, which seems to indicate a kind of tradition that it had been originally derived from that State, though this, at present, is exceedingly indistinct.

At and around Hendersonville, Henderson County, North Carolina, I found the cattle suffering from this disease to a very considerable extent; and as this county lies in the elevated district west of the Blue Ridge, where the winters, if not so long, are nearly as cold as in New Jersey, it affords a most interesting subject of study. Nine years ago, as I was informed by Mr. McDowell, a cow was brought from South Carolina, by A. W. Cummings, from which an outbreak of the disease occurred four miles north of the town, destroying some forty cattle. The trouble is said to have died out subsequently; but of this there is reason for the gravest doubt. In fact, there appears to have been more or less losses every summer within a circle of two miles from the farm originally infected; and during my visit there the last week in October, 1879, I investigated the cases of two cows just dead, and a bull in the first stages of the disease (killed for examination), and none of these animals, so far as known, had been on lands pastured or traveled by cattle from the infected district.

That there have been importations of the malady since the one mentioned, in localities but three or four miles distant, seems to be established. Five years ago a Mr. Price brought cattle from east of the Blue Ridge, and last summer a colored man brought a yoke of oxen; again, it is not uncommon to drive to South Carolina and back with oxen; so that it must be confessed one cannot be certain as to when lands really were infected. The fact, however, that the disease has remained here, being more or less fatal every summer for nine years, while in no other locality west of the Blue Ridge has it existed more than two years successively, unless known to be freshly imported, leads me to fear that it has successfully scaled the great natural barrier which it found in the Blue Ridge, and that from now onward it will not only remain permanently established here, but that the active principle having become acclimated in this colder region will enable it to advance to the north and west with even greater rapidity than before.

In Georgia.—From Georgia the disease was reported to the Department of Agriculture, in 1868, as existing in Pulaski County, which is between the thirty-second and thirty-third parallels of north latitude; in Hall County, between the thirty-fourth and thirty-fifth parallels; and in Chattooga County, in the same latitude*; while, in 1874, a gentleman in Habersham County (just south of the Blue Ridge, in about the same latitude as the two last-mentioned counties), wrote to the Country Gentleman that all their cattle were subject to a disease during the summer months which from its symptoms must be red water, and that those

* Loc. cit., p. 195.

brought from north of them, if only for a few miles, were especially subject to it.* In this State the malady seems to have been always considered as due to change of climate, and there are no laws bearing upon it.†

In Georgia the disease is popularly termed *murrain*, and the infected parts of the State are called *murrain districts*. Colonel Peters, of Atlanta, who has probably had a longer and more varied experience with it than any man in the South, considers Atlanta to be in a murrain distict, while Gordon County, in which his farm is situated, some eighty miles north, is not such a district, and cattle brought from his farm to Atlanta take the disease.

In *Alabama* the disease seems to be equally fatal, though but little information has been received from that State. Colonel Crawford, of Mobile, introduced eight short-horned heifers from Kentucky a number of years ago, and all died the second summer. ‡ Colonel Peters informs me that there is an extensive region east and west of Decatur, Ala., in the valley of the Tennessee River and along the Memphis and Charleston Railroad, where he had noticed very few cattle were kept, and, on inquiring the reason, learned that it was due to the ravages of the disease known as murrain, which was imported a number of years previously and had proved very destructive.

In *Mississippi.*—Mr. W. B. Montgomery, of Starkville, Miss. (latitude 33½ degrees), imported two Ayrshires and twenty-two Jerseys in the spring of 1873. The two Ayrshires died and twenty of the Jerseys were affected, though but seven died. He remarks that the risk of acclimation is not limited to the first summer, but that the fatality is often much greater the second. Mr. Kennedy, in his efforts to introduce cattle from Kentucky to Mississippi, found much the greater fatality occurred the second season. And Mr. Montgomery's neighbor, L. A. Foote, purchased, in the fall of 1871, eight head of cattle from Kentucky; they passed through the summer of 1872 safely, but, much to his surprise, seven of the eight died from acclimation during the summer of 1873.§

In *Arkansas.*—Our only means of judging of the parts of Arkansas already infected is by the sections in which Texas cattle, when driven through the State, as they have been in large numbers, communicate the disease to the native cattle. Thus, in 1858, in Mississippi County (36 degrees north latitude), the natives were not diseased by mixing with large numbers of Texans, nor were they in Drew County (below 34 degrees), nor in Crawford County (35½ degrees,) though this is in an elevated part of the State, the Ozark Mountains being a county line; but Texas cattle have been driven through Crawford County for years, both before and since the war, and thus the disease has probably been firmly established and the natives inured to it. As evidence that this is the correct explanation, we find animals affected by the fever, in consequence of the introduction of Texans, as far south as Arkadelphia, Clark County (near thirty-fourth parallel), which is off of the regular line of travel; also in Washington County (about the thirty-sixth parallel).‖

In *Texas.*—Not very much is known of the disease in Texas, except that it is very difficult to acclimate thoroughbred cattle on account of it. Mr. Mark Huselby, of Fort Elliott, Tex., in a letter to the National

* J. C., in Country Gentleman, 1874, p. 522.
† Information obtained from State department of agriculture, Atlanta, Ga.
‡ W. B. Montgomery, in Country Gentleman, 1874, p. 386.
§ W. B. Montgomery, in Country Gentleman, 1874, p. 394.
‖ J. R. Dodge, loc. cit., p. 185.

Live Stock Journal (1879, p. 258), says: "It is much more extensive than you seem to be aware of, for in the section of the country known as the Pan Handle of Texas I might make a fair estimate by saying that one thousand head of stock die annually from that disease. * * * We suppose it to be contagious in Texas, though cattle seldom die of it. What are known here as American, Colorado, and Mexican cattle, when brought in contact with Texas cattle, almost invariably suffer severely."

Boundary line of the infected district.—If, now, we attempt to trace the boundary line of the infected district as at present constituted, and from all of the information we have been able to obtain, we must commence in Virginia, at about the thirty-seventh parallel of latitude, take a southwesterly direction, and strike the Blue Ridge a little south of the Virginia line, follow the direction and line of this chain of mountains across North Carolina, South Carolina, and Georgia to about 34½ degrees of latitude, and keep this parallel west to the Rocky Mountains. There is one point in North Carolina and two in Arkansas where the fever has advanced beyond this line, and that there are many such would not be surprising. At present it is enough for us to know the general outline of this district, and that this outline is continually encroaching on uninfected lands.

The district overrun by this disease, then, embraces the whole of the vast area lying to the east and south of this line, an area probably containing five hundred and fifty thousand square miles; and this is nearly seven times the size of England and Scotland combined, nearly eleven times the size of England alone, or two and one-half times the size of France; and yet there are those who would have us believe that this is not an epizoötic, but simply an enzoötic affection, depending on the character of the soil!

The disease beyond the permanently infected district.—For the first account of the southern fever in the North I am again indebted to Mr. Dodge.* Dr. Mease, in the lecture already referred to, mentions an outbreak in the month of August, 1796, in Lancaster County, Pennsylvania, which occurred as the result of South Carolina cattle being brought and sold there. At Anderson's Ferry, now Marietta, they were penned overnight in a plowed field, and did not come in contact with the cattle of the farm, yet the latter commenced dying a short time afterwards. In every instance where sold they communicated the disease to the cattle with which they mixed. The symptoms were loss of appetite and weakness of the limbs, amounting to inability to stand; when they fell they would tremble and groan violently. Some discharged bloody urine, others bled at the nose; bowels were generally costive. On being opened, the kidneys were found inflamed and sometimes in a state of suppuration, and intestines filled with hard balls.

The manner of introduction of this disease, its highly contagious character, the bloody urine and disease of kidneys, are sufficient characters on which to make a diagnosis.

That there have been many occasions on which this disease was introduced in a similar manner in Maryland, Virginia, the mountain region of North Carolina and Tennessee, is very certain, but the accounts of these are so meager that I do not reproduce them.

The fever in the West.—By far the greatest losses seem to have been occasioned by driving Texas cattle through Arkansas, Missouri, and Kansas, and distributing them to feeders in various Western States. As long ago as 1852 and 1853 it was destructive in Missouri, and it is

---

* Report, p. 176.

said to have existed two years before it was traced to the Texas cattle. But as it was, from the first, confined to the great roads or highways running through the country from the South, it was definitely decided in 1853 to be due to the Texans, since it was then confined to a single highway over which these had passed. About fifty per cent. of all the cattle along this road died, and persons living near the fording places lost as high as ninety per cent.

A drove of cattle that had even been wintered at Sarcoxie, Mo., (latitude 37½°), and driven north some fifty miles to Vernon County the following June (1853), communicated the disease to native cattle.*

During the summers of 1856 and 1857 Texas cattle were taken into the States of Kansas and Iowa in great numbers, and it is stated that the native stock was swept away by a "dry murrain" that prevailed along various routes traversed by the Southern droves. From 1858 to 1861 the Southern fever prevailed along roads traveled by Texas stock in Missouri, Kansas, and Iowa, and in 1861 laws were framed in Kansas to regulate the movements of herds from the South. Similar laws were enacted in Missouri, Illinois, and Kentucky. The disease ceased in all of these States during the war, to reappear with the first Texas droves.†

Mr. John H. Tice, of Saint Louis, in a letter to the New York cattle commissioners (Report 1869, p. 92), says the first cases he saw were in 1858, five miles west of Saint Louis, in a drove of 180, three-fourths of which were Texans and the remainder natives. The disease appeared on Saturday, a few days after their arrival, and by Monday twelve were dead, all natives.

In 1860 a large drove of Texans stopped opposite the pasture in which those just mentioned had been inclosed. Some of these died, and the malady soon appeared among the neighboring cattle, nearly all of those affected being cows.

About the middle of October, 1866, a cattle-train was delayed three hours opposite Mr. Tice's house. On this were Texas cattle. A cow turned in afterwards to graze along the inclosed railroad track contracted and died of the disease.

In 1866 a drover brought a steamboat load of Texas cattle up the Mississippi and Ohio Rivers, and, landing them at Louisville, drove on to Lexington, which is in the interior of Kentucky. This drove moved to Georgetown, and wherever the native stock of the district chanced to graze upon the roadside or pastures that were thus traversed by the Texans the former were in the course of two months almost entirely swept off by this cattle disease. A drove of fat cattle that chanced to follow close upon the trail of the Texans were all attacked and all died.†

In 1867 Cairo, Ill., had become the chief point of transshipment of cattle from steamboats to railroads as a result of the prohibition against the movement of these animals through Kansas, Missouri, and Iowa. The first lot was landed April 23, the second April 26, and so on, about thirty thousand head having been disembarked during the spring and summer. The first lots seemed healthy, but later in the season it is stated that from twenty-five to thirty died daily in the yards or about the levee. The farmers in the vicinity, suspecting no danger, allowed their stock to mingle freely with the Texans; but the disease appeared

---

* Letter of Dr. Albert Badger, in Diseases of Cattle in the United States, Department of Agriculture, 1871, pp. 187, 188; also, Report of New York State Cattle Commissioners, 1869, p. 91.
† Report of New York State Cattle Commissioners, 1869, p. 92.

early in June, and the fatality increased until it plainly declared the presence of an epizootic.*

The great outbreak of 1868.—Notwithstanding a more or less imperfect knowledge of the ravages of this disease which I have already recounted, the meaning of such facts was disregarded, and there was a general apathy in regard to the matter, just as there is at present in regard to even a worse disease—the imported contagious pleuro-pneumonia—and, to carry the comparison farther, there were not wanting prominent men then, as now, to make light of the dangerous character of the malady, and even to throw doubts on its very existence. But this indifference and scepticism was soon to give place to genuine alarm and consternation, because of the wide distribution of the fever and the remarkable losses occasioned by it. Let us hope that the comparison may stop with the present indifference and scepticism in regard to pleuro-pneumonia, and that we shall not see similar natural causes distribute this plague over a like area before we can realize its extension; for if this happens, we have a disease not destroyed by frost or winters that must remain to decimate our cattle and the cattle of those who succeed us for untold generations—a permanent infliction, a continual loss to the whole country.

In Illinois.—The Texans having been turned from their original course by the late laws of Missouri, Kansas, and Iowa, were shipped by boat to Cairo, and then distributed by rail to various parts of Illinois and Indiana, to be grazed a few weeks before going to market. It was not suspected by the farmers who harbored this dangerous stock that they carried certain destruction to all native cattle wherever they went. The fever first appeared at Cairo about the 10th of June.

At the little town of Tolono, at the crossing of the Illinois Central and Toledo and Wabash Railroads, from fifteen to eighteen thousand Texan cattle were landed. The fever commenced its destructive work about the 20th of July, sweeping away nearly every animal of the bovine race in that section; two hundred and thirty-five cows died before the first of August; nine hundred and twenty-six head of cattle died in that township, which polled but a trifle over two hundred votes; and five thousand head succumbed in the county!

At Broadlands, the great farm of Mr. Alexander, the losses were very severe. The Texans were intermixed in the pastures with about six hundred native animals, all but two hundred and eighty of which were soon sent to Eastern markets. Those left on the farm began to die July 26, and one hundred and ninety-eight were lost, including a yoke of old Texas steers which had been on the farm. Of the cattle sent East, two hundred and twenty-four died before they reached their destination, and the remainder were said to have been sent to the rendering tanks.

F. A. Atkins, writing from Odell, Ill., reported ninety-eight as the loss of his township.

Within two miles of the Chicago stock-yards, according to the report of the medical officer of the city, but one cow escaped, one hundred and sixty-one animals, mostly cows, having perished.

One thousand eight hundred animals were estimated to have died near Loda, Ill., and many others in different parts of the State.

In Indiana.—In Warren County the Texans began to arrive about the first of June, and the natives were attacked with the fever July 25. The losses in this county amounted to about one thousand five hundred head.

---

* Loc. cit., p. 93.

In Jasper County four hundred were lost; in an adjoining county, four hundred; in Marion County, one hundred.

At Cincinnati, Ohio, in a herd of twenty-nine cows exposed, all died, and a few are reported from other parts of the State.

It is useless to enter into details regarding the number of animals found by the New York commissioners on the cattle-trains and in the cattle-yards; between two and three hundred diseased or dead animals were inspected by their officers, and upwards of forty killed to study the progress and effects of the disease.

This brief sketch, the facts for which were mainly gathered from the Department of Agriculture Report on Diseases of Cattle, 1871, and the Report of the New York State Cattle Commissioners for 1869, might have been considerably extended by noticing outbreaks of less extent in various parts of the North from Illinois to Rhode Island; but I trust what has been given is sufficient for the object of this historical review, viz., to give an idea of the destructiveness of this fever in localities where the native cattle have not been inured to its essential cause by long exposure.

As the disease progresses along the so-called "distemper line" the native cattle are swept away in something like the proportion in which we have seen the deaths occurred in Illinois, i. e., from 50 to 90 per cent. There is consequently a considerable area of the country along this great line, stretching from Virginia south and west to the Rocky Mountains, where the people are continually subject to such ravages. I have no means for estimating the annual losses in this large section, but, certainly, they must be very considerable.

The period of incubation.—It is a difficult matter to arrive at the exact period of incubation from the history of these outbreaks, since we cannot tell exactly when the disease germs find their way into the system. We can only take the time from the first chance of exposure till the disease becomes manifest by its apparent symptoms. Thus, in Warren County, Indiana, this was from June 1 to July 25, or fifty-five days; at Cairo, in 1867, this period was six weeks; in Summit County, Ohio, cattle exposed July 5 were affected September 1, or fifty-seven days; at Dayton, Ohio, the first animal sickened in forty-seven days. In Orange County, New York, one cow sickened in thirteen days; a second in fifteen days; a third in nineteen days; an ox in two weeks; and a second ox in twenty-eight days.

Accounts do not agree as to the date of the first case at Tolono. Mr. Hill states that the first Texans arrived April 29, and that the disease broke out June 10, or forty-two days after. Professor Gamgee states that the largest body of Texans arrived toward the end of May, and the disease broke out July 27, or in about fifty-eight days. One gentleman there gave accommodation, on the night of June 25, to three hundred Texan steers, and the disease broke out July 28, or in thirty-three days.

F. A. Atkins, Odell, Ill., observed the incubation to vary from fifteen days to three months. June 25, Texas cattle were herded on a prairie near Odell, where there were native cattle. The disease appeared August 10, or about forty days afterwards.

At Whitehall, Mr. Gregory put two hundred and fifty-five Texans into a pasture June 20, in which fourteen native steers and three cows were grazing; July 18, one hundred Texans were added, and July 29 forty fat natives; August 8, two of the three cows died, or in forty-nine days; August 10, one of the fourteen steers died, or in fifty-one days; August 15, the fat steers that came July 29 began to die, or in seventeen days.

In Warren County, Indiana, ninety-five cattle were grazed from the

19th of June on an infected pasture; July 28, one was sick (thirty-nine days); by August 4 eleven were sick, and three had died (forty-six days); June 12, a drove of Texans were herded on another pasture in the same county; four weeks after (about July 9) twenty-six native cattle broke from their pasture and grazed over the same ground; July 29 (twenty days), one of these was sick, and twenty-two died during August. At the time the Texans were there (June 12), one hundred native cattle were being pastured on the same prairie; August 1, two of these died (forty-nine days), and twenty-five more were sick.

At Farina, Ill., two hundred and fifty Texans were placed with fifty native steers the 10th of May, and the disease appeared among the latter on or about July 15 (sixty-six days). Near Sodorus this period was from the 1st of June to the 28th of July, or fifty-eight days. In Champaign County, Texas cattle were placed on a prairie on the 15th of June, and the natives began to die on the 3d of August (forty-nine days), and in four days twenty out of thirty-eight were dead.

A case is reported from Kansas where a native animal died two weeks after exposure. A gentleman in Bates County, Missouri, says the disease is seen in from ten days to two weeks after the passage of the Texas cattle. A reliable gentleman of experience at Charlotte, N. C., informed me that cattle brought there from the mountain districts usually sickened in from ten days to two weeks. Colonel Peters, in sending stock during the months of September and October from Northeast Georgia and Middle Tennessee to Edisto Island, found that nine out of ten died in a few weeks, and many in ten to fifteen days.

From these facts it would appear that the period of incubation of this disease is exceedingly variable, and that it may range from ten to ninety days. A partial explanation of this variation might be that animals do not happen to take the contagious principle into their system as soon as they are pastured on infected grounds; but there is reason to believe that this is rather owing to the condition and requirements of the contagious principle.

Peculiarities in regard to the incubation.—The New York commissioners remark, with reason, that there are two classes of facts to be considered here: (1) The time from the first exposure to Texas cattle or their excrements to the date of attack, and (2) the time that elapses after exposure to grounds *that are known to be capable of communicating the disease.* In other words, there is a distinction to be made between exposure to the infection itself and the mere exposure to Texas cattle or their trail *immediately* after they have passed.

We have seen that the time that elapsed between the first exposure of native stock at Tolono till the appearance of the fever was at least forty-two and possibly fifty-nine days; but after the cows of the town were all dead, a large herd of native cattle that had been in no way exposed to the Texans were brought into the town and pastured upon the same grounds where the disease had prevailed so fatally. In about *three weeks* they began to die; that is, the incubation was less than half the original period.

At Cairo, in 1867, nearly six weeks elapsed after the landing of the first five hundred Texans before the native cattle began to die; but after the outbreak had declared the presence of the infection, the incubation was scarcely four weeks.

At Whitehall, as already mentioned, the incubation for the cows and steers which were in the pasture when the Texans came was forty-nine and fifty-one days. respectively ; but the fat steers placed in the pasture more than a month after the arrival of the Texans began to die in seventeen days, or one-third of the time of the others.

The case at Warren County, Indiana, is also very instructive. One hundred natives were pasturing on a prairie June 12, when the Texans were herded there, and they continued on the same pasture until August 1 before any died; while the twenty-six native cattle that broke from their own pasture and grazed on this infected one on the 9th of July sickened as early as the others, though they were not exposed till nearly a month later.

There is still further confirmation of this view from what occurred on Mr. Alexander's farm, in Champaign County, Illinois. One hundred native oxen were grazing continually in a pasture which was used as a resting place for the freshly arrived Texans, and in which the latter were placed in the number and for the time mentioned, as follows: May 31, 400, for one day; June 1, 226, for one day; June 18, 496, for three days; June 21, 349, for one day. The one hundred natives were all removed July 1 to a pasture in which no Texans had been; July 14, 27 of these were placed in a pasture that had been grazed by Texans. Two weeks later several of the twenty-seven sickened, and in the course of the month nearly all of them had the disease.

In another lot of twenty-six natives that had grazed with freshly arrived Texans in a pasture from June 20 to July 1, not a case of the disease occurred, and they remained healthy to the end of the season; but these cattle were removed July 1 to a pasture in which no Texans had been. A large herd of the companions of the twenty-six just mentioned continued in the original pasture where they had grazed together for the last ten days in June, a month later, and these were stricken with the fever, and by the middle of August were nearly all dead.

In Buncombe County, North Carolina, a drove of cattle from east of the Blue Ridge remained overnight at Mr. Tom's farm, either late in winter or early in spring, he had forgotten which, and his own cattle remained free from the disease till August, after which nine out of seventeen died.

At one time Colonel Peters, of Atlanta, Ga., imported six Devons from England, which he divided into two lots; part were kept at home and stabled; they were fed on dry food, and kept carefully shaded; the remainder were sent to his farm and pastured. The following August, and at about the same time, both lots sickened and all died.

Finally, we have the facts already stated that susceptible cattle taken from the mountain region of North Carolina to Charlotte during the hot weather sicken in ten to fifteen days, and that other susceptible cattle taken to Edisto Island, South Carolina, take the fever in about the same time. In these cases they are taken on lands known to be capable of communicating the fever, and the incubation is almost invariably seen to be much less than it would appear from cases where animals had grazed continuously on pastures from the time the Texans were first put upon them.

There is one more point in this connection to which I will call attention. It is generally after the 20th of July when the disease appears, no matter how long the cattle have been grazing with Texans or on infected pastures.

The following table shows this very plainly:

Case occurred at—	First exposure.	Disease appeared.	Days of incubation.
Warren County, Indiana	June 1	July 25	55
Do	June 19	July 28	59
Do	July 9	July 29	20
Do	June 12	August 1	49
Tolono, Ill	April 29	June 10	42
Do	May 26	July 27	53
Do	June 25	July 28	33
Summit County, Ohio	July 5	September 1	57
Odell, Ill	June 25	August 10	40
Whitehall, Ill	June 20	August 8	49
Do	July 29	August 15	17
Farina, Ill	May 19	July 15	60
Sodorus, Ill	June 1	July 28	58
Champaign County, Illinois	June 15	August 3	49
Orange County, New York	August 31	September 12	13
Do	do	September 14	15
Do	do	September 18	19
Buncombe County, North Carolina	Spring	August	
Colonel Peters's Devons		do	
Virginia		July and August	

That is, either in permanently infected districts or in others to which the disease is carried it is usually after July 20 that the first cases appear; in exceptional instances, however, the fever may appear early in June, and in Texas even in January. An attempt at partially interpreting these facts will be made in discussing the nature of the disease.

Effect of weather and temperature.—The fact that this fever usually appears at a certain season of the year is of itself an indication that the weather or temperature may have a considerable effect on its development; but, fortunately, we have more definite evidence than this.

The development favored by hot and dry weather.—Major Ragland, of Halifax County, Virginia, writes me that the disease is more likely to appear and is most destructive in hot and dry weather.

It was during the hot and dry weather of August and September, 1879, that the disease appeared and was most fatal in Buncombe and Henderson Counties, North Carolina.

Mr. V. P. Chilton, of Southwestern Missouri, wrote to the commissioners of the Illinois Agricultural Society, in 1868, that the disease is more virulent in some seasons than in others, the excessive heat of this summer causing it to be worse than usual.

Dr. Albert Badger, of Nevada City, Vernon County, Missouri, wrote to the New York commissioners (Report, p. 91) that the fatality is much greater in a warm and dry summer than in a cold and wet one.

Professor Gamgee, though his investigations only extended through a single year, was of the opinion that heat and drought favored the development of the disease.

The disease arrested by cold weather.—In the South, as well as at the North, the disease disappears after a few killing frosts. In Texas, newly imported stock is said to be liable to the fever in January, but I have not been able to find that this is the case in any other Southern State. When the disease has been carried a considerable distance beyond the line of the permanently infected district, the active principle of the affection is destroyed, and the malady does not reappear till freshly imported.

Dr. Albert Badger, in the letter just referred to, says the disease always ceases when the frosts and freezing weather have killed the vegetation.

Professor Gamgee says a few nipping frosts check its ravages anywhere and everywhere.*

Now, what meaning are we to gather from these observations? Does a slight degree of cold below the freezing point suffice to destroy the germs to which this fever owes its origin, or is the result due to some other and indirect effect?

Atlanta, Ga., is in a permanently infected district, and the temperature is not unfrequently as low as 10° F., on one or more occasions during the winter. In a large part of the infected district the germs of the fever must survive a temperature of 5° to 15° F. nearly every winter; and yet we know that this does not destroy them, although the earth is not protected as at the North with snows, and is consequently subjected, at the surface, to frequent freezing and thawing, which is supposed to be a very destructive process to most organisms.

If now we conclude, as I am inclined to, that the germs have survived nine consecutive winters in Henderson County, North Carolina, where the temperature has been at times several degrees below zero, we must admit that it is not alone the destructive effect of a freezing temperature on the germ-particles that stops the disease after the first few frosts; but that this result follows rather from the death of the vegetation on which animals have been grazing. The reason for this conclusion will be more apparent when we come to consider the manner in which the disease is transmitted; nevertheless, it is certain that the germs do not survive the winter when carried a long distance from the infected district.

Is there a probability of the fever becoming permanent in the Northern States?—It is not my purpose to throw any doubts on the observations of those who have seen the disease invariably disappear in all localities at any considerable distance beyond the boundary line of the permanently infected district. But there is another class of facts here that have never received attention, and which deserve most careful consideration.

If the fever has been destroyed by frosts and freezing weather at the North, that is no more than once happened in North Carolina. When carried a considerable distance beyond the boundary line, to which I have so often referred, whether this occurred in North Carolina or Illinois, the fever was invariably ended by the following winter; but, nevertheless, the boundary line has continued to advance; it has embraced places where it formerly could not survive the winter; it has not been permanently destroyed by temperatures equal to the lowest point reached by the mercury in many winters in the Middle States, and this, notwithstanding the absence of a protecting covering of snow; it has crossed the great natural barrier of the Blue Ridge; it has invaded Virginia, and we cannot doubt, from facts I have already given, that with this advance the germs acquire an increased power of resistance to low temperatures.

This is shown (1) by the laws which at first allowed cattle to be driven after the first of November, but which were then changed so that they could only be driven after the first of December, and again changed to prevent infected cattle going into uninfected districts even in winter, although each law applied to a colder section of the State than the preceding one; it is further shown (2) by cattle bringing the infection into a district in the early spring immediately succeeding a cold winter, as oc-

---

* John Gamgee, M. D., Report on Splenic or Periodic Fever of Cattle, in Diseases of Cattle in the United States, Department of Agriculture, 1871, p. 117.

curred in the outbreak in Buncombe County, North Carolina, in 1879; and, finally, it is proved (3) by the germs surviving the winters in sections where they were formerly destroyed.

I learn, also, in a private note from Professor Law, that this disease has at times survived the winter at Cincinnati, Ohio, but this was supposed to be owing to protection of the germs in the manure of the cattle-sheds.

Whether this acclimation of the germs will continue, and the disease advance over States now entirely free from it, may be a point impossible to definitely decide before this has actually occurred, and an irreparable damage has been inflicted; but so far as one can decide from indications, the danger of such a continued extension is a *real* danger, and one which we cannot neglect without rendering ourselves liable to the most disastrous consequences.

The theory of its climatic origin.—Against such a conclusion, will at once be urged the theory generally accepted at the South that this is not a communicable disease; that it arises from a change in climatic conditions, in food or water. Or, the modification of this theory, that it is due to the influence of those conditions which cause intermittent fevers in man, and is, therefore, capable of but a limited extension beyond such districts.

It is not my intention to consider the nature of the malady, at this place, farther than can be done by the aid of historical incidents, and the reader is referred for a more complete discussion to the section devoted to this topic. It is proper to call attention here, however, to the fact that by far the larger part of the great district infected in Alabama, Georgia, South Carolina, North Carolina, and Virginia, is a very different country from that to which its origin has been attributed in Texas. The land is elevated, rolling, and well drained; the soil is sandy, permeable to water, and is almost free from organic matter—the very opposite of the conditions which Professor Gamgee has considered necessary to its origin and permanence. And, more than this, the eastern foot-hills of the Blue Ridge, on which the fever is now firmly established, are as free from malaria and from intermittent fevers of people as are any parts of the Middle or Western States.

The other theory, that it is due to a change in the conditions of life of the animals, will not bear a moment's thought; for certainly, in such case, it would be the animals moved north and not the natives with which they come in contact, that would suffer.

A fact suggestive of its climatic origin, says Mr. Dodge (Report, p. 178), showed its existence in the mountain lands of Georgia, where it was generated by the presence of lowland cattle that had scarcely been removed the distance of fifty miles. If such facts as this are suggestive of the climatic origin of a disease, what contagious maladies are there that would not be ranged in the climatic category? The fact that the animals were removed so short a distance is rather evidence that it is not of climatic origin.

Again, many in Georgia consider that a chief cause of the disease is removal of animals from limestone to freestone lands; and that, hence, cattle brought from New York or New Jersey are less apt to sicken than those brought from Tennessee or Kentucky. Now, aside from this theory ignoring the transportation of a disease from the South to the North, and the continued prevalence of a disease along what I have called the border line of the infected district—in either case a disease evidently not caused by change of climate or conditions of life by the sick animal—I doubt the historical facts on which the conclusion is based.

Some parts of New York and New Jersey are limestone formations, as is well known, and, consequently, cattle from these States are not always from a freestone district, as is supposed. Again, cattle taken from Buncombe County, North Carolina, a freestone country, to any region east which is also freestone, will as surely die as those brought from Kentucky or Tennessee to Georgia. Hence, we cannot doubt that this explanation of the greater loss among the cattle from these neighboring States is an erroneous one; and, if I should venture an opinion, I would be inclined to say it was due to the less susceptibility of the breed of cattle (Jerseys) almost exclusively imported from New York and New Jersey.

Difference in susceptibility.—In all outbreaks of this disease, as is also true of all other contagious diseases, there are some animals which, though exposed to the essential cause, never contract the malady; again, among the adult animals brought South there is a small proportion which are never affected by it. These animals, for some reason which is not understood, are insusceptible to the disease; this is *individual insusceptibility.*

Going a step further we find that different breeds have a special susceptibility or insusceptibility to the destructive action of the peculiar principle which causes the Southern fever. Colonel Peters, whose great experience with cattle of different breeds at the South I have already referred to, found that the short-horns were exceedingly susceptible, and that scarcely a single one would survive being taken into an infected district; while, on the other hand, the Brahmas were absolutely insusceptible to it, and might be sent to any part of the South with the greatest impunity.

He suggeseed that an explanation of this might be found in the lymphatic temperament, the tendency to fatten, the decreased relative size of the lungs, in the short-horns, which would make them less able to withstand the effects of heat; but this suggestion was evidently inspired by a partial belief in the climatic theory already discussed, and so generally believed in Georgia. In the present condition of science, with so little definite knowledge bearing on this point, it is probably best neither to contest nor favor this explanation, farther than to once more call attention to the fact that there is something in this district to be withstood besides the effects of heat, if cattle are to enjoy an immunity from the ravages of Southern fever. That one breed is entirely insusceptible to this disease is certainly an important fact to place on record, and I am glad to be able to credit Colonel Peters with the discovery.

The Jersey cattle seem to be less susceptible to this fever than either short-horns or scrubs, and this is probably one reason why they are so rapidly growing in favor at the South; thus, although Mr. Montgomery, of Mississippi, imported twenty-two Jerseys in the spring, and although twenty of these had the fever, but seven died; if they had been short-horns of the same age, he would have been fortunate if a single one had survived.

In regard to other improved breeds, I have no definite information.

Do native Southern cattle suffer from this fever?—The great mass of Southern people, if asked this question, would undoubtedly say, no; and, yet, we occasionally hear of cases, as from Georgia and Texas, already quoted, where large numbers of native cattle die from it. These cases are generally, however, along the border line of the infected district; and the purpose of this inquiry is to ascertain the susceptibility

8 C D

of those cattle which have been to a degree inured to the causes of this disease by constant exposure to them.

The only investigations on this point were made in 1868, on Texans. It has been already remarked that considerable numbers of these died daily during the hot weather at Cairo. The heat, crowding, fatigue, and suffering of the animals so exhausted their vital power as to allow the disease germs which they carried in their systems to gain the mastery. *Post mortem* examinations were made on Texans sick of Southern fever at Buffalo, which demonstrated that the lesions, as well as the symptoms, corresponded with those of the diseased Northern cattle.[*]

Other examinations made at the Communipaw slaughter-house showed that Texans which did not clearly exhibit symptoms of the affection, but which had a general unhealthy appearance, presented certain lesions which are characteristic of the disease; that is, the spleens were enlarged, and the stomachs presented erosions, sloughs, and extensive cicatrices. Subsequent examination of a large number of apparently healthy Texas bullocks disclosed the fact that without exception the abomasums contained numerous cicatrices and erosions in process of healing, and there was also a general hyperæmic condition of the mucous membrane, showing a state of chronic gastritis.[†]

The weighing of a large number of spleens, under the supervision of Dr. Rauch, medical officer of the city of Chicago, showed that the Texan spleens averaged two and one-half pounds, while those of natives averaged but one and one-half pounds.[‡]

Professor Gamgee went further and made *post mortem* examinations of these cattle at their home in Texas, and he found an abnormal weight of the spleen, coupled with gastric redness, erosions, pale blood and frequently the presence of bloody urine in the bladder. In this connection he remarks that there are few fevers that do not at times attack animals in such a way as to produce so little general disturbance as to prevent their recognition in the living animal. He has seen such cases in the rinderpest, lung-plague, and sheep-pox. In enzoötic maladies, especially in anthrax, it is often found in *post mortems* of animals from districts where such diseases arise that the healthiest and strongest are suffering or have suffered changes which a special systematic vigor or constitutional resistence hides, so long as the animals are in life. In one case, in Kansas, a sudden shock to the system of a steer by being stampeded caused the development of the fever; other animals there apparently fresh and healthy had abnormally high temperature; so that it became evident that a large herd, from the home of this disease, not only carries the active cause of it in the systems of the animals, but also the evidence of a specific disease induced, which remains for an indefinite time latent and unobserved.[§]

The length of time the poison is retained by animals from infected districts.—It was found, in 1868, that a two months' journey from Texas to the Union Pacific Railroad in Kansas was not sufficient to rid the animals of the poisonous principle which they distributed. Professor Gamgee states that without a doubt cattle driven into Kansas, Missouri, or other States, in the summer or autumn of one year, and grazed there till the next spring or summer, failed to retain any deleterious principle.[‖]

Mr. Wilkinson stated that the drive from the thirty-fourth parallel in

---

[*] Report of New York State Cattle Commissioners, 1869, p. 77.
[†] Soc. cit., pp. 71, 76, and 77.   [‡] Gamgee, loc. cit., p. 83.
[§] Gamgee, Report, pp. 86, 87.   [‖] Report, p. 115.

South Carolina to the Virginia line, 250 miles, was sufficient to free them of it.*

On the other hand, we have the case from Southwestern Missouri, where a drove of cattle were wintered at Sarcoxie, and, when driven 50 miles north the following June, conveyed the malady;† also, the statement of Dr. Mease that cattle driven from South Carolina as far as Pennsylvania still infected the native cattle.‡ There is besides the fact observed by me of a drove conveying the disease into Buncombe County, North Carolina, late in winter, which also seems to have a bearing. Altogether, it would appear that at least three to four months in an uninfected district is necessary to make these animals safe, and that they cannot be safely mixed with susceptible animals in the summer if they have left the locality where the disease is indigenous the preceding winter; also, that such animals traveling over ground in winter or early spring, in Western North Carolina, may infect the ground passed over, so that native stock running on it the following July or August will contract the fever.

How the active principle is communicated.—It is not transmitted by sick Northern stock.—One of the most peculiar features of the Southern fever is that while the natives outside of the infected region may be sick and dying of it, they do not communicate it to other susceptible animals. This has been the universal testimony of drovers, dealers, and farmers who have had experience with it.

Dr. Manheimer, of Chicago, reported that in many instances cows were housed in the same stable with others that were sick without contracting the disease.§

Mr. Hosack, cattle inspector at Pittsburgh, says that two diseased droves from Illinois were placed in the stock-yards where they were surrounded by other stock, but in no one instance did the disease show itself but in the two herds mentioned.||

Mr. E. C. Hall, of Onanga, Ill., who saw much of the disease, says: "The fact is, the native stock do not seem to give the native stock disease."¶

Mr. Fred. A. Atkins, of Illinois, wrote:

I have no reason to believe that native cattle, even under circumstances the most favorable for infection, will infect other native cattle. Not one of the many I have seen die of the disease but that was exposed to infection from Texas stock; and not one of those in this vicinity now living but that was exposed to sick native cattle. I have seen a calf, now living and in good health, that was suckled in succession by three different cows, that died of the disease in its most aggravated form, while they were sick, without deleterious effect on its health."**

Professor Gamgee did not find a single instance where it was transmitted by Northern stock.†† On the other hand, Dr. Rauch, the medical officer of Chicago, wrote as follows:

With regard to the transmittal of this disease by native cattle, I must confess that, notwithstanding the weight of testimony against it, I am inclined to believe that such can and does take place. Several instances of this character fell under my observation during the past three months, but the most conclusive evidence I have is that native cattle were purchased at Chicago in August and taken to Lebanon County, Pennsylvania, and that a short time after they died, and that other native cattle on the same farm and in the neighborhood died, and that no Texas cattle had been near the place.‡‡

* Reports on Diseases of Cattle in the United States, p. 172.
† Loc. cit., p. 188.   ‡ Loc. cit., p. 176.
§ Report New York State Cattle Commissioners, 1869, p. 82.
|| Loc. cit., p. 84.   ¶ Loc. cit., p. 85.   ** Loc. cit., p. 85.   †† Report, p. 115.
‡‡ Report New York State Cattle Commissioners, 1869, p. 96.

The New York commissioners also report a case where native cows from Painesville, Ohio, communicated the disease to other native cattle in Orange County, New York, and from which eight animals died. *Post-mortem* examinations were made on some of the cattle dying in Orange County, and the Ohio cattle were traced with much care.*

We may conclude, then, that while in the vast majority of cases the disease is not spread by Northern cattle, there may occasionally be an instance where this is the case.

It is not contracted directly from Southern animals.—In regard to this point there seems to be less evidence on record. The two cases occurring on Mr. Alexander's farm of Broadlands, referred to under the heading " Peculiarities of incubation," are, however, very clear evidence in this regard. Professor Gamgee observed similar instances; and, indeed, is, I believe, the only one who has called attention to the subject. As proof that Texans do not infect other cattle directly, he refers to an observation in Chicago. Mr. Sherman, of the Union stock-yards, had thirty-five cows, which grazed all summer close up to the cattle-pens where thousands of Southern steers were continually inclosed. Of these the majority had been purchased out of the yards at different times, some had been in the cattle-pens with Texan droves. A Texan calf was even placed with the cows, yet no animals could be in better health. †

How communicated.—The fever seems to be only contracted, then, by native cattle grazing on lands which have been previously grazed or traveled by Southern cattle. It has been claimed that simply driving across a road over which Texans had passed was sufficient for natives to contract the disease; but, in view of the facts cited, it may be doubted if the fever would attack animals unless they had eaten grass which had been soiled by the excretions of the Southern animals, or, at least, which had been so near these excretions as to be contaminated by them. In the vast majority of instances a good fence is sufficient protection; indeed, there is but one case on record, to my knowledge, where cattle in a well-fenced inclosure, that Southern cattle had not been allowed to enter, contracted the fever. That case was at Tolono, Ill., in 1868, but the whole country was then so thoroughly infected that it would not be strange if either these cattle had succeeded in getting at the grass beyond the fence, or if the excretions of Texans had been washed by rains into the inclosure.

Its extremely fatal character.—If the destructive work of this malady is limited to a few months of the year, the great susceptibility of most cattle to it, and the large proportion that dies, are more than an offset to this; indeed, it would be difficult to find an equally destructive disease in the whole list of epidemics or epizoötics. Within two miles of the Chicago stock-yards, according to Dr. Manheimer, but one cow escaped the disease, while one hundred and forty-seven died; at Tolono, Ill., but two cows were left within two miles of the town, two hundred and thirty-five having died between the 20th of July and the 1st of August; at Cincinnati, Ohio, twenty-nine cows exposed and all died; at Dayton, 50 per cent. of those exposed were affected; at Odell, Ill., 40 per cent. of the cattle exposed sickened; of the herd of nearly six hundred and fifty cattle exposed at Broadlands two-thirds are known to have died; in Vernon County, Missouri, the loss along the infected roads was 50 per cent. of the whole stock, and at the fording places 90 per cent.; of a herd of forty found September 12 on board the Fah Kee, bound for Bermuda, fourteen had died the previous night, and three more were lost subsequently, and so on, wherever the disease was noticed.

In regard to susceptible animals taken South, we have seen that a large proportion take the disease and die, some the first, but many the second summer. Thus, of twenty-two Jerseys, twenty sickened and seven died; of the eight cattle taken to Mississippi by Mr. Foote seven died; eight Short-horns taken to Mobile all died. By proper selection and management it would seem, from the experience of those who have imported on a large scale, that cattle can be so inured to this disease that thoroughbreds may be taken South with little loss. (See Acclimation.)

Horses affected.—Very little study has so far been given to the disease so often affecting horses taken to the South; it is sometimes referred to as the result of a change of climate, at other times we hear of it as "blind staggers," and still oftener the cases are not reported. Were it not for the fact that in Illinois, where the Southern fever was most fatal in 1868, it was followed by a similar disease among horses, we should hesitate in attributing the losses in acclimating horses in the South to the same disease. As it is, it would certainly be desirable to study the disease in these two families of animals at the same time, and it is possible this would throw light upon the very considerable losses among horses attributed to "blind staggers."

Effect of eating the flesh of diseased animals.—In 1868, the metropolitan board of health found that the use of meat from animals affected with Southern fever caused an enormous increase in the death rate by diarrheal diseases, and especially was there developed an utterly obstinate and incurable class of choleraic diarrheas.[*]

Dr. Mackey, health officer of Buffalo, fed two dogs upon the diseased flesh, but it was soon vomited, and in neither case would their stomachs retain or digest it.[†] In New York three dogs were kenneled and fed sparingly on two enormous livers taken from cattle that had been slaughtered for examination. Two of them were attacked with an obstinate diarrhea and one died at the end of twelve days. The stomach of the latter was ulcerated, but the essential signs of the disease were not found. Rabbits, fed upon bread soaked in the bile of diseased cattle, died in from one to four weeks, according to the amount ingested, with many of the phenomena of the Southern fever.[‡]

SYMPTOMS.

Temperature.—A considerable elevation of temperature is the first symptom of the disease. At Tolono, Ill., Professor Gamgee examined a number of cows, apparently healthy, which gave the following temperatures: 106°, 106.5°, 106.7°, 106.7°, 106.1°, 107.2°, 106.7°, 107.2°, 104.2°, 106.7°, F. At other places similar observations were made (Report, p. 89). A bull in the first stages examined by me, near Hendersonville, N. C., had a temperature of $103\frac{1}{2}^\circ$. It seems, therefore, that a considerable elevation of temperature precedes the other manifestations of the disease by several days.

During the greater part of the course of this fever the high temperature is maintained, as is shown by many observations. Dr. Morris, of the New York commission, records about twenty cases, varying from $101\frac{1}{2}^\circ$ to 110° F., and averaging $105\frac{1}{4}^\circ$ (Report, pp. 56 to 80). Professor Gamgee has recorded sixty observations, varying from 98.6° to 107.4°

* Report of New York State Cattle Commissioners, 1869, p. 98.
† Loc. cit., p. 82.   ‡ Loc. cit., pp. 122, 130.

F., and averaging 105° (Report, p. 90). As death approaches the temperature gradually sinks to, or even below, the normal.

General appearance.—The first symptom is an unmistakable appearance of languor and fatigue; the animal may remain at a distance from its fellows, with drooping head, lopped ears, and an arched back. The eyes are staring, the coat rough, frequent efforts are made to dung, which are often ineffectual. The horns are hot and the nose dry. As the disease advances the drooping of the head is carried to such an extent that the nose nearly, or quite, touches the ground; the hind legs are advanced under the body and placed in a bracing attitude; the fetlocks are bent; the gait is sluggish, and there is a disposition to lie down, particularly in water. Though covered with flies, the animals seem unconscious of their bites.

Pulse and respiration.—The pulse is frequent or rapid; in some cases it is hard and wiry, but in others it is soft and feeble. As death approaches it may be over one hundred per minute.

The respiration is increased in frequency until in hot weather it becomes labored.

Excretions.—The droppings are dry, scanty, and stained with blood; in exceptional cases there is diarrhœa.

The urine is generally colored with blood in the earliest stages of the affection, but occasionally this does not occur till the disease has reached its height; it is passed frequently and in considerable quantities. It coagulates with heat or by addition of nitric acid. In some cases so much of the blood-coloring matter is contained in the urine that this becomes almost black in appearance.

Weakness and partial paralysis.—There is often tremor of the flank muscles; the gait is unsteady and staggering, with swaying of the hind quarters. In a considerable number of cases there is an inability to control the voluntary muscles, from a defective co-ordination of movement. While there is listlessness and indifference to surrounding objects, there is also restlessness from a desire to lie down, though a seeming fear to do this till compelled by loss of strength. At this time they are frequently unable to raise themselves, and if helped to their feet can only stagger a few steps and drop to the ground again.

Delirium.—Very often the animals become delirious, and stand pressing their heads against each other, against a fence, or some other unyielding object, or with their nose resting near or on the ground. Some are frantic, though too feeble to do injury.

The last stages.—The animals do not lose their appetite at once, as occurs with most other acute affections, but, in spite of this being retained, they rapidly waste and become lean, the flanks are drawn in, the abdomen raised, and the animals acquire in a few days that appearance which is seen in other wasting diseases only after weeks of illness.

The indifference and listlessness increase with the progress of the disease, and finally the animal sinks into a profound stupor, stretches out upon the ground, and dies almost without a struggle.

In other cases the febrile phenomena, the indifference and other symptoms gradually subside, and the animal becomes convalescent in from ten days to two weeks; but it is months afterwards, as a rule, before they have regained the perfect health and thrift possessed before the attack.

POST-MORTEM APPEARANCES.

General appearance.—The New York commissioners found that, with animals dead of this disease, the decomposition was so rapid that the special lesions were almost entirely obliterated within three or four hours (Report, p. 80). Professor Gamgee does not record any observation which would lead one to believe that the putrefactive changes are more active in this disease than in many others; and it has certainly seemed to me that blood gathered from animals slaughtered for examination resisted decomposition fully as long as from healthy animals.

Effusions of serum and sometimes of blood have been witnessed under the lower jaw and sternum. On opening the animal the muscular tissue is seen of a dark red color ; the fat is of a deep brownish yellow, having in intense cases a green bronzed tinge, and has not the firm resistance of health. When cut into, the lean meat has a peculiar sickening odor; in some cases the superficial muscles may have a natural color, but the deeper ones, and especially the intercostal muscles (those between the ribs), have the dark brown color and nauseous smell. These appearances were particularly marked in nine out of eighteen cases.

The above observations, made first by the New York commissioners, I have been able to confirm in part, though Professor Gamgee was unable to discover any difference between the color of the muscles in such animals if examined immediately after death and those slaughtered in health.

The subcutaneous areolar tissue may have a pallid appearance or yellowish tinge, but is not congested as in anthrax.

It is not unusual to find petechiæ scattered over the serous and mucous membranes and dotting particular organs.

Digestive tract.—The œsophagus and first three stomachs, as a rule, have a normal appearance. Very often the third stomach is reported dry and obstructed, and though this sometimes occurs as in all fevers, it is not seen oftener than in other febrile affections. Professor Gamgee found no special lesions or abnormal conditions in connection with this organ, and this accords with my own observations.

Fourth stomach.—It is here that the most constant lesions are found. The whole mucous membrane may be more or less congested, but this is most marked at the cardiac end, where there is generally very dark redness. Here, too, are found ecchymosis, yellow, granular elevations and erosions. The pyloric portion, however, is the seat of the pathognomonic lesions. It is more normal in color, but it invariably presents erosions, sloughs, and deep excavated ulcers of various forms and extent; in animals long sick, in those convalescing, and in native Southern cattle some of these will be found healing, while extensive cicatrices mark the positions that have been occupied by others.

Intestines.—The small intestines are often the seat of petechiæ and congestions, while engorgement and blood extravasations are common in the large intestines, particularly in the rectum. The epithelium lining these organs is frequently softened and is easily scraped off by the finger.

Liver and gall bladder.—The liver is often enlarged and increased in weight, generally fatty or waxy, its bile-ducts and radicles fully injected with bile, and its color changed to a yellowish brown. In five cases out of seventeen recorded during the New York investigations it weighed over twenty pounds, in one case reaching twenty-seven and one-half pounds, and in twelve cases it weighed over fifteen pounds. Still, it must be admitted that many times similar enlargements are found in healthy cattle, and that these animals are somewhat inclined to chronic changes in this organ.

The gall bladder is almost invariably distended with an accumulation of dark, thick, flaky, or even tarry-looking bile, in some cases weighing nearly four pounds. In early stages of the disease I have found the bile thin and very fluid, but this receptacle was, nevertheless, distended to its fullest capacity. The lining membrane of this and the hepatic duct is frequently or always found congested and the coats thickened.

Urinary organs.—Kidneys.—These are usually congested, enlarged, and the cortical substance softened. In ten out of fifteen cases they were enlarged; in nearly all dark in color and congested; in one they were pale, and in two perfectly disorganized. In a few cases they appear entirely normal.

Urinary bladder.—This is almost always distended with dark, bloody urine, even though the animal has micturated just before slaughter. In three cases the mucous membrane was thickened, in one it was inflamed, and in several there were numerous petechiæ. The quantity of contents is from one to two quarts.

Spleen.—The spleen is always enlarged, more or less engorged with dark-colored blood, frequently softened, and in some cases is simply a disorganized, structureless, pulpy mass. In one case, recorded by Dr. Morris, it was two feet six and one-half inches long, and weighed nine and one-fourth pounds.*

The following table shows plainly the very considerable enlargement of the spleen even with healthy Texans, also the weights of livers, as obtained by Dr. Ranch, medical officer of Chicago:

	Spleens, average pounds.	Livers, average pounds.
2,607 native Western cattle	1.45	12.263
954 Cherokee (?) cattle	1.94	10.529
782 Texan cattle	2.50	12.236

A marked reduction was found in the average weight of the spleens in Cherokee and Texan cattle from August to September, thus:

	Native.	Cherokee. (?)	Texan.
	Pounds.	*Pounds.*	*Pounds.*
August weights	1.38	2.36	2.83
September weights	1.45	1.94	2.53

Heart and lungs.—The heart is frequently blood-stained both on its inner and outer aspects. The lungs are generally normal, the most frequent lesion being interlobular emphysema; they are sometimes covered with ecchymoses and in other cases collapse imperfectly, though retaining their normal color.

The blood.—Is nearly normal in appearance, and coagulates quickly, firmly, and completely; it is considered by most observers to be more fluid and watery than in health.

HISTOLOGY.

Nervous system.—Professor Gamgee found, in all cases in which partial paralysis in the hind quarters alone was marked, the upper coruna of the gray matter in the lumbar region reddened, with blood extravasations and staining of the nerve cells. This appearance could be traced in all parts of the cord in cases of more general paralysis; and in

---

* Report New York State Commissioners, 1869, p. 84.

one instance, in which it was most general and marked, there was blood extravasation outside of the dura mater, beneath the medulla oblongata. The gray matter of the medulla was itself slightly stained. In one instance he found the inner surface of the dura mater studded with bright red spots, similar to the small ecchymoses seen in the urinary bladder. Beyond a tendency to congestion and occasional blood extravasation, no lesion was discovered in the nervous system; both white and gray matter were usually firm and not softened.—(Report, pp. 96 and 97.)

Dr. R. Cresson Stiles, of the Metropolitan Board of Health, has studied the histology of this disease with great care, and I am indebted to his report for the remaining observations on the microscopical characters.

Blood.—The white corpuscles were in excess only in one case; the liquor sanguinis was yellower than in health and contained minute yellow flocculi of granular matter; the source of the yellow color was indicated by the occasional presence of crystals supposed to be of cholesterine. In several instances, in specimens of defibrinated blood and serum poured from the clot, complete dissolution of the blood globules took place in a few hours and long before putrefaction.

Urine.—This rarely contained blood disks, but coagula of fibrine inclosing *débris* of corpuscles, dark crimson granules, and casts of the *tubuli uriniferi* were of frequent occurrence. The dark red coloring matter of the blood globules was diffused through the urine.

Bile.—Instead of the normal greenish or brownish transparency, it was opaque, thick, and grumous, nor was its opacity lost by filtration. It contained granular flakes and masses of a brilliant yellow or orange tint to transmitted light. These were so abundant in many cases as to give the bile a semi-solid consistence; when most dense they were mingled with crystals of hematoidine; sometimes they were impacted so that they might have offered resistance to the flow of bile into the gall bladder; some of the yellow coagula had been molded into cylindrical casts by the smaller biliary ducts and were found in this shape in the gall bladder. It was evident that the source of the yellow flocculi was an admixture of blood with the bile in the most minute ducts within the substance of the liver; the blood corpuscles being dissolved, the coagulating fibrin imbibed the bright yellow dye of the mingled coloring matter of the bile and red blood corpuscles. Both blood and bile contained micrococci in large numbers, which, when cultivated by Professor Hallier, of Jena, produced a variety of fungus not before described, and which he named the *Coniothecium stilesianum* in honor of the discoverer.

The liver.—Under a power of five hundred diameters, the ultimate biliary radicles were seen to be distended with a bright yellow secretion. About the center of each lobule and shading into it was a zone of fatty degeneration which affected the superficial portion of each lobule. The mucous membrane of the hepatic duct was always of a crimson hue.

Spleen.—The nuclei, or the same with delicate cellular investment, which, in the natural state, filled the closed vesicles of the Malpighian bodies and the trabecular interspaces, were replaced by large cells undergoing fatty degeneration. In some cases yellow flocculi were found free in the splenic pulp; in others, cells filled with black pigment granules and crystals of hæmatoidine were abundant.

Kidneys.—The *tubuli uriniferi* of both the cortical and tubular portion were rendered opaque by a deposit of granules of fat in their epithelium, and their cavity was occupied for the most part by coagula reddened or blackened by *débris* of blood corpuscles and by granules of dark-crimson pigment, with occasionally a recognizable blood disk.[*]

---

[*] Report of N. Y. S. Cattle Commissioners, 1869, pp. 127 to 132.

CHEMISTRY.

According to analyses of Prof. C. F. Chandler, it was found that there was a considerable decrease in the total solids contained in the blood. In every case there was a marked loss of red blood globules, amounting to a loss of more than fifty per cent. of the total quantity that healthy bullock's blood contains, and in the most striking cases the loss of these globules was almost complete. The albuminous state of the urine was in all cases marked, and the liver was found to contain a greatly increased proportion of fat.*

INTERPRETATION OF THE MORBID CHANGES.

Dr. Stiles concluded that the most characteristic phenomena of this fever were explicable as the results of a hepatic affection. The repletion of the ultimate bile ducts induced the absorption of bile into the general circulation; this in turn caused the dissolution of the blood corpuscles and was an active cause of the hæmaturia. The solvent action of the bile upon the blood corpuscles, and the consequent liberation of their coloring matter, can be witnessed, he says, under the microscope, each disk disappearing suddenly like a light blown out, and the liquid assuming an orange tint. In the experiments of Kuhne and Frerichs, the injection of bile or its salts into the blood was followed, in the great majority of cases with the lower animals, by the appearance of blood in the urine. The blood thus altered in character becomes liable to extravasation; other hemorrhages than hæmaturia being frequent attendants upon attacks of jaundice. The yellow flocculi of the bile were attributed to the admixture of blood with this liquid; the corpuscles being dissolved, the coagulating fibrin imbibed the bright yellow dye of the mingled coloring matter of the bile and the red blood corpuscles. The flocculi of the liquor sanguinis and spleen were attributed to a similar source, i. e., a mixture of the bile with the blood.

It must be admitted that this seems a very reasonable theory, and yet when I mixed blood and bile from an animal suffering from this malady with the most pronounced hæmaturia, the blood globules not only refused to disappear like a light blown out, but they were not in the least affected by it. Of course, a single case is not sufficient to found a general principle upon; but still, if the hæmaturia is due to the solvent effect of the bile on the blood corpuscles, then in every case where the hæmaturia exists the bile ought to have the property of dissolving the corpuscles.

I have no doubt there is, in most cases, a mixture of bile with the blood; the yellow flocculi and yellow staining of the tissues, fat, and mucous membranes are sufficient proof of this; nor do I see reason to doubt the admixture of blood with the bile. But that the hæmaturia is the direct result of bile in the blood I am not prepared to accept; it appears to me that it would be more reasonable to attribute it either to changes in the blood from other causes or to lesions of the kidneys.

There seems to be a tendency to hæmaturia in many contagious fevers and allied maladies; it occurs in scarlatina, in typhus fever, in smallpox, in malignant measles, purpura hæmorrhagica, and yellow fever in man; and in malignant anthrax, typhoid of horses, swine plague, septicæmia, ergotism, and poisoning by cantharidin and other agents in animals, besides being an accompaniment to inflammation of the kidneys in whatever way produced. Various changes of the blood are,

* Loc. cit., pp. 123, 124, 137, 138.

also, followed by hæmaturia ; for instance, anæmia and plethora both give rise to it ; and in the *wood evil* of Europe it is supposed to be caused by astringent food.

Now, that the blood is greatly altered in Southern fever is proved not only by the ecchymoses occurring in nearly every organ at times, but also by direct chemical analysis. Again, the kidneys are found congested and engorged with blood, and casts of the tubuli are found in the urine, while in no case did Professor Chandler find bile in the urine, which would lead us to believe that the local changes in the kidneys may partly explain the passage of blood with this excretion.

All the changes, however, were considered due, either directly or indirectly, to the effects of the micrococcus found in the blood and bile; and the inclination was to accept this as the essential cause of the disease. In the single case which I had the opportunity to study, before the disease was arrested by the approach of cold weather, I was unable to find either micrococcus or cryptococcus cells in the bile; the reason may have been that the animal was slaughtered in the first stages of the affection, before these became numerous, or the germs may have existed and were overlooked, as so often happens in such cases.

The distention of the gall bladder with viscid bile is by no means confined to this fever; it is also usual in rinderpest, and is explained in our case by the inflammation and thickening so common in the mucous membrane of the hepatic duct. The enlargement of the liver, which is by no means general even in Southern fever, the increased size, engorgement, and disorganization of the spleen, the congestion and engorgement of the kidneys are all common in different contagious fevers, and various enzoötic affections, particularly anthrax, and in poisoning by those agents which act principally upon the blood.

If we confine our theory, then, to what strictly and legitimately follows from the lesions mentioned, we can only conclude that the exciting cause of the malady acts primarily and principally upon the blood, and that the disease consists essentially in physical, chemical, and possibly vital changes in this liquid, which is so necessary to the existence of the animal.

NATURE OF THE DISEASE.

No one can realize or regret more than myself the obscurity still surrounding the real nature and cause of many of the epizoötic diseases, and particularly the one under consideration, in spite of the many facts already discovered in regard to this and analogous affections which throw light upon it. And though we may not be able to reach conclusions entirely satisfactory, even to ourselves, it is yet of the greatest importance to review our knowledge, and to bring out the points that still need clearing up; for, as in all discoveries, a clear statement of the question is half the difficulty.

The tick theory.—One of the most widely-spread opinions in regard to the causation of Southern fever, is the pathogenic influence of the ticks with which Southern cattle are generally covered, and which migrate in large numbers to the bodies of other cattle with which they mix. But the acceptance of this view is simply an evidence of the desire of the human mind to explain the origin of mysterious phenomena. The same principle is exhibited in the popular views regarding the pathogenic nature of *hollow-horn, hollow-tail, wolf-teeth, black-teeth, hooks,* &c., none of which have the least foundation in fact or reason. The tick theory scarcely explains a single one of the many peculiar phenomena of the disease. Ticks are not kept from cattle by a fence. They are found everywhere, but are simply more numerous at the South. Their attacks are not confined to the latter half of the summer; nor would they be likely to remain on a pasture from spring till August without doing harm, and then suddenly cause an outbreak of the disease. Again, the *post-mortem* examination plainly indicates the cause of the disease to be an agent taken into the circulation, and causing the most important changes in the composition of the blood.

The theory of Professor Gamgee.—This gentleman, whose great reputation and attainments are so well known, spent ten months in the study of this disease, and his views certainly cannot be dismissed without a careful consideration. We will therefore examine them at considerable length.

This disease is enzoötic, he says, where the prevailing influences are such as to favor development of intermittent fever in man. The bilious remittent and intermittent fevers of man are represented in animals by the deadly *charbon* or anthrax, as also by the marked form of splenic fever now being discussed. He did not fail to find gastric erosions in Texas cattle, though he has opened as many as twenty-six per day. From the earliest date the calf feeds on grass, to the oldest day the bullock attains, the lesions are found; and during summer a few bullocks in a herd may be seen to drop behind the others, and to void bloody urine.

It is important to show how it differs from maladies which spread from country to country, devastating broad tracts of land, and calling for the most decisive and energetic means for their suppression. It is not an epizoötic, and not propagated by contagion. The true plagues or epizoötics spread by direct or indirect transference of an animal poison or virus from sick to healthy animals. These poisons take effect without regard to seasons, and are alike developed in the systems of sick animals. It is not contact with Texan cattle that induces the malady; the animals affected do not propagate it to other animals. Even coming in contact with Texans through a fence, drinking the same water, or being housed in sheds with sick natives does not induce it. We must therefore distinguish it from contagious maladies, and refer it to another group.

This is an enzoötic originating in various parts of the Gulf States. Florida cattle are as dangerous as Texans, deriving the same deleterious properties from the soil on which they are reared, and, in all probability, from the vegetation on which they feed. It is indigenous to the South, and, as far as Texas is concerned, is universally prevalent. Southern cattle become charged with principles which are thrown off in the excretions for many weeks, and they necessarily deposit a large amount of whatever they excrete; thus, pastures are contaminated, the grasses of which prove deadly poison to susceptible cattle. It is certain that feeding on land over which Texans have passed is the ordinary and probably invariable cause of splenic fever.

Southern animals become affected in consequence of the nature of the soil and vegetation on which they feed and the water which they drink; and with Northern animals it is a modification—a poisoning of the food, and possibly of the water, tainted by the manure of the Southern cattle. It is thus with human cholera. There is something tangible and ponderable, which some future chemist may reveal, that renders the grasses and perhaps the waters of the South so deleterious. It is not a contagious plague, and will probably cease when the agriculture of the South is fairly and fully developed.

There are two instances where enzoötic diseases may spread a certain distance by contagion—*i. e.*, in malignant anthrax and milk-sickness. There is another striking similarity between milk-sickness and splenic fever: the animal poisoned in the disease-producing district may show no signs of disease unless subjected to a definite exciting cause, such as being driven or frightened. In classifying milk-sickness or trembles, it should undoubtedly be placed among the effects of vegetable poisons. He would be disposed to deal with splenic fever in the same way.

The circumstances under which it manifests itself favor the view that it is allied to the numerous forms of anthrax which prevail in hot countries, usually in low lands, and especially in wet seasons. Splenic fever is propagated most readily by heat and drought; it is indigenous where vegetation is rank. The condition of the spleen would induce many a pathologist to classify it unhesitatingly with anthrax; but there is a line of demarcation. Southern cattle do not die in large numbers, as is always the case where anthrax originates, on the lands which foster the development of the subtle poison. Their systems are not charged with an inoculable poison, as in anthrax.

The darn of Aberdeenshire presents most of the characteristics of splenic fever—bloody urine, weakness of gait in hind quarters, associated in some cases with intense fever, in others with anæmia, bladder distended with bloody urine, liver light color, spleen congested, and blood extravasations. The opinion formed on the spot was that the darn of Aberdeenshire was caused by the cattle eating the young shoots of oaks and other astringent plants.

Since the introduction of turnip husbandry in Great Britain a malady has arisen among cows, after calving, which is usually known as "red water," due to the condition of turnips grown on ill-drained lands. This is a distinct form of disease due to modification in the character of the root grown on damp and retentive soils. It is therefore proved that the conditions of the soil may injuriously affect domesticated animals, and produce a definite and distinct disease through foods that are usually wholesome.—(Report, pp. 106–129.)

Professor Gamgee might also have added to his list of somewhat similar enzoötic affections the disease known in Italy as *gastro-hepatic fever*, in Spain as *hiel*, and mentioned by Gellé as *ictère des bêtes bovines*— a disease characterized by the same yellow colorations, listlessness, stupor, a weak, staggering gait, and delirium—and supposed to be due to feeding in wet, swampy pastures, to the pernicious and irritating action of putrid water, to sudden changes from moisture to drought, to severe work, fatigue, &c.*

It will be seen that Professor Gamgee has made an able presentation of his views; that he has drawn arguments to support them from his great experience in different parts of the world; but I think I may safely say that the investigations that have been made in the last few months in regard to the history and present distribution of the disease, render these views, which I have endeavored to fully and fairly present, entirely untenable. The great importance of this subject leads me to discuss with considerable detail the points just enumerated.

1. *The disease is not confined to the localities to which its origin was attributed.*—As I have already shown, this disease is now distributed over an immense area where the soil is elevated, rolling, sandy, well drained, and containing but the very smallest quantity of organic matter; where, consequently, the vegetation is sparse, and should be free from the poisons to which Professor Gamgee refers. This district

---

* Lafosse, Traité de Pathologie Vétérinaire, iii, p. 777.

comprises a great part of the northern half of Alabama and Georgia, the northwestern part of South Carolina, a belt in North Carolina extending from the Blue Ridge more than two hundred miles eastward, and the Virginia district, embracing Halifax and adjoining counties as far as known—a district in itself nearly or quite as large as England and Scotland combined.

2. *It exists permanently where intermittent fevers are unknown.*—If, now, we examine the district of the eastern foot-hills of the Blue Ridge, where the Southern fever is now recognized to be a permanent institution, we find the disease existing where, I believe it is justly claimed, there is perfect freedom from intermittent fevers among people; certainly in a region as free from them as the Middle or Western States, and where the heat is no greater in summer. Here, then, we have complete evidence to prove that this disease does not require a peculiar soil, or a particular quality of vegetation for its development, and that its permanence is entirely independent of miasma or any other conditions supposed to develop intermittent fevers; and with the demonstration of this fact must disappear all hopes of the disease ceasing with any possible development of Southern agriculture, since it must depend upon conditions over which this has no control.

3. *It is an epizoötic.*—It is difficult to see how it can be claimed that this disease is not an epizoötic. An epizoötic is a disease existing over a large extent of country, and capable of indefinite extension; an enzoötic is a disease attacking a considerable number of animals in a particular locality, and depending for its existence on local conditions. That a disease is an enzoötic affection is no guarantee that it may not become epizoötic, nor can we distinguish between such diseases by peculiarities of origin or transmission alone. It is the extent of country which is involved and its capabilities of extension that are to be considered. Cholera, for instance, is endemic in India, but that does not prevent its spread as an epidemic. The Southern cattle fever may, therefore, be properly recognized as enzoötic in the South, but it certainly possessed all the essential features of an epizoötic in 1868.

4. *It differs essentially from the enzoötics to which it was compared.*— Milk-sickness, or trembles, does not resemble the Southern fever in any important particular. It is not confined to the Southern States, but occurs in Ohio, Indiana, Illinois, and perhaps other Western States. Its origin is plainly limited to certain very small areas of land. In the mountain districts of the South, where it is so common, such localities are well known and their limits quite accurately defined. Such districts are seldom more than a few miles in extent, and are located in valleys or coves on the north side of a mountain; the sun shines on these but a few hours in the day, if at all; the soil is rich and full of moisture; the trees gigantic and interlacing overhead; the atmosphere damp and chill, like that of a deep cellar—the very conditions, one would suppose, for the production of the poison referred to by Professor Gamgee; but certainly the opposite of what we find in much of the country in which the cattle fever is enzoötic.

In milk-sickness the flesh and milk contain a poison capable of producing the disease in animals or persons who eat them. This poison affects dogs, sheep, and horses as well as cattle. It has an affinity for oil, so that it is chiefly removed from the milk in the butter. The poison is generally supposed in the South to be of a mineral nature, and is considered by some to be a form of arsenic. In Ohio there are those who claim to have produced it by feeding the white snake-root (*Eupatorium ageratoïdes*), but as this grows in many localities where the disease is unknown, that fed in these instances must have contained a special poison. Still others suppose they have found the true cause of the affection in cryptogamic vegetation; but, in reality, nothing definite is known as to the nature of the poison, nor have any systematic investigations ever been made in regard to it.

The fact that localities where this disease is produced are numerous, that they are scattered over a large part of the country, makes it very desirable that an effort should be made to learn in what the essential cause consists. It is not so much the loss of the animals, though this in the aggregate must be quite serious, as the effect on human health and life. for a certain number of people accidentally contract it every year near the home of the disease, and it may be produced by butter sent long distances to unsuspecting consumers, and may cause death at once or leave the victim but a wreck of his former self, with shattered nervous system and permanently impaired digestion.

According to Professor Gamgee, the bodies of animals affected with Southern fever contain no principle which can be inoculated, or which is injurious to man. Here, again, there is a wide difference.

Milk-sickness is transmitted in no other way than by eating the flesh or drinking the milk of sick animals. It cannot be conveyed, as far as known, from one living animal to another, except through the milk as just stated. It is, consequently, impossible for it to become an epizoötic. Southern fever, on the contrary, is a disease that is readily transmitted by means of pastures or roads, and it may extend, as we have seen, indefinitely, if the conditions of traffic are favorable for its transmission.

The only point of resemblance, therefore, is that the apparent symptoms of the affections may be developed as a consequence of fatigue or fright; and this is also claimed to be true of malignant anthrax and rinderpest in certain cases, and of the gastro-hepatic fever of Italy. This comparison, then, neither throws light upon the nature of Southern fever nor favors the view that this latter disease in an enzoötic and due to a similar poison.

Comparison with malignant anthrax.—We have seen that a mistake was made in supposing that the Southern fever originates exclusively on low, wet lands, and we will not dwell upon it now longer than to remark that as to origin there is consequently no resemblance in the character of the soil to which they are at present indigenous. Anthrax, besides, is a common disease at the North.

If our fever is propagated most readily during heat and drought, these are not the conditions most favorable to anthrax, for with this disease it has been observed that heat and moisture are most favorable to its development.

Again, anthrax is an inoculable disease; it is occasionally transmitted in stables, and to no great extent in pastures, and in none of these points is there a resemblance.

It is only the enlarged spleen, the congested and engorged kidneys, the enlarged, congested liver, and the hæmaturia that the diseases have in common; and these features occur, as has been mentioned, in a number of diseases, and in poisoning by various agents.

Compared with darn or enzoötic hæmaturia.—This disease was referred to by Professor Gamgee to strengthen two points: first, the similarity of symptoms was intended to impress the idea of the Southern fever being also an enzoötic; and, secondly, to "prove that the conditions of the soil may injuriously affect domesticated animals and produce a definite and distinct disease through foods that are usually wholesome."

European writers distinguish two kinds of hæmaturia, one characterized by anæmia, the other caused by feeding on astringent and irritating agents, and it is said that the food which causes the latter form will cure the former. In symptoms our fever most resembles the form caused by astringent and irritating food, i. e., by the buds and young shoots of oak, beech, ash, and willow trees, when cattle are turned into the woods in early spring. In each disease there is fever, a weak, unsteady gait, bloody urine, and a malady running its course in a few days. On *post-mortem* examination there is also some similarity: the spleen is engorged, the kidneys enlarged, congested, and of a dark color, and there are blood extravasations—lesions found in a variety of diseases, as we have seen. But, going further, there is considerable difference; the *mal de brou* of the French shows its origin by signs of inflammation in the entire digestive tract,* while the splenic or Southern fever is seldom accompanied by active inflammation of these organs.

The other form of hæmaturia is not only caused by turnips grown on wet, undrained land, but even by grass grown under such conditions; more than this, it may be caused by any food that contains a large quantity of water, such as green oats or mustard that has grown rapidly in spring.† It is probably, therefore, not due to a special poison, unless we consider water to be such, but to anæmia from the great quantity of water passed through the system. In this form there is little fever, and with the exception of the hæmaturia there is scarcely a point of resemblance with the disease we are studying.

In neither of these forms of hæmaturia is there any poison capable of transmitting the affection to other animals by any means whatever.

In a careful examination, then, of the different diseases to which this gentleman refers for support to his hypothesis that Southern fever is a strictly enzootic malady, capable of but limited extension, and caused by a poison developed in consequence of the nature of the soil on which it grows, we find no sufficient resemblance in the phenomena of the diseases to enable us to draw any such inferences, nor, leaving resemblance aside, are there any facts in regard to them favoring such views.

THE NATURE OF THE CONTAGION.

It is not a chemical poison.—By "chemical" is meant, in this connection, any poison derived from either the organic or inorganic kingdoms of nature that is not endowed with life. The peculiarities in the action of chemical as opposed to living poisons are (1) their effect is produced almost as soon as they enter the circulation; (2) the effect is in proportion to the quantity introduced; (3) the amount entering the body cannot increase itself there at the expense of totally different substances.

The period of incubation.—All observers admit that there is a period of incubation of certainly ten days, generally two to six weeks, and often of longer duration. Now, this is opposed to the idea of a chemical poison, because with these there is no such incubating period; the admission of an incubation is equivalent to admitting the cause to be a virus; and in the present state of science this is equivalent to admitting a living contagion.

If it is claimed that this period of incubation is simply the time required to take a sufficient amount of the poison into the system to pro-

* Traité Pratique des Maladies de l'Espèce Bovine, J. Cruzel, p. 93.
† Nouveau Dictionnaire Pratique de Méd., de Chir., et de Hyg. Vét., ix, p. 15. Lafosse, Traité de Pathologie Vétérinaire, iii, p. 761.

duce decided effects, we oppose to this the fact of the twenty-six cattle in Indiana which broke from their pasture and grazed on the infected prairie but a single night, or the other fact of the drove of fat cattle in Kentucky that simply traveled the road passed over by the Texans, and of which not one recovered from the disease. There are a vast number of similar facts on record that might be referred to if necessary.

The peculiarities of the incubation.—If cattle may graze with impunity on pastures at certain seasons of the year side by side with freshly arrived Texans, as seems to be shown by the observations at Broadlands; if the incubation may be long or short, as it is early in spring or not till summer when the exposure occurs; if pastures infected in spring or early summer may be grazed by susceptible animals as late as the first of July without deleterious effects; if hot and dry weather is most favorable to the development of the disease; if pasturing a single night on a prairie on which Texans had been but one night, and that four weeks before, is as fatal as pasturing a month, continually mingling with Texans, all of which points are well established, the evidence is certainly against the supposition that it is a chemical poison.

Resembles some of the characters of yellow fever.—While we know very little of the real nature of yellow-fever virus, it seems doubtful if the disease is ever communicated direct from the bodies of persons affected; but all agree that it may be carried in vessels, or in packages of merchandise or clothing, and is doubtless taken into the system with the inspired air.* Here, there is no poisoning of the food, as in the cattle fever; but in the amount required to produce the disease; in its suspension in the atmosphere; in its wonderful effects in producing a fatal, specific disease, we have the characters of other communicable fevers. Both yellow fever and Southern fever stop their ravages at the approach of cold weather; but with neither does the patient apparently give rise to a virus. The active agent seems to be transported without change from the section to which it is indigenous, and to be capable either of at once entering the system of susceptible individuals, or of existing and multiplying outside of the body.

In the one case, the poison is transported in the confined air of vessels or packages; in the other, it would seem to be carried in the vast reservoirs of the digestive tract, in which it probably finds the conditions necessary for its growth and multiplication. Here, again, we fail to find any resemblance to those chemical poisons which may be isolated and which are capable of solution, precipitation, crystallization, &c.

It encroaches on new territory.—We have proved, by reference to the laws of North Carolina, and by the testimony of reliable men, that the northwestern part of South Carolina and the whole of the central and western belts of North Carolina were but a few years ago free from the malady, except as imported from year to year. We have also proved by the laws of the State, and the present condition of nearly the whole of that part of the State east of the Blue Ridge, that the great central belt has become infected; that taking cattle there from further south no longer causes outbreaks of the disease, but that driving cattle from east of the Blue Ridge into any of the counties west of it does cause such outbreaks.

We have even pointed out the counties of Burke, Rutherford, and Polk, as instances of the progress of the disease within the last twenty

---

* A Treatise on the Principles and Practice of Medicine, by Austin Flint, 1868, p. 884. Traité des Maladies Épidémiques, Origine, Évolution, Prophylaxie, par Leon Colin, Paris, 1879, p. 844.

9 C D

years, and have shown that it was introduced into Virginia sixty or seventy years ago from Eastern North Carolina, and that it has remained there till this time. Now such facts are not true of chemical poisons; they are not true of diseases caused by modifications of the food and water in consequence of the character of the soil; and consequently we conclude from all these facts that we are dealing with a disease not produced by such a poison, and that is not due to the character of the soil and vegetation.

A living poison the probable cause.—To me, at least, the conclusion seems unavoidable that the poison is something that lives, grows, and multiplies itself, and that in this respect its characters are common with those of all communicable epizoötics.

Dr. Stiles discovered in the blood and bile of Northern cattle affected with the disease, a micrococcus which was only visible under a high magnifying power (about 1,000 diameters); this appeared when so magnified simply as minute points or granules, which under favorable conditions increased in size. They were considered as minute vegetable germs, and when cultivated by the accomplished Professor Hallier, of the University of Jena, they developed into a distinct fungus plant, of the genus *Coniothecium*, which he named, in honor of the discoverer, *Coniothecium stilesianum*. He says: "Perhaps you may succeed in finding out the places where this *Coniothecium* grows in nature. At all events *it is a parasitic fungus growing on plants, and to be looked for in the food of the wild bullocks.*"

Now it is true that Mr. H. W. Ravenal, who accompanied Professor Gamgee to Texas in the spring of 1868, failed to find such a fungus growing on the forage plants of that State; but the time of year was at least unfavorable, and no positive conclusions can be formed from only negative results.

The same may be said of the cultivations of the blood of Texas bullocks, made by Drs. Billings and Curtis, in connection with Professor Gamgee's investigations, and in which they did not succeed in obtaining any fungi except such as appeared under similar conditions from blood of healthy animals. This is also true of my own microscopic studies of the blood and bile of an animal slaughtered in the first stages of the disease. The difficulties of these investigations are so great and there have been so many failures in the past for every success, that we should be sufficiently warned not to lay too much stress on negative experiments or observations.

The virus of anthrax or charbon.—There is no disease to which we can refer, in support of our views regarding the living nature of the poison causing this fever, with such satisfaction as to malignant anthrax, thanks to the persistent scientific investigations of Davaine, Delafond, Koch, Pasteur, and Toussaint. Malignant anthrax is an enzoötic, and although communicable by inoculation was known to develop on low, moist, imperfectly-drained lands; it was not contagious to any great extent except by inoculation, accidental or otherwise; and, consequently, it might very reasonably be considered due to variations in the character of the food and water, induced by the nature of the soil. The points wherein it resembles the Southern fever, as well as those by which it differs from this, have been already pointed out.

As long ago as 1850, MM. Rayer and Davaine, of France, discovered microscopic filiform bodies, twice the length of a blood globule, in the blood of animals spontaneously affected or inoculated with anthrax.[*]

* Recherches Expérimentales sur la Maladie Charbonneuse par H. Toussaint, Paris, 1879, p. 20.

It was the organism which has since received the name of *bacillus anthracis*. Since then, thousands of observations have been made in different countries, and although it was long ago admitted that this organism was constantly present, and although it was almost accepted as a pathognomonic sign of this specific affection, nothing but the most convincing and overwhelming proof was sufficient to overcome the determined opposition to its acknowledgment as the essential and only cause of the malady. But this proof has been obtained by the investigations of the last few years; and we can now state with the same certainty that the bacillus is the essential cause of anthrax that we can refer to the acarus as the essential cause of the itch, mange, or scab.

The investigations necessary to determine this fact were long, tedious, delicate, and most difficult to accomplish; the way was a new one, and had to be marked out by careful soundings, very much as a vessel is conducted when entering an unknown harbor full of hidden rocks and reefs. Besides this, the arguments brought against the pathogenic action of the organism were such as to make the task almost a hopeless one.

It was claimed, and at the time apparently with the greatest justice, that specific fevers were not caused by bacteria, because the germs of these were universally present in the atmosphere; because they multiplied in vast numbers in the mouth and digestive organs of animals in perfect health; because solutions containing them might be injected into the blood even, without producing disease. In anthrax it was further shown that the blood became virulent before bacteria could be discovered with the microscope; and it was claimed that these multiplied and grew in the blood of anthrax patients, simply because the peculiar changes of the blood induced by the disease made this a favorable medium for their development. Certain microscopists went even beyond this, and pointed out white granules, similar to the white blood globules and lymph granules in appearance, and supposed by some to be degenerated *bioplasm* of the body, which were considered to be the true pathogenic agents.*

Gradually, however, it became apparent that there were different varieties of bacteria, and fortunately, after a quarter of a century of contest, Dr. Koch was able to discover points which made the pathogenic action of this parasite a matter of certainty.†

He discovered that, in the blood and juices of living animals, the bacillus threads multiplied rapidly by lengthening and dividing, and that though he inoculated them for twenty generations in mice no other forms were produced. In the blood and juices of dead animals, or other favorable nutritive fluids, with admission of air and between certain extremes of temperature, they grew to extraordinarily long threads, with the formation of numerous spores. These spores, in a proper liquid, with the conditions just mentioned, germinate and produce again the bacillus threads. Now the bacillus threads are easily destroyed; a few days in an unfavorable temperature, an equal time deprived of air, diluting the fluid containing them with a large quantity of water, or drying the substances containing them, was sufficient to destroy their life; but their spores, on the other hand, when once formed, are capable of existence under the most unfavorable conditions.

Here, then, was the clue by which Koch was able to demonstrate the agency of this particular organism in causing the disease known as anthrax. When he preserved fresh anthrax blood without contact with

---

* Disease Germs: Their Nature and Origin; Beale, London, 1872; pp. 232 to 251. The Microscope in Medicine; Beale, London, 1878; pp. 309 to 321.
† Beiträge zur Biologie der Pflanzen, B. I; H. I, 1876; pp. 277 to 308.

air till the threads died, the fluid at once lost the power of communicating the disease; on the other hand, with the admission of air, under otherwise identical conditions, the organism grew, formed spores, and the liquid retained its virulence.   Again, by admitting air, but by keeping the fluid in a temperature of 8° R. (50° F.), which prevented the formation of spores before the organism died, the active properties were again lost.   So, by drying rapidly or by diluting with much water, the death of the *bacilli* resulted, and the activity of the liquid disappeared.

This was of itself good evidence, but Koch showed further that when the spores had once formed neither cold, nor deprivation of oxygen, nor drying, nor dilution with water any longer destroyed the virulence of anthrax fluids.   This was scientific demonstration and proof positive of the pathogenic action of the *bacillus anthracis.*

How different is this from the supposed demonstrations of other investigators in regard to this and similar organisms.   For instance, one man filters the *bacilli* from a virulent fluid, and as it no longer retains its activity he asks us to accept it as a demonstration that the *bacilli* are the cause of that contagious disease.   Another inoculates a few drops of a nutritious fluid with a virulent liquid and sees a form of bacteria multiply in it; he inoculates again and again into fresh fluid, new generations of bacteria being constantly produced, and when he has ascertained the virulence of a tenth or a twentieth generation he triumphantly points to this as a demonstration that the bacterium which he has seen develop in each must be the virulent principle—the cause of the particular disease.

If scientific demonstrations could be made with this facility, we should not have been a quarter of a century discovering the connection between the *bacillus anthracis* and malignant anthrax, nor would it now require much time to investigate either swine plague or Southern fever.   But, unfortunately, these are not demonstrations; they are entirely unreliable and worthless towards proving that a particular organism is the source of the activity of the liquid; for, in the first case, our experimenter has forgotten that when he filters out the bacillus he at the same time removes every other organism, including the granules of bioplasm, from whatever source derived.   He shows that the active agent is a solid and not a liquid; that it is something that has size and form, but he goes no further; and this was demonstrated years ago by Chauveau, and since by Klebs, Pasteur, and others.   So, in the second case, when an inoculation for a cultivation experiment is made with a particle of virulent matter, we have no guarantee that other particles, besides a particular bacterium, are not transferred at the same time; that these are not seen is no proof that they do not exist, since the highest powers of the microscope still show particles which appear as points, and which are less than the $\frac{1}{100000}$ of an inch in diameter, and other particles so clear and structureless that they are scarcely visible—and who can say what is beyond our vision?

But even Koch's demonstrations were not sufficient for all the doubters, and Pasteur, who has gained a world-wide reputation for his brilliant investigations, took up this subject.

The first objection was that when the blood of an inoculated animal first becomes virulent there are no bacteria visible under the microscope.   This Pasteur said was owing to the difficulty of finding one or two bacteria in such a large surface as is made by a drop of blood under the higher power lenses.   For instance, a drop of blood pressed flat between the object glass and thin cover presents a surface of one-half inch in diameter; this magnified only five hundred diameters gives a surface with a diameter of 20 feet. And this is sufficient to fill the field of vision nearly eighteen hundred times. But Pasteur was not the man to stop at an explanation; he proved the presence of bacteria by planting a drop of this blood in a suitable liquid, in which, if present, they would develop in such numbers as to be easily recognized. In this way he demonstrated that as soon as the blood became virulent it contained bacteria or their germs. This seemed to be a fitting climax to a long series of investigations ; but new difficulties were to arise.

The distinguished investigator, M. Paul Bert, read a paper before the *Académie des Sciences* in which he showed that compressed oxygen, which was supposed to kill all living things, did not destroy the anthrax virus, and that blood treated with it would still form an *alcoholic* extract capable of producing anthrax. Now alcohol, as well as compressed oxygen, had been supposed to be fatal to all life; it was a standard disinfectant for destroying bacteria or their germs in apparatus used for cultivation experiments. If, then, the virus survived the action of both compressed oxygen and solution in alcohol, reasoned M. Bert, it was because it was not a living organism, but a chemical agent allied, perhaps, to diastase.*

But once more the genius of Pasteur was equal to the occasion, and he demonstrated, to the satisfaction of M. Bert, that although the *bacilli* were destroyed by the agents used, their germs were not, and that these were even capable of resisting absolute alcohol. He subjected M. Bert's alcoholic extract to microscopic examination and proved the presence of large numbers of these spores. M. Bert afterwards found that these spores, preserved for five months in ordinary alcohol, not only retained their vitality, but were as active as at first.†

Finally, Pasteur showed that fowls, owing to their high temperature, were not susceptible to anthrax, and would not contract it if inoculated, but by placing the lower third of their bodies in cold water their temperature was reduced to a point favorable to the development of the *bacilli*; and, if inoculated under such circumstances, they contracted the disease and died; but if, after the disease made its appearance, the fowl was removed from the bath and the temperature allowed to rise the malady was cured.

The smoke of this thirty years' contest is just clearing from the field, and one of the most important victories is won in the name of modern science. For years the obstacles seemed insurmountable, but they have at last disappeared, and medical science takes a most important stride in advance.

Other late discoveries.—The good results of these investigations and discoveries are not confined to a single disease, but are felt in our whole theory of epizoötic and enzoötic maladies; long before these investigations were completed, Pasteur had shown that the *pébrine* and *flacherie* of silk worms were not due to miasma, but to microscopic organisms; Obermeyer and Heydenreich had shown the presence of a *spirillum* in all cases of relapsing fever; and, within the last year or two, Toussaint has found an organism in the disease known as cholera of fowls; Pasteur has found another in septicæmia; and Klein has discovered the bacillus of swine plague.

Hallier goes beyond most other specialists and claims that the micrococci and bacteria are formed from the elementary granules of fungus spores, and believes these, with proper culture, will reproduce their parent forms and may be thus identified. Having demonstrated, as he thought, the fungi which thus produced a number of contagious dis-

* Recueil de Médecine Vétérinaire, 1877, p. 695.
† Loc. cit., p. 919.

cases as long ago as 1868, and finding his views were not generally accepted, he has gone back to the study of the diseases of plants and the
lower forms of animal life. He now claims to have demonstrated that
the potato-rot is not due to the mycelium of the *peronospora infestans*,
as has been supposed, but to the formation, within the cells of this
fungus, of micrococci and bacteria which then become free and are the
destructive agents. He also shows that the disease called *gattine* of the
cabbage worm (*pontia brassicæ*) is caused in the same way.*

Pasteur has also shown, within the last few months, that when the
bodies of anthrax animals are buried the *bacilli* multiply in the earth
and form spores which, notwithstanding extremes of moisture and
drought, remain active a year afterwards.† He and Toussaint have further shown that the so-called spontaneous cases of anthrax are almost
invariably caused by the germs being taken in by way of the mouth.‡
so that now we begin to understand some important points in the causation and manner of preventing this fatal malady.

The bearing of these discoveries on Southern fever.—The present tendency
of science is, then, to consider these diseases to be due to parasitic organisms of a vegetable nature; and Pasteur, who has done so much
work in this field, is but giving expression to the most advanced scientific thought when he says that the contagious and communicable fevers
so destructive to both man and animals may, by the use of proper means,
be swept out of existence.

We accept this statement as a probable fact, in respect to most of
these diseases, because we see no reason for believing the other theory
which regards contagion as degenerated bioplasm of the animal body;
and claims that, if all contagious diseases were exterminated, they would
soon reappear by a new formation of the virus. I have found the bio-
plasts of Beale in swine plague side by side with the micrococci; but I
have also found them in inflammations, no matter how produced; indeed,
their presence in simple inflammations is admitted by the author of the
theory.

If there were any truth in the theory of the spontaneous evolution of
these diseases, why is it that we have no rinderpest, sheep-pox, or venereal
disease of solipeds in the United States? Why is it that contagious
pleuro-pneumonia and epizoötic aptha have never appeared in this
country except as introduced? For me the answer is plain: the organisms which cause these diseases have never been introduced and scattered over the country, and we will be free from such diseases till these
are so introduced or so scattered; and it seems equally plain that if
such organisms are once completely destroyed the diseases which they
cause will no more reappear than do the extinct races of animals which
inhabited this world in prehistoric times.

I have shown that no theory but that of a living organism explains
the phenomena of Southern fever; and I see no reason to doubt that
future investigations will not only discover this, but will demonstrate its
action as clearly as the connection of the *bacillus anthracis* has been
demonstrated with malignant anthrax. According to Hallier, it is a
fungus growing on plants, and, from the period at which the disease
usually appears, it would seem that such plants do not reach a proper
stage of growth to be suitable for its full development before the first of
July.

* Die Parasiten der Infectionskrankheiten bei Menschen, Thieren, und Pflanzen.
Buch I, Die Plastiden der niederen Pflanzen, ihre selbständige Entwickelung, ihr Ein-
dringen in die Gewebe, und ihre verheerende Wirkung. Leipzig, 1878.
† Recueil de Médecine Vétérinaire, 1879, pp. 1026, 1027.
‡ Recherches Expérimentales sur la Maladie Charbonneuse, 1879, pp. 111 to 122.

MEANS OF COMBATING THE DISEASE.

Medical treatment.—In most cases, we must admit that medical treatment has been anything but flattering to the resources of our profession; but this experience has been simply that of the medical profession during its whole history in treating this class of affections; consequently, we were not to expect success at the first trial. Professor Gamgee says : " Page after page might be filled with notes on the administration of nitrate and chlorate of potash, iodide of potassium, quinine, salts of iron, sesquicarbonate of ammonia, epsom or glauber salts, sulphur, ginger, calomel, soap and oil, &c. None of these agents have affected the steady progress and fatality of the disease." And he concludes that the great majority of epizoötic and enzoötic maladies never can and never will be arrested by the medical treatment of the sick ; and that this fever may be classed among the incurable maladies, inasmuch as we know of no antidote to the mysterious poison inducing it.

For our part, we are not disposed to take quite such a gloomy view of the resources of therapeutics, and we are in hopes that a series of careful investigations will yet show that there are agents which may be relied upon to the same extent in the contagious fevers as in other affections. Since some animals without such medication prove unfavorable to the development of the virus, it is evident that but a slight assistance is needed for the others to resist it, if this assistance is only in the proper direction.

By a long series of laboratory experiments, Davaine found that iodine destroyed anthrax virus when diluted with twelve thousand parts of water, and this agent has since been used with considerable success in the treatment of the disease.

The New York commissioners treated five cattle successfully which were affected with Southern fever, with carbolic acid. A mixture was made of twelve ounces of carbolic acid crystals, four ounces glycerine, and twelve ounces of bicarbonate of soda; one ounce of this was then dissolved in three or four gallons of water, and the animals given no other drink. This was readily drank by the steers. The surface of the ground was liberally sprinkled with heavy oil of coal-tar mixed with sawdust. The next day the animals were evidently improving, and the second day after they were put on treatment they were convalescent (Report, p. 60). Under this treatment it was claimed the fungus spores rapidly and completely disappeared from the blood.

Whether the early recovery of these animals was due to the carbolic acid or not can only be conjectured till more experiments have been made; but considering the large proportion of animals that died from this affection, the recovery of every one of a lot of five must be looked upon either as a very singular coincidence or as the result of a very efficacious remedy.

Carbolic acid was also considered by the English commission which investigated the cattle plague to have a remarkable influence in preventing that malady; but in other contagious fevers it has signally failed. The truth seems to be that an agent which has a remarkable effect on one virus may be almost harmless to another; and, hence, there is a necessity to test the whole list of these agents with each virus, in order to discover the most useful remedies in that particular affection.

Col. D. D. Fiquet, of Texas, who holds the obviously erroneous theory that the malady is due to digestive derangement, is reported to treat it very successfully as follows: One tablespoonful of powdered charcoal and one teaspoonful of powdered ginger is given as soon as sickness is discovered; in one hour he gives one quart of marsh-mallow infusion and one quart of camomile tea. If not relieved in eight or ten hours, this is repeated; as the appetite returns, he gives plenty of green food. Ninety-five per cent. of the cases thus treated are said to recover.*

It is evident to any one of experience that the cases giving such favorable results have been very mild ones, for this is simply palliative treatment—not curative; it seeks to relieve an effect of the disease, but cannot modify the disease itself. The fact that the attacks under some circumstances are so mild gives hope that a combination of carbolic acid, or, perhaps, some more active agent, with stimulants and palliatives, varied to suit the particular case, may prove very useful.

Disinfectants.—The New York commissioners were very enthusiastic over the favorable results obtained with carbolic acid. They had the yards, to be disinfected, sprinkled with heavy oil of coal tar, which contained about 10 per cent. of carbolic acid; they recommended its use in stables and they administered the pure acid to sick animals. Still it was not tried on a sufficient scale nor in such cases as to allow an opinion of its effects. At that time this agent was supposed to have wonderful virtues in destroying the different varieties of virus. The English cattle plague commission supposed they had demonstrated that it was remarkably efficacious in rinderpest.

Thus, air from a shed containing sick animals was filtered through cotton wool; one piece of this wool was exposed to carbolic acid vapor for half an hour. The disinfected wool was placed, by an incision, under the skin of a healthy calf; another calf was inoculated in the same manner with a piece of the wool that had not been disinfected. The one inoculated with the disinfected wool remained perfectly well; the other took the cattle-plague and died in a few days.

A healthy animal was put in a shed with a sick one, and was given an ounce of carbolic acid a day; the floors were sprinkled daily with the acid, and the shed whitewashed once a fortnight with a mixture containing the same agent. The shed was so small that the animals touched each other at times. The sick animal died in six days and was replaced by another. It was a month before the healthy animal showed signs of sickness, and then only in a mild form, recovering in a few days. The ordinary period of incubation being nine days, was extended to three times this length, and the disease was deprived of its malignant character.

On one farm where the disease existed the cattle were divided into two lots, forty-five in disinfected houses, twenty-eight in undisinfected open sheds. The disease was communicated to each lot by inoculation. Only those actually inoculated of the disinfected animals sickened, while the entire lot not protected in this way were quickly destroyed. After a few weeks the remainder of the forty-five disinfected animals were turned out to grass and all died within a few days. †

Since then carbolic acid has lost much of its reputation. Pettenkofer has shown that carbolic acid simply arrests development of certain germs, but does not destroy their vitality,‡ which seems to be confirmed by the experiment just quoted. Dr. John Dougall exposed vaccine lymph for thirty-six hours under a bell-glass to air saturated with the

* National Live-Stock Journal, 1878, p. 541.
† Beale: Disease Germs, pp. 276 to 279.
‡ H. Letheby, M. D.: M. A. Prof. of Chemistry in the College of London Hospital; Medical Officer of Health; Public Analyst for the city of London; President of the Society of Medical Officers of Health. Paper read at meeting of Society of Medical Officers of Health, October 18, 1873.

vapors of carbolic acid without destroying its activity; and in one case he mixed lymph with a 2 per cent. solution of this disinfectant, and when inoculated ten days later it still retained its vitality. (Quoted from *Lancet* in *Veterinarian*, 1873, p. 837.)

Davaine, in his numerous experiments on anthrax virus, found that this was destroyed by solutions of disinfectants of the following strength: ammonia, 1.150; carbolic acid, 1.150; ordinary vinegar, 1.150; caustic potash, 1.500; permanganate of potash, 1.1250; sulphuric acid, 1.5000; solution of iodine (compound), 1.12000 * Here carbolic acid certainly showed no special activity, being no more useful than common vinegar.

Other experiments have shown that some bacteria are apparently destroyed instantly by a one-half per cent. solution of carbolic acid, while others are lively after being in such a solution for twenty-four hours.†

From these facts we cannot hesitate in our conclusion that a disinfectant cannot be considered efficacious in any disease until its effect has been carefully tested with that particular virus. An agent that is very destructive to certain germs scarcely affects others in the least. Hence the necessity of isolating and studying the germs of each contagious disease from different standpoints if we would hope to control these plagues. Strange as it may appear, notwithstanding the great number of human lives and large amount of property swept away annually by these diseases, there never has been a systematic and complete trial of all the different agents likely to prove useful in the case of a single one of the contagious fevers. There is here a vast field for work that promises the most important results.

Acclimation.—A writer in the National Live-Stock Journal (1879, p. 16) claims that thoroughbreds may be imported into Texas with a loss of not exceeding ten per cent.; and that of one hundred and four thus imported in the preceding fourteen months, but three had died. Although all danger could not be considered over, the results were so favorable and the recommendations so judicious that I condense them, the more so since I know they correspond with the precautions taken by successful stock-raisers in other parts of the South.

The animals must be of sound parentage; under eight months old; they should have been dropped early in spring; must be of strong, robust frames, and may be shipped in November or December. They should be well fed to date of shipment on milk, bran, oats, &c., and should be turned on good grass pasture, at least of nights; they should be made comfortable on the trip, and receive the same kind of food after their arrival as before starting. When the railroad journey is completed they should have a week's rest in dry lots, and then, if driven any distance, it should be by slow and easy traveling. If troubled either with purging or costiveness, these should be corrected by simple remedies. Should have free access to running water, and to an oat or barley patch sowed in the fall for their benefit. They should be suitably protected from the weather, and receive the charcoal and ginger treatment of Colonel Fiquet as soon as symptoms of disease are noticed. If December or January are safely passed, one can feel safe till the heat of June. Avoid exposing them to either rain or sun. In a word, they must have generous feed, comfortable quarters, and kind treatment.

I would suggest as modifications of these rules, (1) that cisterns be constructed where practicable, adapted to keeping the water pure, and that the imported animals obtain their drink from these alone; (2) that

---

* Recueil de Médecine Vétérinaire, 1874, p. 565.
† Billings and Curtis in Report on Diseases of Cattle in United States, p. 169.

the oat or barley patches for green food be sowed on land to which no cattle have had access for several years; (3) that the hay, if any, which they are fed, be gathered from land from which cattle have been excluded for an equal period; (4) that for the first two years they be pastured on land from which natives have been excluded an equal time: (5) that in case of sickness, carbolic acid be tried in one or two drachm doses, two or three times a day, in connection with the other remedies proposed. The first four suggestions are intended to prevent animals from taking the germs of the disease into their systems during the first two years after importation.

Prevention.—The preventive measures necessarily include every means of transportation of the to disease to uninfected districts, of stopping its encroachment on adjoining healthy sections, and even of combating it in the districts which have been already overrun by it. The measures which are to be recommended in each of these cases vary, because of the peculiar phenomena which we have already discussed at length.

Preventive measures at a distance from infected districts.—Cattle from the infected parts of the South must not be allowed to come north during the spring or summer months. If, as has been the custom in the West, they are taken to those localities beyond the infected district where there is little danger of their coming in contact with native cattle, either on the route or after their arrival, it is advisable to allow such driving even in summer, providing they are prevented from leaving such new locality till the following winter or later; since experience has shown that after a five or six months' sojourn in a healthy district these cattle are no longer able to communicate the disease. At present there is no reason for preventing Southern cattle going north *by rail* in winter, since, as far as I have been able to learn, there are no instances of the disease having been carried north of the thirty-ninth parallel of north latitude at this season. But the traffic cannot be safely extended as late as the last frosts of spring, if these cattle are to be herded on grounds on which natives are to be pastured or allowed to run during the succeeding summer; because, as I have shown by an outbreak in Buncombe County, North Carolina, the germs may be deposited on the ground early in spring, and will survive frosts, rains, and drought, and cause the appearance of the disease the following July or August. The Southern fever may, then, be easily prevented from causing any losses in the greater part of the Northern or Western States.

Prevention near the border line of the infected district.—We have here a much more difficult problem, one that may not be solved with entire satisfaction until more is known in regard to the essential cause of the affection. There is one rule, however, which should be enforced to the letter, and to which no exceptions whatever should be allowed, viz., a line should be drawn around the permanently infected district, and no cattle, under any circumstances, allowed to be driven from within that district across the line at any season of the year.

If railroads are safely fenced, so as to prevent cattle from grazing along the track, then cattle might be transported by rail or by boat. The great danger from this disease, under present circumstances, is its character of gradually and slowly, but surely, extending itself over hitherto healthy districts, from which, when once infected, there seems no possibility, with our present knowledge, of ever removing it. It is a matter of such great importance to absolutely prevent this gradual extension of the infected district, that I fear I am unable to use language that will convey an adequate idea of the real danger and vast amount of damage that is being caused, and which must increase from year to year as the great stock-raising districts of the country are en-croached upon. The danger principally arises from, (1) herds of cattle driven across the above line to Northern markets; (2) those driven to be slaughtered but a short distance beyond the infected district; (3) those driven but a short distance beyond, to be grazed a certain time before going to market; (4) milch cows and breeding animals bought by farmers, usually in small numbers, but no less dangerous; (5) oxen driven backward and forward by teamsters, farmers, and others. By these means the roads for a certain distance beyond the infected districts are kept loaded with the germs of the disease; many pastures, fenced and unfenced, are likewise infected; and as the difference in the climatic conditions of localities but a few miles distant is so slight, the organism causing the disease, whatever its natural history may be, gradually inures itself, by a kind of natural selection, to the slight change in conditions of life, and thus it is able to slowly extend its habitat towards the colder parts of the country. Nor is there anything wonderful in an organism thus becoming able to withstand greater extremes of temperature by such a gradual change; for do we not have varieties of wheat and oats which are unable to withstand the cold of winter and other varieties which do this perfectly? And is not the difference due to gradually changing the conditions of life to which these were exposed, and by a survival of those plants capable of enduring the most? Indeed, summer wheat or oats may be changed even now, by careful selection, into the winter varieties.

The researches of the last few years have made it probable that the cause of this class of diseases is a parasite of a vegetable nature, and, consequently, it is not surprising that such parasites, like other members of the vegetable kingdom, may by natural selection, under suitable conditions, become able to withstand a climate that at first proves destructive to them. And no demonstration is needed that the most favorable conditions for such natural selection at present exist along the whole border line of the infected district.

It is true both North Carolina and Virginia have laws intended to prevent the extension of this disease, but they were enacted in ignorance of the most important phenomena connected with it, and the many exceptions allowed make these laws almost useless. For instance, there is no attempt to prevent oxen from being driven backward and forward across the line. Even fat cattle are allowed to be driven to Danville, Norfolk, and Portsmouth for slaughter; and the ordinary traffic between farmers along the line is scarcely interfered with. The extension of the disease will probably never be checked without a national law, operating uniformly along the whole line of this district, and strictly enforced.

Exterminating the disease within the infected district.—With the knowledge we now possess it is impossible to make any recommendations looking to the annihilation of this fever in those districts which are at present overrun by it.

If we accept much that is now believed concerning it, its extinction would be almost a hopeless task; but, as is always the case with other diseases of this class, much of the mystery concerning it may turn out to be entirely imaginary. With so many cattle running at large, it would not be surprising to find almost any contagious disease distributed to the same extent. In a country so thoroughly infected as the Southern States, not only the roads but water-courses are probably infected; even the food grown on such lands is probably contaminated, or the attendants might carry the contagion to where imported cattle were stabled; so that its spontaneous origin (origin *de novo*), even in the

South, is not to be accepted until more is known concerning it; nor would it be wise to decide, in advance of investigation, that it can never be exterminated.

The success that has attended the late investigations of anthrax and some other diseases in Europe, which enables us to isolate and study the organisms that are the essential cause of these diseases, leads us to anticipate that the observations of Dr. Stiles and Professor Hallier on our Southern fever may be confirmed, or, at least, that the exciting cause may be discovered. With a firm faith in the ability of modern science to conquer, by persistent and systematic investigation, every difficulty connected with the control of communicable diseases, I believe in the possibility of freeing our land completely from this scourge.

INVESTIGATIONS NEEDED.

The suspension of the disease during the late fall and winter months has prevented direct investigation of the germs causing the malady, but a study of the bacterium of swine plague is being carried on, and it is hoped that the experience thus gained, together with a familiarity with other organisms likely to be encountered in this part of the country, will be of great assistance when the summer outbreaks again occur.

One great difficulty has been that we have had no trained specialists for carrying on such work; indeed, taking the world over, we may count such specialists on our finger ends. It is not the work of a week or a month to make such important discoveries as have lately been given to the world by Chauveau, Cohn, Davaine, Hallier, Klein, Koch, Pasteur, and Toussaint. These men are all specialists and have devoted years to the work; and while the facts they have discovered are not numerous, the effects of their work on medical science and on the well-being of humanity may be far greater than can at present be estimated.

Until the appropriations of 1878, there had been few inducements for men to prepare themselves by special training for such delicate investigations as are needed in an attempt to reach the ultimate facts concerning the communicable diseases; and for a man in general veterinary or medical practice to at once take up such investigations and carry them on successfully, when these noted specialists were so long baffled by less difficult problems, is very much like the ordinary farmer analysing his own fertilizers and calculating to a fraction of a per cent. the valuable ingredients.

The delicacy and difficulty of this class of investigations cannot be overestimated if substantial and reliable results are to follow. Not only must we seek to become acquainted with the character of a considerable number of different organisms so small that in some cases fifty thousand placed side by side would not cover the space of an inch, but in cultivating them the germs of other forms, *with which the air is loaded*, must be excluded; not an easy task, when we consider that distilled water, even in a well-closed bottle, swarms with life within a few hours after it has been thoroughly boiled; and that some germs resist a boiling temperature from thirty to sixty minutes, and the action of ordinary or absolute alcohol for an indefinite time.

The difficulty in deciding if a given organism is the cause or only the result of a disease is not to be overlooked; it can be appreciated when we consider that it was a quarter of a century after the discovery of the *bacillus anthracis* before positive proof could be obtained that it was the cause of malignant anthrax. Then, in searching for the habits and vulnerable points of these organisms, we must feel our way in the dark, and can only be certain of a point after it has been confirmed many times; and in determining the effect of the considerable number of agents destructive to the lower forms of life, we must test their effect in solutions of different strength; and here we are often impeded in our work by our animals becoming infected by means of the attendants, and by our cultivating apparatus being invaded by other organisms, destroying the work of whole days or weeks. New forms of apparatus and new methods of cultivation must be invented, and everything must be done near the section where the disease exists, and at a distance from the source of supplies, so that the delays, the difficulties, the discouragements to the worker can scarcely be appreciated except by those who have had the experience.

Although it is only through the most delicate and difficult investigations yet undertaken by science that we can hope to obtain such information as is necessary to completely check the progress of this disease and to destroy it where it already exists, I cannot believe that the American people are to be discouraged from continuing such an important research; for is a vast nation to be frustrated by such obstacles, and to be subject to such losses, when there is good reason to believe that a persistent investigation may soon reveal wherein lies the power of the hidden foe, and indicate the means by which it can be destroyed?

I would mention the following as points to be decided at the first opportunity, and as a guide to future investigations:

1. Can the disease be inoculated by any fluids or excretions of either healthy Southern cattle or of susceptible animals which have contracted the fever of them? (I have made some inoculations with blood and bile from diseased animals so far without result.)

2. Is there any peculiar organism in the tissues, fluids, or contents of digestive organs of diseased animals, or of apparently healthy Southern cattle?

3. If such organism exists, is it the essential cause of the disease?

4. If the virus can be isolated, what are the effects upon it of the various agents supposed to be effective in destroying the lower forms of life?

5. Are such agents useful when administered to sick animals?

6. Determine, if possible, the conditions of life necessary to the existence of the pathogenic germ, its habitat outside of the body, the changes which it undergoes, and the manner in which it enters the system.

7. Study more completely the changes produced in the tissues and fluids of the body by the disease.

8. Trace the boundary line of the infected district throughout its whole extent.

9. Determine if lands in the infected district can be freed from the disease.

Respectfully submitted. ·

D. E. SALMON, *D. V. M.*

THE SOUTHERN CATTLE "DISTEMPER."

Mr. James P. Phillips, Clarksville, Habersham County, Georgia, writes as follows concerning the ravages of the disease known in the south as "cattle distemper." The disease referred to is splenic or Texas fever of cattle, and is treated of at length in this report by Dr. Salmon, of North Carolina. Mr. Phillips says:

The only disease, or *the disease*, among cattle, in this and all the mountain counties on either side of the Blue Ridge, from Alabama into Virginia, is "distemper." This term means anything or nothing. The disease is not malarial; it is not contagious through the air, as a board or rail fence quarantines against it. Soquee, a small river 30 feet wide, insures freedom from it on the opposite side. But one diseased animal ranging in the neighborhood will infect hundreds of animals, 90 per cent. of which will die. During the past season twenty-two milch cows died within a space of one mile square in Fair Play district—a mountain district noted for fat beef, golden butter, and juicy mutton, and where the Indian custom of burning the woods in chestnut time (Indian summer) is kept up.

The disease is usually first noticed by the animal seeking seclusion in a branch or ditch, where it stands with head lowered, ears drooping, and tail often a little up from its rest. It dislikes sunshine or heat; the heat of the body is greatly increased and the pulse is high. Sometimes, when not standing in the water, the feet and the legs, as high up as the hock, are cold. I have never found the brain affected by infiltration, thickening of the membrane, or softening. In two cases, last summer, I found the lungs hepatized. The omassum (manifold) was hard and impacted with dry food. Every drop of juice of the grass or food seemed to be absorbed. In every case examined I found the kidneys more or less diseased. I have never found the small intestines diseased.

The diseased cow usually dies within forty-eight hours after the first symptoms are observed, but sometimes the animal may linger for two weeks. Should the cow recover, she is never herself again. The remedies used have been innumerable—all the mineral and vegetable purgatives—and time and again a remedy boasted of as *the cure* fails in the next case. In my opinion, a certain preventive is "soiling" and complete isolation from range stock. Our small farmers, however, cannot or will not adopt this plan, but depend upon the "range." A remedy for this would be found in the "no fence law." This being impossible now, Congress should enable the Commissioner of Agriculture to send a competent veterinarian next summer to the district where the disease is most prevalent, which I think will be Batesville, in this county. Here he can study the disease, and give us a remedy, if there be one. In the absence of proper instruments, our *post-mortem* examinations do not amount to much.

CONTAGIOUS PLEURO-PNEUMONIA OR LUNG-PLAGUE OF CATTLE.

Prof. JAMES LAW, V. S., of Cornell University, New York, has issued a valuable work of about one hundred pages, entitled "The lung-plague of cattle—contagious pleuro-pneumonia." This work is all the more valuable from the recent experience of Dr. Law in the treatment of this disease, for it will be remembered by many of the readers of this brief review that he was last spring appointed by the governor of New York to act as chief of a commission of veterinarians to assist the State authorities in devising and carrying out such measures as it was hoped would result in the complete suppression of this deadly malady among the cattle of that State. He states many facts connected with the history of the disease in this country not heretofore generally known, and also corrects some errors and misapprehensions touching the disease itself which English veterinarians have fallen into. He says that the name of the disease (*pleuro-pneumonia*) has been largely misapprehended by the medical mind, and that there is no proof that the malady, like other imflammations of the organs within the chest, is caused by exposure, inclement weather, changes of climate or season, imperfect ventilation, &c. Other names have been, at different times, employed; for instance, *Peripneumonia, Peripneumonia pecorum enzootica or epizootica, Peripneumonia exudativa enzootica or contagiosa, Peripneumonia pecorum epizootica typhosa, Pleuro-pneumonia interlobularis exudativa, Pneumonia catarrhalis gastrica asthenica, Pleuritis rheumatica-exudativa.* But Dr. Law regards all of these terms as objectionable, and says if the term *contagious* (*contagiosa*) be added to any of these definitions it only removes the difficulty a short step, "for the physician still concludes that the affection is due to local or general causes, and that if it arises in one animal under such circumstances it may in one million, subject to the same conditions; that its general prevalence, at any time or place, may be altogether due to the environment, and that the doctrine of contagion is either founded on insufficient data or true only in a restricted sense and entirely subsidiary to the generally acting causes. But the malady, as known to veterinarians of to-day, is always and only the result of contagion or infection." Therefore, a name better adapted to set forth the character of the disease without the risk of misleading should be chosen, and for this reason Dr. Law has adopted that of *contagious lung-plague of cattle*, the new counterpart of the *Lungenseuche*, by which it has long been known in Germany. He regards the old term, *pulmonary murrain*, as equally good. The German *Lungenseuche* is especially apposite, the real meaning being *lung contagion*, which conveys the idea of transmission by contagion only. He therefore gives, as a definition of the malady, a specific contagious disease peculiar to cattle, and manifested by a long period of incubation (ten days to three months) by a slow, insidious onset, by a low type of fever, and by the occurrence of inflammation in the air passages, lungs, and their coverings, with an extensive exudation into lungs and pleuræ.

After reciting the history of the malady in the Old World, in which

the statement is made that Great Britain alone has lost not less than
$10,000,000 per annum by the ravages of the disease since the year 1842,
the following brief history of its invasion and continous existence in this
country is given:

Into Brooklyn, Long Island (New York), it was introduced in 1843 in the system of
a ship cow, purchased by Peter Dunn from the captain of an English vessel. From
Dunn's herd it spread to others adjacent and speedily infected the whole west end of
the island, as will be noticed later at greater length.

Into Massachusetts the plague was introduced on the 23d of July, 1859, in the bodies
of four Dutch cows, imported by Winthrop W. Chenery, of Belmont, near Boston.
These cows were procured from Purmerend and the Beemster, and were kept in stables
for several days at the port of Rotterdam, an infected city, before being put on board
the vessel. They were shipped April 6, passed forty-seven days at sea, and were ill
during the last twenty days, one of the number having been unable to stand. On
landing, two were able to walk to the farm, while the other two had to be carried in
wagons. The worst cow was killed May 31, and the second died June 2; the third
did well till June 20, when she was severely attacked and died in ten days; the fourth
recovered. On August 20 another cow, imported in 1852, sickened and died in a few
days, and others followed in rapid succession. In the first week of September, Mr.
Chenery isolated his herd, and declined all offers to purchase, being now convinced
that he was dealing with the *bovine lung-plague* of Europe.

Unfortunately, on June 23, he had sold three calves to Curtis Stoddard, of North
Brookfield, Worcester County, one of which was noticed to be sick on the way to
Curtis' farm. Several days later Leonard Stoddard (father of Curtis) took this calf to
his farm to cure it, and kept it in his barn with forty cattle for four days, when he
returned it to his son. It died August 20. Curtis Stoddard lost no more until November 1, when he sold eleven young cattle to as many different purchasers, and
wherever these went the disease was developed. In one case more than 200 cattle
were infected by one of these Stoddard heifers. Of the nine cattle which he retained
seven were killed and found to be badly diseased.

An ox of L. Stoddard's sickened two weeks after he had returned the diseased calf
to his son, and soon died. Two weeks later a second was taken sick and died; then
a dozen in rapid succession. From this herd were affected those of the following:
Messrs. Needham, Woods, Olmsted, and Huntingdon. Olmsted sold a yoke of oxen
to Doane, who lent them to assist with twenty-three yoke of cattle in removing a
building in North Brookfield. These belonged to eleven different herds, all of which
were thereby infected.

This will suffice to show how the disease was disseminated. In the next four years
it was found in herds in the following towns: Milton, Dorchester, Quincy, Lincoln,
Ashby, Roxborough, Lexington, Waltham, Hingham, East Marshfield, Sherborn, Do-
ver, Halliston, Ashland, Natick, Northborough, Chelmsford, Dedham, and Nahant,
and on Deer Island.

By the spring of 1860 the State had been aroused to its danger, and in April an act
was passed "to provide for the extirpation of the disease called pleuro-pneumonia
among cattle," which empowered the commissioners to kill all cattle in herds where
the disease was known or suspected to exist. With various intervals this and succeeding commissions were kept in existence for six years, and the last remnants of the plague
having been extinguished, the last resigned definitely in 1866. The records show
that 1,164 cattle were slaughtered by orders of the commissioners, in addition to others
disposed of by the selectmen of the different towns in 1863, when the commission was
temporarily suspended. The money disbursed by the State was $67,511.07, and by
the infected towns $10,000, making a grand total of $77,511.07, in addition to all losses
by deaths from the plague, depreciation, &c. Dr. E. F. Thayer, Newtown, was the
professional commissioner who brought this work to a successful end.

An importation into New Jersey in 1847 is recorded, to check which the importer,
Mr. Richardson, is said to have slaughtered his whole herd, valued at $10,000, for the
good of the State. Unfortunately, all New Jersey men were not so public spirited, and
subsequently importations from New York and mayhap also from Europe have since
spread this pestilence widely over the State. From New Jersey it spread to Pennsylvania and Delaware, and thence to Maryland, District of Columbia, and Virginia, in
all of which it still prevails.

Of the progress of the disease southward from New York the records are somewhat
imperfect, yet sufficient to show a steady advance. Robert Jennings records its existence in Camden and Gloucester Counties, New Jersey, in 1859, and its introduction into
Philadelphia in 1860. It spread to "The Neck," in the southern part of the county,
killing from 30 to 50 per cent. of infected herds, and spread in 1861 into Delaware and
into Burlington County, New Jersey. In 1868 Mr. Martin Goldsborough assured Professor Gamgee of the extensive prevalence of the disease in Maryland, infection having
been introduced by cattle from the Philadelphia market. The professor personally

traced the disease in New Jersey, Pennsylvania, Maryland, District of Columbia, and Virginia, and makes the following assertions:

"That the lung plague in cattle exists on Long Island, where it has prevailed for many years; that it is not uncommon in New Jersey; has at various times existed in New York State; continues to be very prevalent in several counties of Pennsylvania, especially in Delaware and Bucks; has injured the farmers of Maryland, the dairymen around Washington, D. C., and has penetrated into Virginia."

He adds a table compiled by Mr. G. Reid, Ingleside farm, Washington, D. C., showing that in an average of 471 cows, kept in Washington and vicinity, 198 had died of lung plague since its introduction; 39 head perished in 1868 and 16 in 1869, up to date of report.

More recently illustrations of the existence of the disease in these States have been frequent, and among comparatively recent cases the author has been consulted concerning a high class Jersey herd near Burlington, N. J., in 1877, and a herd of imported Ayrshires in Staten Island later in the same year.

In 1878, the town of Clinton, N. J., was invaded, the infection coming through a cow that had strayed for some days in New York City. This was alleged to be an Ohio cow, but had strayed long enough in New York to have contracted the affection.

After showing that the disease is a purely contagious malady, and cannot arise *spontaneously*, Dr. Law gives the following brief history of the introduction, progress, and continual presence of the affection since its introduction among the cattle in and near the city of New York.

From different old residents (including Wm. Geddes, of Brooklyn, and Hugh T. Meakim, of Flushing) who were in the milk business in Brooklyn at the time of the importation, the following facts have been obtained:

The first cow was introduced from England, on the ship Washington, in 1843, and was purchased by Peter Dunn, a milkman, who kept his cows in a stable near South Ferry. This cow soon sickened and died, and infected the rest of his cows. From this the disease was speedily conveyed into the great distillery stables of John D. Minton, at the foot of Fourth street, and into the Skillman-street stables, Brooklyn, through which my informant, Fletcher, showed the Massachusetts commission in 1862. In this long period of nineteen years, the plague had prevailed uninterruptedly in the Skillman-street stables, and the commissioner reported that they "found some sick with the acute disease," and having killed and examined one in the last stages of the affection, stated that it showed a typical case of the same malady which existed in Massachusetts.

As dealers found it profitable to purchase cheap cows out of infected herds, and retail them at a round price, the malady was soon spread over Brooklyn and New York City. One or two cases will enable us to trace one unbroken chain of infection down to the present time.

In 1849, William Meakim, of Bushwick, Long Island (New York), kept a large dairy, and employed a man with a yoke of oxen in drawing grain from the New York and Brooklyn distilleries. A milkman on the way, who had lung fever in his herd, persuaded the man to use his oxen in drawing a dead cow out of his stable. Soon after the oxen sickened and died; and the disease extending to his dairy cows, Mr. Meakim lost forty head in the short space of three months. The stables having thus become infected, Mr. Meakim continued to lose from six to ten cows yearly for the succeeding twenty years, or as long as he kept in the milk business. This, which is but one instance out of a hundred, covers fifteen years of the plague in the Skillman stables, and brings the record down to 1869. It will be observed that this was the first occurrence of any such sickness in Mr. Meakim's herd; it commenced, not among the cows cooped up in hot buildings and heavily fed on swill, but in the oxen that were almost constantly in the open air, but which had been brought in contact with a dead and infected cow; the infection of the cows followed, and for twenty long years no fresh cow could be brought into these stables with impunity.

Dr. Bothgate, Fordham avenue and Seventeenth street, New York, informed us that twenty years ago (1859) his father kept a herd of Jerseys, which contracted the disease by exposure to sick animals, and that all efforts to get rid of it failed, until when, several years later, the barns were burned down. The devouring element secured what the skill of the owner had failed to accomplish—a thorough disinfection.

For some time so prevalent was the disease that Dr. Bothgate did not dare to turn his cattle out in the fields, lest they should be infected by contact with cattle over the fence. Since the period of the infection in his own herd, he knows that the pestilence has been constantly in many of the dairies around him. This bridges over the time from the Skillman-street and Meakim cases down to the present day.

Twenty years ago (1859) Mr. Benjamin Albertson, Queens, Queens County, Long Island (New York), purchased four cows out of a Herkimer-County herd which had got belated and had been kept over night in a stable in Sixth street, New York, where the cattle market then was. These cows sickened with lung fever and infected his large herd of 100 head, 25 of which died in rapid succession and 10 more slowly. He was left with but 60 head out of a herd, after the purchase of the four, of 104 animals, and honorably declined to sell the survivors at high prices to his unsuspecting neighbors, but sold a number at half price to a Brooklyn milkman, who already had the disease in his herd and knew all the circumstances.

Twelve years ago (1867) Lawrence Ansert, Broadway and Ridge Street, Astoria, (New York), bought of a dealer two cows, which soon after sickened and died, and infected the remainder of his herd of 18. Eight of them died of the disease, and he fattened and killed the remaining ten, and began anew with fresh premises and stock. He has lost none since.

The next case, like the last, affords a most instructive contrast to the first two, as showing how the disease may be permanently eradicated by proper seclusion. In 1872, Frank Devine, of Old Farm-House Hotel, West Chester, purchased from a dealer a cow which soon sickened and died. The disease extended to the rest of his herd, and in seven months he lost thirty-six cows. He appreciated the danger of contagion, and began again with new stock, keeping them rigidly apart from the infected beasts and premises, and from that time onward avoided all dealers and bred his own stock, with the happy result that in the last six years he had not had a single case of lung fever in his herd.

The virulence and infectious nature of the disease does not seem to have been lessened by its transplantation to this country. Many instances are given which show conclusively that it is equally as fatal to-day in those localities in the United States in which it exists as it is in its home in the far east, or in those nations of Europe which it has invaded. Speaking of the contagious and infectious nature of the malady, Dr. Law says:

No one who has studied the plague in Europe can truthfully claim that it is less infectious here than in the Old World. What misleads many is, that during the cooler season many of the cases assume a sub-acute type, and others subside into a chronic form with a mass of infecting material (dead lung) encysted in the chest, but unattended by acute symptoms. But this feature of the disease renders it incomparably more insidious and dangerous than in countries where the symptoms are so much more severe, that even the owners are roused at once to measures of prevention. Then, in moderating the violence of its action, the disease does not part with its infecting qualities, but only diffuses them the more subtilely in proportion as its true nature is liable to be overlooked. A main reason why unobservant people fail at first sight to see that the lung fever is contagious is, that the seeds lie so long dormant in the system. A beast purchased in October passes a bad winter, and dies in February, after having infected several others. She has had a *long period of incubation*, and when the disease supervenes actively, she has passed through a chronic form of illness, so that when others sicken, people fail to connect the new cases with the infected purchase. Then, again, in an ordinary herd of 10 or 20 head the deaths do not follow in rapid succession, but at intervals of a fortnight, a month. or even more, and those unacquainted with the nature of the disease suppose that it cannot be infectious, or all would be prostrated at once.

The disease may be communicated by immediate contact, through the atmosphere for some considerable distance, by the inhalation of pulmonary exudation when placed in the nostrils, from impregnated clothing of attendants, through infected buildings, infected manure, infected pastures, infected fodder, &c. Healthy cattle have been contaminated after being lodged in stables that were occupied by diseased ones three or four months previously. Hay spoiled by sick cattle has induced the disease after a long period, and pastures grazed upon three months before have infected healthy stock. The flesh of diseased animals has also conveyed the malady; and it is recorded by Fleming that the contagion from cattle buried in the ground infected others 50 or 60 feet distant.

There seems to be much difference of opinion with regard to the power of the virus to resist ordinary destructive influences. Under ordinary circumstances, it will be preserved longest where it has been dried up and covered from the free access of the air. In close stables and buildings having rotten wood-work, or deep dust-filled cracks in the masonry, and in those with a closed space beneath a wooden floor, it clings with the greatest tenacity. Again, in buildings which contain piles of lumber, litter, hay, fodder, or clothing, the virus is covered up, secreted, and preserved for a much longer period than if left quite empty. In such cases it is preserved as it is in woolen or other textile fabrics when carried from place to place in the clothing of human beings. As carried through the air the distance at which the virus retains its infecting properties varies much with varying conditions. Dr. Law states that he has seen a sick herd separated from a healthy one by not more than fifteen yards and a moderately close board fence of 7 feet high, and in the absence of all intercommunication of attendants, the exposed herd kept perfectly sound for six months in succession. At other times infection will take place at much greater distances without any known means of conveyance on solid objects. Röll quotes 50 to 100 feet, while others claim to have known infection transmitted a distance of from 200 to 300 feet. But the author questions whether, in such cases, the virus had not been dried up on light objects, like feathers, paper, straw, or hay, which could be borne on the wind.

Because the lesions are concentrated in the lungs, and begin with cloudiness and swelling of the smaller air tubes and surrounding connective tissues, the presumption is favored that the virus is usually taken in with the air breathed. Its progress and the results of all attempts at inoculation would seem to confirm this. The exudation into the interlobular tissue, the congestion of the lung tissue itself, and the implication of the lung covering, are regarded as secondary phenomena, or, in other words, the disease begins where the inspired air must lodge the germs. The inoculation of the virulent lung products on distant parts of the body transfers the seat of the disease to the point inoculated, and in such cases the lesions of the lungs are not observed, or at least are not greatly marked.

A diseased animal is more likely to infect a healthy one at that period when the fever runs highest and the lung is being loaded with the morbid exudation. Proof appears to be wanting as to the infecting nature of the affection during the incubation stage, but it must not be inferred that with the subsidence of the fever the danger is removed. It is a matter of frequent observation that animals which have passed through the fever, and are again thriving well and giving a free supply of milk, and to ordinary observers appear in perfect health, retain the power of transmitting the disease to others. This may continue for three, six, nine, twelve, or, according to some, even fifteen months after all signs of acute illness have disappeared.

The number of animals infected by contact or exposure to the contagion is somewhat irregular, as is also the virulence and fatality of the disease. The French commission of 1849 found that of 20 healthy animals exposed to infection 16 contracted the disease, 10 of them severely. Dr. Lindley gives examples from his South African experience in which whole herds of 80, 130, and even of several hundred died without exception, showing that in warm climates the mortality is greatest. Dr. Law found the disease much more virulent and fatal during the hot summer months in New York, and says that during the winter season it is far less violent in its manifestations, and a great number of animals resist it.

Lung plague (pleuro-pneumonia) confines its ravages entirely to the bovine genus, and no race, breed, or age is exempt from its attacks. Sex gives no immunity; bulls suffer as much as cows; and oxen and calves, if equally exposed, furnish no fewer victims than bulls and cows. As in rinderpest, measles, scarlatina, and the different forms of variola, an animal once afflicted with lung plague is usually exempt or impervious to a second attack. Only occasional instances are given where an animal has suffered from a second attack. The losses caused by the plague ranges all the way from 2 to 63 per cent. of all the animals in the country or locality in which it prevails, the losses varying according to climate, surroundings, condition of stock, &c.

The period of latency, that is, the time that elapses between the receiving of the germs into the system and the manifestation of the first symptoms of the disease, varies greatly. Veterinarians differ as to their experience and statements, and set this period at from five days to three months. Dr. Law has seen cases in which cattle have passed three or four months after the purchase in poor health, yet without cough or any other diagnostic symptom, and at the end of that time have shown all the symptoms of the lung plague. It is this long period of latency that renders the disease so dangerous. An infected animal may be carried half way round the world before the symptoms of the malady become sufficiently violent to attract attention, and yet all this time it may have been scattering the seeds of the disease far and wide. The average period in inoculated cases is nine days, though it may appear as early as the fifth, or it may be delayed till the thirtieth or fortieth day. In the experimental transmission of the disease by cohabitation, under the French commission, a cough (the earliest symptom) appeared from the sixth to the thirty-second day, and sometimes continued for months, though no acute disease supervened. Hot climates and seasons abridge the period of latency, as the disease has been found to develop more rapidly in summer than in winter, and in the South than in the North. A febrile condition of the system also favors its rapid development. Of the symptoms of the disease, Dr. Law says:

These vary in different countries, latitudes, seasons, altitudes, races of animals, and individuals. They are, *caeteris paribus*, more severe in hot latitudes, countries, and seasons, than in the cold; in the higher altitudes they are milder than on the plains; in certain small or dwarfed animals, with a spare habit of body, like Brittanies, they appear to be less violent than in the large, phlegmatic, heavy-milking, or obese short-horn Ayrshires and Dutch. A newly-infected race of cattle in a newly-infected country suffer much more severely than those of a land where the plague has prevailed for ages; and finally certain individuals, without any appreciable cause, have the disease in a much more violent form than others which stand by them in precisely the same conditions.

Sometimes the disease shows itself abruptly with great violence and without any appreciable premonitory symptoms, resembling in this the most acute type of ordinary broncho-pneumonia. This, however, is mostly in connection with some actively exciting cause, such as exposure to inclement weather, parturition, overstocking with milk, heat, &c.

Far more commonly the symptoms come on most insidiously, and for a time are the opposite of alarming. For some days, and quite frequently for a fortnight, a month or more, a slight cough is heard at rare intervals. It may be heard only when the animal first rises, when it leaves the stable, or when it drinks cold water, and hence attracts little or no attention. The cough is usually small, weak, short and husky, but somewhat painful and attended by some arching of the back, an extension of the head upon the neck, and protrusion of the tongue. This may continue for weeks without any noticeable deviation from the natural temperature, pulse, or breathing, and without any impairment of appetite, rumination, or coat. The lungs are as resonant to percussion as in health, and auscultation detects slight changes only, perhaps an unduly loud blowing sound behind the middle of the shoulder, or more commonly an occasional slight mucus rattle, or a transient wheeze. In some cases the disease never advances further, and its true nature is to be recognized only by the fact that it shows itself in an infected herd or on infected premises, and that the victim proves dangerously infecting to healthy animals in uninfected localities. It may be likened to those mild cases of scarlatina which are represented by sore throat only, or to the modified variola known as chicken-pox.

In the majority of cases, however, the disease advances a step further. The animal becomes somewhat dull, more sluggish than natural, does not keep constantly with the herd, but may be found lying alone; breathes more quickly (20 to 30 times per minute in place of 10 to 15); retracts the margins of the nostrils more than formerly; the hair, especially along the neck, shoulders, and back, stands erect and dry; the muzzle has intervals of dryness, and the milk is diminished.   The eye loses somewhat of its prominence and luster; the eyelids and ears droop slightly, and the roots of the horns and ears and the limbs are hot or alternatively hot and cold.   By this time the temperature is usually raised from 103° F., in the slightest or most tardy cases, to 105° and upward to 108° in the more acute and severe.   Auscultation and percussion also now reveal decided changes in the lung tissue.

The ear applied over the diseased portions detects in some cases a diminution of the natural soft-breathing murmur, or it may be a fine crepitation, which has been likened to the noise produced by rubbing a tuft of hair between finger and thumb close to the ear.   Where this exists it is usually only at the margin of the diseased area, while in the center the natural soft murmur is entirely lost.   In other cases a loud blowing sound is heard over the diseased lung, which, though itself impervious to air and producing no respiratory murmur, is in its firm, solid condition a better conductor of sound and conveys to the ear the noise produced in the larger air-tubes.

Percussion is effected by a series of taps of varying force delivered with the tips of the fingers of the right hand on the back of the middle finger of the left firmly pressed on the side of the chest.   Over all parts of the healthy lung this draws out a clear resonance, but over the diseased portions the sound elicited is dull, as if the percussion were made over the solid muscles of the neck or thigh.   All gradations are met with as the lung is more or less consolidated, and conclusions are to be drawn accordingly.

In other cases we hear on auscultation the loud, harsh, rasping sound of bronchitis, with dry, thickened, and rigid membranes of the air-tubes, or the soft, coarse, mucus rattle of the same disease when there is abundant liquid exudation, and the bursting of bubbles in the air passages.   In others there is a low, soft, rubbing sound, usually in jerks, when the chest is being filled with or emptied of air.   This is the friction between the dry, inflamed membrane covering the lungs and that covering the side of the chest, and is heard at an early stage of the disease, but neither at its earliest nor its latest stage.   Later there may be dullness on percussion up to a given level on one or both sides of the chest, implying accumulations of liquid in the cavity, or there is a superficial dullness on percussion, and muffling of the natural breathing sound with a very slight, sometimes almost inaudible, creaking, due to the existence of false membranes (solidified exudations) on the surface of the lung or connecting it to the inner side of the ribs.   This is often mistaken for a mucous rattle that can no longer take place in a consolidated lung in which there can be no movement of air nor bursting of bubbles in breathing.   The mucous rattle is only possible with considerable liquid exudation into the bronchial tubes, and a healthy, dilatable condition of the portion of the lung to which these lead.   In rare cases there will be splashing sounds in the chest, or when the patient has just risen to his feet a succession of clear ringing sounds, becoming less numerous and with longer intervals until they die away altogether.   These are due to the falling of drops of liquid from shreds of false membrane in the upper part of the chest through an accumulation of gas into a collection of liquid below.   It has been likened to the noise of drops falling from the bung-hole into a cask half filled with liquid.   Peculiar sounds are sometimes heard, as wheezing, in connection with the supervention of emphysema, and others which it is needless to mention here.

In lean patients pressure of the tips of the fingers in the intervals between the ribs will detect less movement over the diseased and consolidated lung than on the opposite side of the chest where the lung is still sound.

As seen in America, in winter, the great majority of cases fail to show the violence described in books.   The patients fall off rapidly in condition, show a high fever for a few days, lie always on the same side (the diseased one) or on the breast, and have a great portion of one lung consolidated by exudation and encysted as a dead mass, and yet the muzzle is rarely devoid of moisture, the milk is never entirely suspended, and may be yielded in only a slightly lessened amount as soon as the first few days of active fever have passed.

During the extreme heats of summer, on the other hand, the plague manifests all its European violence.   The breathing becomes short, rapid, and labored, and each expiration is accompanied by a deep moan or grunt, audible at some distance from the animal.   The nostrils and even the corners of the mouth are strongly retracted.   The patient stands most of its time, and in some cases without intermission, its fore legs set apart, its elbows turned out, and the shoulder-blades and arm-bones rapidly losing their covering of flesh, standing out from the sides of the chest so that their outlines can be plainly seen.   The head is extended on the neck, the eyes prominent and glassy, the muzzle dry, a clear or frothy liquid distils from the nose and mouth, the back is slightly raised, and this, together with the spaces between the ribs and the region of the breast-bone, are very sensitive to pinching; the secretion of milk is entirely arrested, the skin becomes harsh, tightly adherent to the parts beneath, and covered with scurf, and the arrest of digestion is shown by the entire want of appetite and rumination, the severe or fatal tympanies (bloating), and later by a profuse watery diarrhea in which the food is passed in an undigested condition. If the infusion into the lungs or chest is very extensive, the pallor of the mouth, eyelids, vulva, and skin betrays the weak, bloodless condition. The tongue is furred, and the breath of a heavy, feverish, mawkish odor, but rarely fetid. Abortion is a common result in pregnant cows.

During the summer the disease shows its greatest violence, and it is then that its mortality is not only high but early. The great prostration attendant on the enormous effusion into the organs of the chest, the impairment of breathing, and the impairment or suspension of the vital functions in general, causes death in a very few days. In other cases the animals die early from distention of the paunch with gas, while in still others the profuse scouring helps to speedily wear out the vital powers. In certain severe cases the rapid loss of flesh is surprising. Dr. Law says that in such cases a loss of one-third of the weight in a single week is by no means uncommon, and even one-half may be parted with in the same length of time in extreme cases. In fatal cases all symptoms become more intense for several weeks, the pulse gradually becomes small, weak, and accelerated, and finally imperceptible; the breathing becomes rapid and difficult, the mucous membranes of the mouth, eyes, &c., become pale and bloodless, emaciation goes on with active strides, and death ensues in from two to six weeks. Sometimes, in cold and dry weather, a portion of dead lung may remain encysted in the chest, submitting to slow liquefaction and removal, and such animals will go on for months doing badly, at last to sink into such a state of debility that death ensues from exhaustion and weakness. In still other cases the retention of such diseased masses, and the consequent debility, determines the appearance of tuberculosis, from which the animal dies. Purulent infection and rupture of abscesses into the chest are also causes of death, but the author states that no such cases have come under his observation.

Dr. Law gives the following description of the *post-mortem* appearances :

If the disease is seen in its earliest stages, the changes are altogether confined to the tissue of the lung. From the examination of the lungs of several hundred diseased animals, I can confidently affirm that the implication of the serous covering of the lung (pleura) is a secondary result. In all the most recent cases we find the lung substance involved and the pleura sound, while in no one instance has the pleura been found diseased to the exclusion of the lung tissue, or without an amount and character of lung disease which implied priority of occurrence for that. Yet, in all violent attacks the disease will have proceeded far enough to secure implication of the pleura as well, and hence we may describe the changes in the order in which they are usually seen when the chest is opened. The cavity of the chest usually contains a quantity of liquid varying from one or two pints to several gallons, sometimes yellowish, clear, and transparent, at others slightly greenish, brownish-white, and opaque, or even exceptionally slightly colored with blood. This effusion contains cell-forms and granules, and gelatinizes more or less perfectly when exposed to the air.

On the surface of the diseased lung, and, to a less extent, on the inner side of the ribs, is a fibrinous deposit (false membrane), varying from the merest rough pellicle to a mass of half an inch in thickness, and, in the worst cases, firmly binding the entire lung to the inside of the chest and to the diaphram. These false membranes are usually of an opaque white, though sometimes tinged with yellow, and, in the deeper layers, even blood-stained, especially over an infarcted lung. A noticeable feature of these false membranes, and one that serves to distinguish them from those of ordinary pleurisy, is that they are commonly limited to the surface of the diseased portion of lung, or, if more extensive, that portion which covers sound lung-tissue is much more recent, and has probably been determined by infection from the liquid thrown out into the chest.

In the lung itself the most varied conditions are seen in different cases and at different stages of the disease. The diseased lung is solid, firm, and resistant, seems to be greatly enlarged, because it fails to collapse like the healthy portion when the chest is opened; is greatly increased in weight, and sinks in water. When cut across it shows a peculiar linear marking (marbling) due to excessive exudation into the loose and abundant connective tissue which separates the different lobules of the ox's lung from each other. This exudation is either clear, and therefore dark, as seen by reflected light, or it is of a yellowish-white, and when filled with it the interlobular tissue appears as a network, the meshes of which vary from a line to an inch across, and held in its interspaces the pinkish-gray, brownish-red, or black lung tissue.

When only recently attacked the lung may present two essentially different appearances.

1. Most frequently the changes are most marked in the interlobular connective tissue, which is the seat of an abundant infiltration of clear liquid, a sort of dropsy, while the lung-tissue, surrounded by this, retains its normal pinkish-gray color, and is often even paler, and contains less blood than in health. It has, in short, i: : ae compressed by the surrounding exudation, and air and blood have been alike in great part expressed from its substance. (See Plate I.) This extreme change in the tissue surrounding the lobules and the comparatively healthy appearance of the lobules themselves, have led many observers to the conclusion that the disease commenced in the connective tissue beneath the pleura and extended to the proper tissue of the lung. There is, however, as pointed out by Professor Yeo, a coexistent disease of the smaller air-tubes corresponding to the lobules that are circumscribed by this infiltration, and there is every reason to believe that the infiltration in question is the result of antecedent changes in the air-tubes.

2. Less frequently we find the lobules of the lung-tissue presenting the first indications of change. The lobules affected are of a deep red, and more or less shining, yet tough and elastic. They do not crepitate on pressure, yet they are not depressed beneath the level of the adjacent healthy lung-tissue as they would be if collapsed. The interlobular connective tissue, devoid of all unhealthy exudation, has no more than its natural thickness, and reflects a bluish tint by reason of the subjacent dark substance of the lung. Here the lung-tissue itself is manifestly the seat of the earliest change—congestion—and the interlobular exudation has not yet supervened. Specimens of this kind may be rare, but a number have come under the writer's observation, and in lungs, too, that presented at other points of their substance the excessive interlobular exudation.

Both of these forms show a tendency to confine themselves to particular lobules and groups of lobules of the lung. They correspond, in short, to the distribution of particular air-tubes and blood vessels, as will be explained further on. The fact, however, is noteworthy as characteristic of the disease, that it attacks entire lobules, and the limits of the diseased lung-tissue are usually sharply marked by the line of connective tissue between two lobules, so that one lobule will be found consolidated throughout, and the next in a perfectly natural condition.

The two forms just described differ also in cohesion and power of resistance. The lung saturated with the liquid exudation has its intimate elements torn apart, and is more friable, giving way readily under pressure, while that in which there is red congestion but no extensive exudation retains its natural elasticity, toughness, and power of resistance.

Another condition of the diseased lung-tissue, more advanced than either of those just described, is the granular consolidation or hepatization. In this condition the affected regions of lung are as much enlarged as in the dropsical condition, but they are firmer and more friable, and on their cut surface present the appearance of little round granules. These granules are not peculiar to the lung-tissue proper, though most marked on this; they characterize the interlobular connective tissue as well. They consist mainly of lymphatic cell growths, filling up the air-cells, the smaller air-tubes, the lymph spaces, and the meshes of the connective tissue. The color of these portions varies from a bright reddish-brown to a deep red, according to the compression to which the lung-tissue has been subjected by the exudation in the early stages. (See Plate I.)

Another form of lung consolidation is of a very dark red or black, and always implies the death of the portion affected. The dark aspect of the diseased lobules forms a strong contrast with the yellowish-white interlobular tissue, excepting in cases where that also becomes blood-stained, when the whole presents a uniform dark mass. This form has the granular appearance of that last described, and on microscopic examination its minute blood-vessels are found distended to their utmost capacity with accumulated blood globules. This black consolidation is always sharply limited by the borders of certain lobules or groups of lobules which are connected with a particular air-tube and its accompanying blood-vessels, and the artery leading to such lobules is as constantly blocked by a firm clot of blood. The mode of causation is this: the artery, being in the center of a diseased mass, becomes itself inflamed. As soon as the inflammation reaches its inner coat, the contained blood coagulates; the vein is usually blocked in the same way. The blood formerly supplied by the artery to certain lobules is now arrested; that in the capillary vessels of these lobules stagnates; nutrition of the walls of the capillaries ceases, and these, losing their natural powers of selection, allow the liquid parts to pass freely out of the vessels, leaving the globules only in their interior. More blood continues to enter them slowly from adjacent capillaries supplied from other sources, and as this is filtered in the same way by the walls of the vessels, these soon come to be filled to repletion by the globules only; and hence the intensely dark color assumed. The color is often heightened by the escape of blood from the now friable vessels into the surrounding tissue, and it is by this means that the interlobular tissue is usually stained. (See Plate I.)

This black hepatization, or, as it is technically called, *infarction*, is an almost constant occurrence in the disease as seen in New York, and the death and encysting of large portions of lung is therefore the rule. If too extensive, of course, the patient perishes, but not unfrequently a mass of lung measuring four or six inches by twelve is thus separated without killing the animal.

If at a later stage we open an animal which has passed through the above condition, the following may be met with: A hard resistant mass is felt at some portion of the lung, usually the lower and back portion, and on laying it open it is found to consist of dead lung-tissue in which the hepatized lobules and interlobular tissues, the air-tubes, and blood-vessels are still clear and distinct, but the whole is separated from the still living lung by a layer of white pus-like liquid, outside which is a dense, fibrous sac or envelope, formed by the development of the surrounding interlobular exudation. From the inner surface of this dense cyst, the firm, thick bronchial tubes and attending vascular systems project in a branching manner like dirty white stalactites, and these, with the interlobular tissue thickened by its now firmly organized exudation, may form bands extending from side to side of the cavity.

At a still more advanced stage the dead and encysted lung-tissue is found to have been entirely softened, and the sac contains but a mass of white liquid *débris*, or, still later, a caseous mass of its dried, solid matters, upon which the fibrous covering has steadily contracted, so as to inclose but a mere fraction of its original area. In hundreds of *post mortems* we have only once seen the dead and encysted lung the seat of putrid decomposition, and never found the cavity opening into a previous air-tube.

There remains to be noticed the condition of the air-tubes and accompanying vessels in the diseased lungs. In all cases where we see the starting point of the disease we find in the small tubes leading to the affected lobules a loss of the natural brilliancy of the mucous membrane, which has become clouded and opaque, and the tissue beneath it infiltrated and thickened. In more advanced cases, and above all in those showing the dropsical condition of the interlobular tissue, we find a similar infiltration into the connective tissue around the air-tubes and their accompanying vessels, and in the hepatized lung this is always seen as a thick, firm, resistant, white material, having the compressed and contracted and often plugged air-tubes and vessels in the center. (See Plate I.) These thickened masses have already been referred to as standing out in stalactite form from the inner wall of the sac in which the dead (necrosed) lung is undergoing solution.

As to the nature of the plague, Dr. Law states that there can be no doubt but it is determined by an infecting material conveyed in some manner from one beast to another. The intimate nature of this material has never been determined. No special anatomical element, no specific organism of animal or vegetable origin, has been detected as constant in the diseased organ and peculiar to it, yet the presence of a specific *contagium* has been fully demonstrated in all the experience of the disease by the author and others. This infecting material, as shown by the records of inoculation, rarely affects the lungs when first lodged on a raw surface of some other part of the body, differing in this essentially from most other specific disease poisons, which have a definite seat of election in which their morbid processes are invariably established, no matter by what channel they may have been communicated. Since this *contagium* does not usually affect the lungs when introduced by some other channel, it follows of necessity that when it does attack the lungs it must have been introduced directly into them. If inhaled in the air breathed, it will fall upon one of two points—the air-tubes or the air-cells—and there begin its baleful and destructive course. This is exactly in accordance with the early lesions of the disease as found by Dr. Law in his *post-mortem* examinations.

In treating of preventive measures, Dr. Law quotes an article prepared by him and published in the National Live-Stock Journal for March, 1878. This valuable paper was afterwards transferred to the pages of special report No. 12 of this department, issued in September last. Following this is a brief summary of the work of the New York commission in its efforts to stamp out the disease in that State; but as the department has later advices from the author in regard to the work actually accomplished by this commission, extracts from Dr. Law's letter are given in preference to quotations from this monograph work. The letter bears date of New York City, December 9, 1879, and contains, among other things, the following:

To place our work in a "nutshell," I would say that in the past ten months the inspectors in New York have examined 40,000 head of cattle, many of them several times; that we have slaughtered and indemnified the owners for 500 head of diseased cattle, and that we have all but exterminated the plague from seven of the counties in which we found it. At present the main center of the plague is in Kings County and the adjacent border of Queens County.

In all country districts, where the cattle are kept on inclosed farms, and where the people heartily co-operated, the work has been easy and in every case speedily crowned with success. In the cities and suburbs, on the other hand, where cattle had been accustomed to graze on open lots, where interchange between different herds was frequent, and where the facilities for secret slaughter favored the covering up of the disease, the greatest difficulties had to be overcome. In New York City we secured the hearty co-operation of the police, and effectually arrested all movement between city stables, allowed only sound animals from healthy counties to enter these stables, and none to leave save to immediate slaughter, and, finally, promptly slaughtered all acute and chronic cases of the disease and saw to the disinfection of the premises, and the most gratifying success crowned our efforts.

In Brooklyn, on the other hand, where our work was systematically opposed, where the aldermen defied the State law by passing an ordinance authorizing the pasturage of cattle on open commons and unfenced lots, and some of them signed special permits for the movement of cattle in defiance of General Patrick's authority, and where magistrates dismissed offenders who were brought before them and reprimanded the policemen who had made the arrests, we soon lost the assistance of the police, which was at first all we could wish, and we naturally failed to meet with the splendid success seen in New York.

It became evident early in the work that unless we could establish special inspection yards under our own control, and abolish the system of distributing cows and other store cattle from dealers' stables, our success would be very partial and slow. In New York we were enabled to do this through the liberality of the Union Stock Yard Company, who built new yards for this purpose, which we opened July 1. In Brooklyn no such favor awaited us, and as the appropriation made by the legislature would not meet the needful outlay and enable us to hold what we had gained until the legislature should again meet, we had to be content with a system which was confessedly ineffective. By the end of August the approaching exhaustion of the appropriation compelled the dismissal of one-half of our veterinary force, and soon after we had to stop nearly all indemnities and consequently nearly all killing. Fortunately, New York City was now so nearly sound that we could continue the work there with but one inspector in addition to the one in attendance at the Union Stock Yard, and wo could still kill and indemnify for all sick cattle in the city. Brooklyn, still widely infected, and with authorities still somewhat inimical, could only have her infected herds quarantined, and in her the scourge is but very partially abated.

In certain outlying districts most gratifying results have been secured. In May we learned that animals from an infected herd had been turned on the Montauk pasture on the east end of Long Island. The range was visited and eighteen animals killed to save the 1,100 that remained. Later, two other cases developed in animals that had been in infected herds and had been overlooked at the first visit. Fortunately, for some months at first the cattle turned on this immense range kept apart from each other in small groups, composed of such only as had herded together prior to their coming on the range, and this most fortunate condition, coupled with the prompt disposal of each animal as it sickened, secured the escape of 1,100 animals. Had the occurrence been later in the season, when the cattle had learned to come together into one great herd, the results must have been most disastrous.

A second case is that of Putnam County, in which the plague had been smouldering since 1878, but was only discovered in September last. The State appropriation would not warrant us to offer indemnities, but the county authorities promptly assumed the responsibility, and every herd in which infection was found to exist was at once exterminated. In this way six herds have been disposed of, consisting of about 100 head, and a seventh, where sickness has existed for months but where it has only just been discovered, will be attended to to-morrow.

As regards the future, I would strongly urge the National Government to assume not only the direction but the execution of this work of stamping out the plague. The following among other reasons require this:

1. The disease is an exotic, and if once suppressed could only reappear in America as the result of importation.

2. It is gradually extending, and if neglected must lay the entire continent under contribution.

3. If it reached our unfenced ranges in the West it would be incradicable, as it has proved in the European Steppes, in Australia, and in South Africa.

4. As the seeds remain latent in the system for three months, infected cattle may be moved all over the continent, from ocean to ocean and from lakes to gulf, and live for a length of time in a new herd before they are suspected.

5. Old cases with encysted masses of infecting matter in the lungs may show no obvious signs of illness, and may be bought and sold as sound and mingle with many herds in succession, conveying infection wherever they go. There is, therefore, the strongest temptation for the owner to seek to secure a salvage by the sale of apparently sound but really infecting animals. There is further the strongest probability that in a new locality these cattle would not be suspected until one or more herds had been irretrievably ruined.

6. The infection of the South and West would inevitably spread the infection over the whole Middle and Eastern States, as infection would pour in continuously through the enormous cattle traffic, and all rolling-stock, yards, &c., of railways would become infected.

7. The live stock bears a larger proportion to the State wealth West and South than in the East, hence the West has most at stake in this matter, and should bear its share in the work of extermination.

8. The plague is more violent in proportion to the heat of the climate, so that it will prove far more destructive in the semi-tropical summers of the South and West than on the Atlantic seaboard.

9. No State can be rendered secure unless all States are cleared of the pestilence. One remaining center of infection on the continent is likely to prove as injurious as the one infected cow landed in Brooklyn in 1843, the sad fountain of all our present trouble.

10. It has been decided by a United States Supreme Court in Illinois, that a State law forbidding the introduction of cattle from a neighboring State, because it is feared they may introduce disease, is unconstitutional. Therefore each State must keep a guard along its whole frontier, with quarantine buildings, attendants, and inspectors, and must quarantine all cattle as soon as they shall have crossed.

11. Smuggling is inevitable so long as there are distinct authorities in two adjacent States. Rascally dealers have repeatedly run cattle into New York from New Jersey, sold them and returned with their money before the matter could be discovered and the law officers of New York put on their track. Were the law and execution one for all the States such men could be apprehended and punished wherever found. In Europe it is found that an armed guard with intervals of 200 yards patrolling the whole frontier day and night is not always sufficient; how much less, therefore, with us a law that can be evaded with such impunity.

12. Finally, there is little hope of Delaware, Maryland, and Virginia stamping out the plague at their own expense, so that unless the United States takes the matter up, the work of New York, New Jersey, and Pennsylvania will be but money thrown away.

This is a matter which threatens with dire disaster the inter-State live-stock trade of the future, and the National Government is called upon to stamp out the scourge with the view of protecting the trade between States.

As respects the organization that should be charged with the work, it certainly ought to have a responsible head, and while the live stock interests should be represented, it should not be made too unwieldy to act at a moment's notice in any emergency. The conditions of success are well enough understood, and while special adaptations would be demanded in many localities, yet the work should be carried out actively without the necessity of calling together a large and unwieldy committee before anything can be done.

Another point of vital importance is that a sufficient sum of money should be appropriated for this exclusive purpose, to obviate the necessity of stopping the work or giving it a material check before success shall have been accomplished. Any material arrest or any entire cessation of the work and a renewed spread of the disease will bring the whole question of veterinary sanitary work into disrepute, and may be the means of indefinitely and fatally postponing further action. While a large sum should be appropriated, its expenditure may be sufficiently guarded, but above all it should not be a common fund to be devoted to this and other objects.

Aside from the moral question, this is of far more immediate importance than even yellow fever, the germs of which are destroyed by frost, and the neglect of which for one year places the sanitarian in no greater difficulty for the next. With a disease like the lung plague, which is favorably affected by no change of climate nor season, and the germs of which survive all extremes of heat and cold, the loss of a year, a month, or even a day, may make the difference between an easy success and disastrous and irremediable failure—a live-stock interest which can supply the world with sound beef, and a general infection of the continent, and continuous embargo on the foreign trade.

REPRESSIVE MEASURES ADOPTED IN PENNSYLVANIA.

The department is indebted to Mr. Thomas J. Edge, secretary of the Pennsylvania State Board of Agriculture, for advance sheets of his forthcoming report on the subject of contagious lung plague of cattle. After citing the history of the disease in Europe and in this country, and alluding to its long presence in Pennsylvania in a malignant and destructive form, Secretary Edge states that finally, but not until after the farmers of the State had sustained heavy losses, a meeting of the dairymen of Delaware, Montgomery, and adjoining counties was called. This meeting was held in Philadelphia in March last, and before its adjournment a committee was appointed to wait upon the secretary of the board of agriculture and urge the importance of legislative action. The veterinary surgeon of the board, in company with this committee, visited herds supposed to be infected. Surgeons who had had years of experience with the disease in Europe and elsewhere were also called in; *post-mortem* examinations were made, and the existence of the malady established beyond a doubt. The legislature being in session, the secretary of the board laid all the evidence before the joint committee of agriculture, and, after discussion and mature consideration, it was decided that the State should adopt a line of precautionary and preventive action, not only for the benefit of its own citizens, but also out of respect to the action of adjoining States. A subcommittee was, therefore, appointed to consult with the governor, and, if deemed expedient, they were instructed to draft an act providing for the suppression of the disease. After consultation, the following resolution was offered and adopted by both branches of the legislature:

WHEREAS, The States of New York and New Jersey, by recently enacted laws to prevent the dissemination among live stock of the disease known as pleuro-pneumonia, now invite this State, by a concert of action, to assist them to eradicate this contagion: Therefore,

Resolved by the Senate (if the House of Representatives concur), That the governor be, and he is hereby, authorized to take such preliminary action as may be necessary to prevent its further spread.

This resolution was approved by the governor March 27, 1879. At the same time, an act previously adopted by the committee was introduced, which, after amendment, passed both branches of the legislature, and was approved by Governor Hoyt May 1, 1879. The enactment bears the title of "An act to prevent the spread of contagious or infectious pleuro-pneumonia among the cattle in this State," and is as follows:

SECTION 1. *Be it enacted, &c.,* That whenever it shall be brought to the notice of the governor of this State that the disease known as contagious or infectious pleuro-pneumonia exists among the cattle in any of the counties in this State, it shall be his duty to take measures to promptly suppress the disease and prevent it from spreading.

SECTION 2. That for such purpose the governor shall have power, and he is hereby authorized, to issue his proclamation, stating that the said infectious or contagious disease exists in any county or counties of the State, and warning all persons to seclude all animals in their possession that are affected with such disease, or have been exposed to the infection or contagion thereof, and ordering all persons to take such precautions against the spreading of such disease as the nature thereof may, in his judgment, render necessary or expedient; to order that any premises, farm, or farms where such disease exists or has existed be put in quarantine, so that no domestic animal be removed from said places so quarantined, and to prescribe such regulations as he may judge necessary or expedient to prevent infection or contagion being communicated in any way from the places so quarantined; to call upon all sheriffs and deputy sheriffs to carry out and enforce the provisions of such proclamations, orders, and regulations, and it shall be the duty of all the sheriffs and deputy sheriffs to obey and observe all orders and instructions which they may receive from the governor in the premises; to employ such and so many medical and veterinary practitioners and such other persons as he may, from time to time, deem necessary to assist him in performing his duty as set forth in the first section of this act, and to fix their compensation; to order all or any animals coming into the State to be detained at any place or places for the purpose of inspection and examination; to prescribe regulations for the destruction of animals affected with the said infections or contagious disease, and for the proper disposition of their hides and carcasses, and of all objects which might convey infection or contagion (provided that no animals shall be destroyed unless first examined by a medical or veterinary practitioner in the employ of the governor aforesaid); to prescribe regulations for the disinfection of all premises, buildings, and railway-cars, and of objects from or by which infection or contagion may take place or be conveyed; to alter and modify, from time to time, as he may deem expedient, the terms of all such proclamations, orders, and regulations, and to cancel or withdraw the same at any time.

SECTION 3. That all the necessary expenses incurred under the direction, or by authority, of the governor in carrying out the provisions of this act shall be paid by the treasurer, upon the warrant of the auditor-general, on being certified as correct by the governor: *Provided*, That animals coming from a neighboring State that have passed a veterinary examination in said State, and have been quarantined and discharged, shall not be subject to the provisions of this act.

During the passage of this act the existence of the disease in the State had been denied. Hence, immediately after its approval, his excellency Governor Hoyt appointed a commission to "examine and determine whether infectious or contagious pleuro-pneumonia exists among cattle in any county or counties of this commonwealth, and report the same to the governor without unnecessary delay." This commission consisted of Hon. Samuel Butler and Hon. H. C. Greenawalt, on the part of the legislature; Thomas J. Edge and C. B. Michener, on the part of the board of agriculture: Hon. John C. Morris and George Blight, on the part of the Pennsylvania Agricultural Society; and George S. Garret, on the part of the dairymen of Philadelphia and vicinity. At the first meeting of this commission Hon. John C. Morris was elected president, and Thomas J. Edge secretary.

At a meeting held in Philadelphia May 16, 1879, a large number of practical dairymen and veterinary surgeons were examined, and their evidence taken down by a stenographic reporter. As a result of this meeting, Messrs. Morris, Butler, and Greenawalt were appointed a committee to report to Governor Hoyt, on behalf of the commission, that the disease did exist in at least two counties of the State, and that the decision of the commission was unanimous.

Under authority of the act before quoted, and based upon the report of the commission, his excellency Governor Hoyt appointed a special agent to take charge of the matter, to whom he issued the following commission :

It having been ascertained that an infectious and contagious disease of neat cattle, known as pleuro-pneumonia, has been brought into and exists in certain counties of this State, I hereby appoint you as my assistant to carry out the provisions of the acts of 1866 and 1879, for the prevention of the spread of this disease. As such assistant you are hereby authorized—

To prohibit the movement of cattle within the infected districts, except on license from yourself, after skilled veterinary examination under your direction.

To order all owners of cattle, their agents, employés, or servants, and all veterinary surgeons, to report forthwith to you all cases of diseases by them suspected to be contagious; and when such notification is received you are directed to have the case examined, and to cause such animals as are found to be infected with said disease to be quarantined, as also all cattle which have been exposed to the infection or contagion of said disease, or are located in any infected district; but you may, in your discretion, permit such animals to be slaughtered on the premises and the carcasses to be disposed of as meat if, upon examination, they shall be found fit for such use.

You may prohibit and prevent all persons not employed in the care of cattle therein kept from entering any infected premises. You may likewise prevent all persons so employed in the care of animals from going into stables, yards, or premises where cattle are kept, other than those in which they are employed. You may cause all clothing of persons engaged in the care, slaughtering, or rendering of diseased or exposed animals, or in any employment which brings them in contact with such diseased animals, to be disinfected before they leave the premises where such animals are kept. You may prevent the manure, forage, and litter upon infected premises from being removed therefrom; and you may cause such disposition to be made thereof as will, in your judgment, best prevent the spread of the disease. You may cause the buildings, yards, and premises in which the disease exists, or has existed, to be thoroughly disinfected.

You are further directed, whenever the slaughter of diseased animals is found necessary, to certify the value of the animal or animals so slaughtered, at the time of slaughter, taking into account their condition and circumstances, and to deliver to their owner or owners, when requested, a duplicate of such certificate. Whenever any owner of such cattle, or his agent or servant, has willfully or knowingly withheld, or allowed to be withheld, notice of the existence of said disease upon his premises, or among his cattle, you will not make such certificate.

You are also directed to take such measures as you may deem necessary to disinfect all cars or vehicles or movable articles by which contagion is likely to be transmitted. You will also take such measures as shall insure the registry of cattle introduced into any premises on which said disease has existed, and to keep such cattle under supervision for a period of three months after the removal of the diseased animal and the subsequent disinfection of said premises.

You are further authorized and empowered to incur such expenses in carrying out the provisions of the foregoing orders as may, in your judgment, appear necessary, and see to it that all bills for such expenses be transmitted to this department only through yourself, after you have approved the same in writing.

The agent appointed by the governor at once issued the following notice:

To all owners of cattle, their agents, servants, or employés; to all common carriers by land or water; to all veterinary surgeons; and to all others whom it may concern:

His excellency Governor Hoyt having decided to co-operate with the executive officers of the States of Massachusetts, Connecticut, New York, and New Jersey in a united effort to eradicate the disease known as pleuro-pneumonia from the herds of this State, it becomes my duty, under the foregoing commission, to request that you will promptly report to me all cases among neat cattle by you suspected to be contagious or infectious. Without your co-operation and assistance this attempt can only result in partial success; with it the result can scarcely be doubtful, and the work thus far accomplished gives us assurance of good results.

His excellency is anxious that all owners of cattle and others interested should be fully impressed with the belief that this commission, as well as the laws of 1866 and 1879, for the prevention of the spread of the disease, are in their interest as well as that of the State. It is also the wish of his excellency that while the provisions of these laws are fully enforced and made most effective, and their purposes promptly and fully accomplished, it shall at the same time be so managed as to cause the least possible inconvenience and injury to all concerned, and with a minimum of expenditure to the State.

I would particularly call your attention to the language used by his excellency in relation to the line of action to be pursued when interested parties have concealed the existence of the disease in their herds. This provision is very important not only to the stock owner, but also to the State, for while the concealment of the existence of the disease will result in pecuniary loss to the owner of the stock, it, at the same time, greatly increases the danger of infection and the subsequent expense to the State. With your active co-operation in this respect we may hope for the prompt suppression of a disease which, while it has already caused a great loss to our stock owners, will, should it become established in our Western States, inflict an incalculable and lasting injury to the stock-raising interests of the whole nation. So far as known, all infected herds in this State have been quarantined and all diseased animals promptly isolated or killed. In the future, as in the past, it will be our duty to cause as little injury and inconvenience to the owners of stock as is consistent with our duties to the State, and to carry out, to the full letter, the directions of his excellency relative to the valuation of all stock condemned and killed.

All reports of supposed infection should be made direct to the office, and all interested are requested to accompany the report with a correct and full account of the location of the herd and the symptons, in order that all unnecessary expense to the State may be avoided. No special line of action has yet been marked out for application to cattle in motion from one portion of the State to another, or to those in transit to other States, but it will be the duty of those in charge to cause the least possible inconvenience consistent with the best interests of the State.

Up to November 1, 1879, the agent of the governor quarantined twenty-seven herds, including four hundred and eight animals liable to infection, and distributed in the following counties: Adams, one; Lancaster, four; York, one; Bucks, one; Delaware, four; Montgomery, five; and Chester, eleven. Of these herds, eight (one in York, three in Montgomery, and four in Chester) were afterwards released from the quarantine and pronounced safe from another outbreak, except from a fresh infection from outside sources.

As soon as the supposed existence of the disease is reported, each animal in the herd is inspected by a veterinary surgeon in the employ of the State, and if the disease is found to exist is promptly quarantined to prevent its spread to adjoining herds; in order, and if possible, to prevent further contagion in the same herd, all diseased animals are appraised and killed.

The individual history of these herds is given as follows by the secretary:

No. 1.—In *York County*, infected by steers bought in Baltimore market. Six head were either lost by death previous to quarantine, or were killed for the purpose of stopping the disease. The whole herd were more or less affected, though a number had a very light attack, and when released from quarantine, September 4, were as well as they probably ever will be. A rigid quarantine, which was very much assisted by the local surroundings, and the prompt support of neighboring stockowners, prevented the disease from infecting other stock; and the killing of diseased animals and the use of disinfectants prevented further loss.

No. 2, containing twenty cows, two bulls, and ten calves, was quarantined June 12. Previous to quarantine four head had died, and after the enforcement of the quarantine fourteen head were killed. With one possible exception, *all* the animals were affected, and a number of them are now in a condition in which they are worse than useless to the owner. In this case the evidence is strongly in favor of the theory that the owner conveyed the disease to his herd by assisting in the care of another infected dairy. No spread of the disease to adjoining farms; but it is quite probable that the disease was carried from this herd to herd No. 8 in the clothing or on the person of the owner, who administered medicine to both herds. This herd has furnished an illustration of the disease in one of its worst forms, but is now believed to be clear, but not beyond the danger of infecting other stock.

No. 3, in *Delaware County*, contained fifty head of stock, and previous to quarantine a number had died. The probability is that the disease was introduced by purchase. After passing into the charge of the State authorities, eleven of the herd were killed. This herd, with Nos. 2 and 7, furnish by far the most stubborn cases we have yet met with. In all three cases every animal had been repeatedly exposed to infection before the existence of the disease was reported; and we may here state that when the first sick animals were promptly isolated, and the case reported, the loss by death has been very slight. By allowing the sick and well to run together, all are infested before the assistance of the State is asked.

Nos. 4, 5, and 5½ adjoin one another, and are all traversed by the same small stream. The disease seems to have originated on the upper farm, where the first sick animal died in the stream and was buried close to its banks. In No. 4, one animal died and one was killed; and in No. 5, the first one was promptly killed. In both cases, the importance of immediate isolation was understood and put in practice. Nos. 4 and 5 have been released from quarantine, but No. 5½ is still infected. Whether, in these cases the stream was the vehicle of contagion or not, we cannot say; but the almost simultaneous outbreak on the three farms can be accounted for on no other hypothesis.

No. 6 had lost eleven head previous to being reported and quarantined. With one doubtful exception, every animal had shown more or less of the effects of the disease, and its owner fully appreciated its contagious nature. Has been released from quarantine.

No. 7, in *Lancaster County*, was composed of forty-two animals, and when reported had been thoroughly infected by two sick animals running with the herd. In this herd, seven animals were killed in one day, and seven placed under quarantine; fifteen have been killed, and to all appearance the disease has been checked. The infection, no doubt, came from an adjoining herd, which in turn had been infected by stock from New York.

No. 8 is supposed to have been infected by the owner of No. 2 administering medicine to the animals after attending to his own. Seven head have been lost in this herd, and the others are not clear of the danger.

In No. 9, containing thirty-one head, the disease seems to have been checked by the prompt isolation and killing of one animal, and has since been released from quarantine.

Nos. 10, 11, and 12 are small herds, in two of which every animal exibited symptoms of the disease; but by rigorous care on the part of the owners, by isolating and the prompt death of infected animals, the loss has been small.

Nos. 13 to 20, inclusive, are herds which have been recently reported and quarantined, and, thus far, the losses in them have been slight. By the prompt action of the veterinary surgeon, assisted by care and co-operation on the part of the owner, it is hoped that most if not all of them have passed the worst point, and that some of them may be released from quarantine as soon as the proper time has elapsed.

No. 21 was infected by six cows purchased in the Philadelphia market, and showed itself ten days after the purchase. Of the six, five have died or been killed, and others are affected. The purchased cows have been traced to the Philadelphia drove yard, but here all further clue to the origin of the disease was lost.

No. 22, in *Montgomery County*, was quarantined October 24, and was infected by a cow purchased from a drover. At the time of purchase she was coughing, and when examined by our surgeon, a week afterwards, she was so far gone that the owner was willing to have her killed as worthless and without a value.

No. 23 was infected by contact with the animals in herd No. 16, previous to quarantine. At the request of the owner, who has insured complete isolation, they are being treated by our surgeon. In this case the fumes of burning sulphur seem to have been effective in preventing further trouble, but all the herd were or are more or less affected.

No. 24, containing thirty-three head, were, no doubt, infected by contact with herd No. 7, as, by accident they were in the inclosure containing the former herd for a short time.

No. 25 was allowed by its owner to graze alongside of No. 7, with no separation other than that afforded by a creek and common fence. Before the infection was reported, most of the steers (fat) were sent to market, but one left on the farm has shown all the symptoms of the disease.

In addition to the care and supervision of cattle already in the State, the agent of the governor was given control of all stock brought in from Europe and not quarantined by the national authorities or those of other States. Under the regulations established all cattle must present a certificate of clearness from any contagious or infectious disease at the point of shipment in Europe. They must also be inspected on their arrival in the State by a veterinary surgeon in the employ and under the control of the governor's agent, must be quarantined closely at the expense of the importer, under the supervision of the State surgeon, and must be again examined at the close of the quarantine. If then found uncontaminated, a certificate is granted which will authorize their removal to any point in the State. In enforcing this quarantine, care is taken to consult the interests of the importers so far as it is consistent with the interests of the State. These precautionary measures are made necessary in order to protect the stock of the State, not only from infection by pleuro-pneumonia or lung plague, but also from rinderpest and the hoof and mouth disease, all of which are now prevalent in Europe.

The report of Secretary Edge concludes with a brief pathological history of the disease—its nature, symptoms, and lesions as shown in *post-mortem* examinations. It is accompanied by colored lithocaustic plates prepared by Prof. J. W. Gadsden, M. R. C. V. S., formerly of England, but now a resident of Philadelphia. These illustrations are given elsewhere, and are marked respectively Plates Nos. II and III.

Mr. Edge closes his report as follows:

In our dealings with the disease under the immediate direction of the government, we find many points upon which scientists differ, and which it would be impolitic for laymen like ourselves to endeavor to settle; but of one point we feel certain, and in which we have the indorsement of *every* practical man who has had the disease among his stock, and this is the contagious and dangerous nature of the disease. Whether the disease can only be conveyed from animal to animal by actual contact, or whether it can or cannot be conveyed in the clothing, by the excretion, breath, or animals of another tribe; whether the disease is of ancient or of comparatively recent origin; whether it can be carried from herd to herd by a stream of water; whether it can be intensified in its ravages by bad ventilation or bad treatment; whether a complete separation of a certain specified number of feet of space will or will not prevent infection; whether in its first stages it is or is not contagious; whether it will or will not affect sheep, are all questions for scientists to determine, and which are all lost sight of in the one great question in the solution of which we are engaged—can the disease be eradicated by prompt and rigid action in the manner proposed? If so, all these questions can be solved in the future; if not, then the future of our stock breeders is indeed precarious. In defense of the propriety of the action of the joint committee of the legislature, and of the legislature itself, as given in the foregoing pages, we have nothing to say, except that the end in view justifies the means. If by the expenditure of a thousand dollars by the State we can save hundreds of thousands to her stock breeders and stock owners, and as many millions to the country at large, then we think no one will complain. If the result in New York, New Jersey, and Pennsylvania shall demonstrate that this cannot be done, we may still point with pride to the fact that this action has saved more thousands than it has cost hundreds; has demonstrated to other States that when Pennsylvania is appealed to for co-operation in a good cause, she is not slow to respond; and that when so important an interest is in danger, the State is not slow in her attempt to extend a helping hand.

Professor Gadsden, of Philadelphia, Pa., writing under date of January 21, 1880, says:

The authorities in this State are still at work in stamping out the disease of pleuro-pneumonia among cattle. There are now but three infected herds left, and they are in three different counties of the State, viz., Delaware, Lancaster, and Lehigh, which are giving the authorities any trouble. Each herd will be kept in quarantine three months after the last trace of disease has disappeared. The owners of diseased cattle complain of the small sum paid per head by the State, but Secretary Edge is afraid to pay more for fear the good work will have to stop for want of means. I notice that the authorities in the State of New York have ordered all work suspended until the legislature shall have appropriated more money.

I inclose you a letter received this morning from J. C. Michener, a veterinary surgeon employed by Secretary Edge. It contains a "bundle of facts" proving the contagiousness of this disease, and shows the great folly of allowing this nefarious traffic in diseased animals. Many such cases as the one alluded to could be traced out in this State. Secretary Edge, in a conversation with me yesterday, said that he was satisfied the disease could be stamped out in this State if diseased animals were prohibited from entering it. He has spent only about one-half of the $3,000 appropriated for the purpose of eradicating the disease. At first he was obliged to pay extravagant prices for some of the diseased animals in order to satisfy the owners. Now he pays but $5 per head for animals suffering with the disease.

The following is an extract from Professor Michener's letter, alluded to by Professor Gadsden:

Your letter making inquiry in regard to the herd of Mr. C. Krauss, of Lehigh County, is at hand. This herd is affected by contagious pleuro-pneumonia beyond the possibility of a doubt, and it is equally as certain that the disease was brought here by a heifer that came from Baltimore through the Philadelphia cattle-yard. Mr. Jonas Graber, who sold the heifer to Mr. Krauss, has traced her back as far as Baltimore, and has kindly furnished me with the names of all the parties to the transaction. Krauss bought the heifer September 11. Within from two to three weeks his attention was attracted to her by her making a grunting noise. She was being fed for a family beef, and was stabled and pastured with the milch cows. She gradually pined away and died in about four weeks. She was examined, and one lung was found swollen solid and adhering to the ribs. A few days after she died other animals commenced to show symptoms, and the local cow doctors suspected pleuro-pneumonia. They had never had the disease in that locality. Pleading ignorance of the law and of parties intrusted with its execution, the disease was allowed to have its own way until December 13, when its existence was made known to the governor's special agent, Edge. Under his orders I visited the herd forthwith, found that two animals had already died, and that twelve more were sick. We killed ten of these and made *post-mortem* examinations of eight. We found them badly affected; all had the characteristic swollen, hard, marbled appearance of lung, and the adhesion that belongs only to contagious pleuro-pneumonia. The diseased lungs weighed from 25 to 45 pounds, and the healthy ones from 3½ to 4½ pounds. We have since killed one more, making thirteen in all which have been lost out of this herd. We have reason to hope that the disease is now under control, although ten more are slightly affected and twenty others have been somewhat exposed to the disease, yet still remain comparatively healthy. No pains have been spared to carefully isolate the sick from the well. Disinfectants have been liberally used and a rigid quarantine enforced. The disease has not spread from the Krauss farm, the location of buildings and surroundings all tending to prevent this.

In all of the seven herds that I have heretofore managed under direction of Mr. Edge, we have been eminently successful in stamping out the disease, and the owners are all well satisfied with the result of our management. Concert of action on the part of the States, with the hearty co-operation of the national government, will effectually rid the country (if not too long delayed) of this most insidious and dangerous enemy to our vast cattle interests. I have lost all patience with those who advocate other means than those now being employed by our commonwealth for the eradication of the disease. When men talk of the disease being curable, and not even a contagious malady, they only show their ignorance. The disease entirely destroys the functions of lung tissue, and can only be cured by the removal of diseased organs and the insertion or substitution of new ones—a feat the best surgeon would hardly undertake to perform.

REPRESSIVE MEASURES ADOPTED IN NEW JERSEY.

In compliance with the provisions of an act entitled "An act to prevent the spread of contagious or infectious pleuro-pneumonia among cattle," passed by the legislature of New Jersey during the session of 1878–'79, Governor McClellan appointed General W. H. Sterling as the head of a commission to form rules and regulations for the proper enforcement of the law. He commenced operations at Trenton, on March 15, 1879; and in order to determine the extent and location of the disease, he caused circular-letters to be addressed to the assessors of each township, to postmasters, farmers, and other prominent gentlemen throughout the State, requesting such information as they could give as to the existence and extent of the malady in their respective and more immediate localities. From the answers returned, he found that the disease was prevailing to a considerable extent in various sections of the State, and that there was, therefore, necessity for immediate and decisive action, if the plague was to be arrested.

In April, General Sterling found it necessary to move his headquarters to Jersey City, in order to meet the requirements of moving cattle to and from the State of New York, from whence the New Jersey dairymen largely draw their supplies of fresh milch cows.

Dr. Holcombe was appointed as surgeon-in-chief, and Dr. Corlies designated as inspector of the abattoir at Jersey City. Four other veterinarians were appointed, and on the 1st of April were commenced regular inspections of the herds in Bergen and Hudson Counties. These counties were quarantined, in order that proper restrictions might be placed upon the movements of cattle in the absence of proper permits. All the ferries and boats on the eastern border engaged in carrying stock into New Jersey were prohibited from landing cattle, unless accompanied with a permit issued by General M. R. Patrick, of the New York State Commission, who had issued similar orders relative to the landing of stock in New York. Thus the carrying trade between the two States at this important point was effectually controlled.

From reports received from the western and southwestern portions of the State, General Sterling was convinced that the disease was being

11 C D

imported from Pennsylvania. Therefore, on the second day of August he appointed Mr. J. W. Allen an inspector, gave him written instructions, and dispatched him to Camden for the purpose of consummating arrangements with the different ferries plying between Philadelphia and the different ports of entry lying between Salem and Phillipsburg, for the transportation of cattle from Pennsylvania, and the inspection of the same on their arrival in the State. His efforts were successful, and the ferry companies caused to be erected on the New Jersey side a sufficient number of sheds and pens to hold all cattle crossing from Pennsylvania until after a thorough inspection had been made. The officers of the Pennsylvania, New Jersey Central, and other railroads cordially co-operated with the State authorities, and soon all the principal avenues for the ingress of the disease were effectually closed. General Sterling closes his brief report as follows:

When we consider the number of cattle in this State, and estimate their value, the importance of this subject will be apparent. The number of cattle in the State on January 1, 1879, was 236,700, valued at $7,828,922. With a knowledge of the past history of the disease in this and other countries, and the difficulty of eradicating its as well as legislative enactments and precautionary measures hitherto adopted for it, prevention elsewhere, a grave responsibility will attach to those in power if the disease be allowed to obtain a foothold, destroying our best stock, checking one of the great interests of the State, and entailing losses appalling to contemplate.

The following is a brief summary of the results accomplished during the year:

The number of cattle found sick with the disease was 572. There were inspected 2,663 herds, containing 40,309 head of cattle.

Many cattle showing symptoms of disease were placed in quarantine and held until the incubative period had passed, being carefully watched during the mean time. When no other symptom than bronchial trouble was manifested the order of quarantine was removed.

The number of cattle found necessary to destroy in order to prevent the spread of the disease was 315, at an average cost to the State of $11.85 per head. There are now in quarantine 99 herds, containing 865 head of cattle, of which number 257 head have been condemned as suffering with contagious pleuro-pneumonia. The total expenses of the commission will aggregate about $19,000.

CONTAGIOUS LUNG PLAGUE OF CATTLE.

Report Commissioner of Agriculture for 1879.

Plate I.

SECTION OF DISEASED LUNG; recent case of Lung Plague.
Thin end showed black hepatization; the centre, red
hepatization; the thick end, interlobular infiltration.
Several blocked vessels are shown.

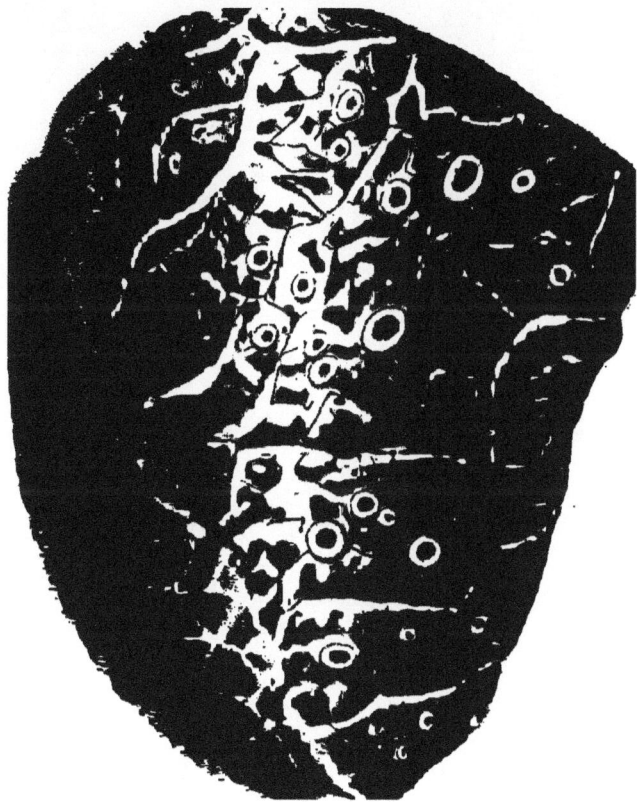

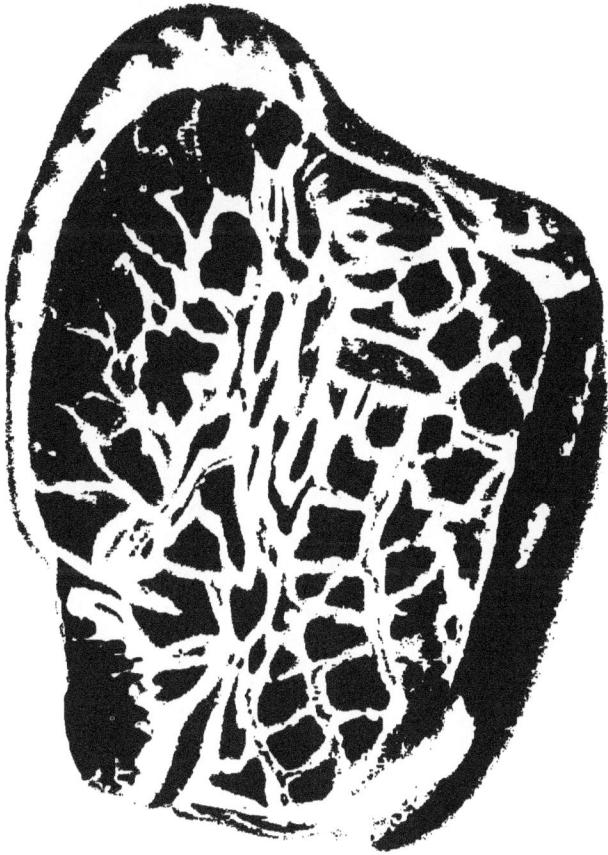

Section of the left lung of a Cow in an advanced stage of Contagious Pleuro-pneumonia showing the characteristic marbled appearance formed by the exudation and consolidation of lymph into the interlobular cellular tissue

Weight of left lung 37 pounds; right lung 5 pounds.

CONTAGIOUS PLEURO-PNEUMONIA—EXTENT OF ITS PREVALENCE.

REPORT OF DR. CHARLES P. LYMAN.

Hon. WILLIAM G. LeDUC,
Commissioner of Agriculture:

SIR: In compliance with the instructions contained in your letter of appointment, dated January 29, 1880, I left Washington on the 29th day of January last, for New York City, where I proposed to commence an investigation for the purpose, if possible, of determining the character and extent of the prevalence of the disease known as contagious pleuro-pneumonia or lung plague of cattle. On my arrival in New York I visited Dr. Liautard, from whom I learned that the disease still prevailed to some extent in Eastern New York and on Long Island, and that there was a reported outbreak at Haverhill, N. H. The New Hampshire State commissioner had pronounced this outbreak as of a sporadic character, yet the circumstances attending it were of a suspicious nature, at least sufficiently so as to throw doubt on the decision arrived at by the State commission, and I regarded a further investigation necessary in order to positively determine the matter. While in New York I gained some valuable information in regard to the disease in the adjoining State of New Jersey, which I propose to make use of on my return to that State.

I arrived in Boston on the 3d day of February, where I met Dr. Thayer, a member of the Massachusetts commission for the suppression of contagious diseases of cattle. He had made some investigations of the New Hampshire outbreak, and gave it as his opinion that the disease prevailing there was not contagious. However, he did not regard his *postmortem* examinations as satisfactory, as he was in no case furnished with whole lungs. I also saw Dr. Billings here, who informed me that he had examined portions of diseased lungs of some of the affected cattle at Haverhill, N. H., and from the appearances he did not regard the disease as that of contagious pleuro-pneumonia. He did not regard his examination as satisfactory, however.

I left Boston on the 5th day of February for Concord, N. H., for the purpose of seeing Dr. A. H. Crosby, chairman of the State commission. He regarded the Haverhill outbreak as of a suspicious character, and advised me to visit that place at once and thoroughly examine the affected herd. He gave me an order for the slaughter of such animals as I might deem necessary for examination, and also a letter to the chairman of the board of selectmen for the town.

I arrived in Haverhill on the morning of the 6th, and in company with Mr. Parker at once proceeded to the farm of Mr. Merrill, the owner of the suspected herd. I found the animals suffering in various degrees from respiratory troubles. As the herd was supposed to have been infected by a drove of cattle from Canada, I asked Mr. M. when this drove stopped with him. He answered—

On the 11th day of September; the first case of sickness occurred about October 2 the animal died on the 2d of November, having been sick only about one week.

163

Which was the second animal attacked, and when?

The light-colored or Fisher cow. She was taken somewhere between the 7th and 14th days of November. The calves first showed symptoms of sickness about the 14th of November. They were taken sick one by one with an interval of about seven days between each, except in one instance, when two were attacked at the same time.

Mr. Merrill described the symptoms as follows:

Coughing is the first symptom. The nose is dry, and the animal stands with its back arched and elbows turned out. If forced to move briskly about it will cough and pant. The disease seems more severe during a thaw than when the weather is colder. There is some running from the eyes. The appetite is invariably good up to about twenty-four hours before death. From six to twenty-four hours before death occurs, the animal is not able to stand. As death approaches, the animal groans quite loudly, the breathing becomes accelerated, and the cough seems to come from a more or less solid body. In coughing, a calf will extend its nose on a level with its neck. This symptom has not been observed to so great an extent among the cows.

How many cows, calves, and yearlings did you have in the fall? How many of each have been sick, and how many have died?

In the fall I tied up the following named animals:

Cows	11 head.	Sick, 4 head.	Died, 1 head.
Calves	23 head.	Sick, 10 head.	Died, 6 head.
Yearlings	13 head.	Sick, 3 head.	Died, 0 head.
Total	47	17	7

How have these animals been housed, and how have they been fed?

The cows and yearlings were tied up on the same side of the barn in a long row, but the yearlings were kept at the further end of the building by themselves. The calves have all been kept together in a separate pen 18 × 12 × 7 feet. During the day they were all turned out into the barn-yard. The cows have been fed on roots and hay, the yearlings on hay alone, and the calves on hay and roots.

Mr. Merrill continued:

The cows from this strange herd were put into the barn-yard, and the calves into a little pasture adjoining. About a week after this herd left, my own calves broke into this pasture. That would make the date about the 18th of September. October 20th I brought home a drove of calves myself from the north, and the first calves taken sick were some of these driven calves.

This being the statement, I regarded a *post-mortem* examination necessary in order to settle the question as to whether this outbreak was occasioned by contagious pleuro-pneumonia. For this purpose I thought it best to take the "Fisher," or light-colored cow, as she was the first attacked and had been the sickest animal of the lot. She was, therefore, slaughtered. I found the lungs in a *perfectly healthy condition.* The pleura of the ribs still showed plain traces of previous inflammation, but she had so far regained her health as to commence to again lay on healthy fat. This cow *never* had contagious pleuro-pneumonia.

I found a calf quite sick, evidently in an almost dying condition. This was next killed, and an examination revealed the fact that it had been suffering from a clear and unmistakable attack of bronchitis. This I demonstrated to the satisfaction of the medical representative of the New Hampshire commission, Dr. Watson. There had been preserved a pair of lungs taken from a calf which had died a few days previously. These showed the lesions of sporadic pneumonia, with some bronchitis. All the specimens were sent to New York for the inspection of the profession.

CAUSE OF THE OUTBREAK.

In looking about for the cause of this outbreak, the buildings and the lay of the land in the immediate vicinity of the premises were thoroughly examined.   The homestead is a meadow farm, lies well, and is inclosed by small hills, with a brook running a crooked course near to the buildings—a place that, in the fall of the year, would retain the fogs rising from the water for a considerable length of time.   Further investigation proved this theory correct.

The calves, when removed from the meadow, where one or two of them had taken cold, were, about the 10th of November, put into a close shed 18 feet long, 12 feet wide, and 7 feet high.   Here they were tied up in two rows, and were so close together that they completely packed the pen.   This huddled condition, to my mind, furnished ample cause for the outbreak, for I do not think that a pen of such dimensions, with so many animals confined in it, could be sufficiently ventilated to preserve health in the absence of mechanical means.   I advised the erection of another pen and a division of the herd.

I found that the cows had caught cold from being constantly subjected to a draft of cold air, so applied as to keep their bodies constantly bathed in a cold current.   The arrangement of the barn is given in the following diagram:

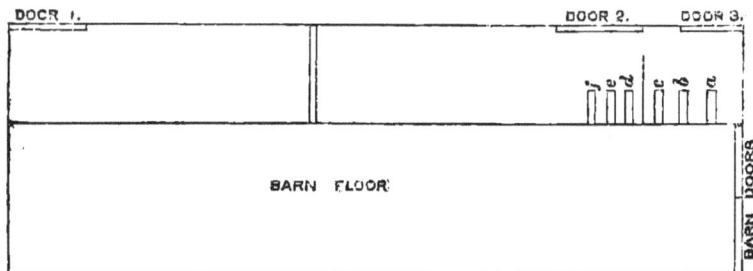

The yearlings were turned in through door No. 1, and divided from the cows by a partition.   This door was then closed, and they were left to themselves.   The cows were turned in at door No. 2, which, together with the large barn-door, was open a considerable portion of the time. Door No. 3 was seldom used.   Between the cows and the barn-floor was a board partition, with the board at the bottom *fixed to lift up*, thus leaving an open space fifteen inches wide directly in front of the cows and down at the floor.   Except in *very* cold weather, this novel ventilating device was left open all the time.   The air rushed in through the wide open door and the opening in front of the cows, passing over and bathing their bodies, and especially the under part of their breasts, chests, and abdomens, on its way out at door No. 2, which, by the way, is a little larger than any of the doors on this side.   This cause I regarded as sufficient to give the toughest animal a cold.

In order to prove this theory correct, it is only necessary to state the following facts: Cow *a* in the diagram, a small and nearly black one, stood in the corner against the partition, *just* out of the line of draught, *and has never even coughed.*   Cow *b* was the first animal taken sick, and the only one that died.   Cow *c* was sick, but not so bad as either cows *b* or *e*.   Being next the partition, even if on the cold side of it, may account for this in a measure.   Cow *d* had been but very little troubled. She coughed slightly, but nothing more; the partition may have protected her.   Cow *e*, the light-colored or Fisher cow, was the second one taken sick, and was more seriously affected than any of the others except the one that died.   Of the four animals which were called sick, although the whole herd except cow *a* were more or less affected with coughs, cow *f* came third. The yearlings were turned in and tied up without any regard to regularity or place. Several of them coughed. Three of them were sick, *i. e.*, the respiration was considerably accelerated, but none of them died. It was advised to close up the feeding space next the floor and put it up in the partition 3½ or 4 feet from the floor, so that the draught through would be over the heads of the cattle.

It may be objected that this cause has been in operation for years, and no such trouble has before occurred. The only answer to this objection is that the past season has been remarkable for its sudden changes. I am told that it has not been uncommon here for the thermometer to vary from 30° to 40° in the course of twenty-four hours. These sudden changes are as liable to affect cattle as human beings, and where exposed as these were, without artificial covering, they could hardly be expected to remain exempt from serious colds. Another thing should be remembered: the past winter has been so mild that attendants no doubt became more careless than usual, and often neglected to close the doors and feeding-troughs.

CONNECTICUT.

In the course of my investigations in Connecticut the following facts were gleaned:

Statement of Hon. E. H. Hyde, chairman of commission.

An outbreak of contagious pleuro-pneumonia had occurred at Greenwich, occasioned by exposure to a calf which had been brought from New York and placed in the herd of Mr. B. Livingstone Mead. This farm is located on the State line, a part being in the State of New York and a part in that of Connecticut. The buildings are in Connecticut. This herd consisted of 20 head. From 7 to 9 animals have died, the last one about the 18th of March, 1879. The remainder are unaccounted for. These animals were at one time examined by Professor Law.

The herd of Daniel M. Griffin, on an adjoining farm, contracted the disease from Mr. Mead's herd. He had 27 head, 8 of which died. With the exception of one animal, Mr. G. sold the remainder of his herd to dealers in New York for slaughter. The one he retained remains with his tenant, and will soon be slaughtered on the place.

Joseph B. Husted, of Greenwich, took some cattle to New York for slaughter, among them two cows. They were all landed at the infected Sixtieth street yard. The cows were not sold, and after some hesitation on the part of the New York commission they were allowed to be returned to Connecticut, the commissioners of the last-named State being notified of the fact. The State authorities at once ordered them quarantined, but before the letter reached Mr. Husted he had sold them, and they are still untraced. They were taken away from Greenwich on or before July 11, 1879.

Mr. Curtis Judson, of Watertown, near Waterbury, keeper of the Gramercy Park Hotel, bought two cows from Hedge, a dealer in New York, and placed them in an excellent herd of his own at Watertown. They proved to be affected with contagious pleuro-pneumonia, and soon infected the herd with which they had been placed. The herd was quarantined by order of the State commissioners, but the owner, on the 8th of March, 1879, broke quarantine and took them to New York. This fact coming to the knowledge of the authorities in time, they were enabled to be in New York on the arrival of the animals, where they were at once killed by order of the New York commission.

Mr. David D. Hawley, of Danbury, had an outbreak of disease in his herd on October 27, 1879. They were visited by Dr. Hopkins, of New York, who made an autopsy of a calf and pronounced the disease tuberculosis. The calf came from New York, and had been with the herd but a month.

Mr. Porter, of Waterbury, had an outbreak among his cattle on the 13th of November, which the attending veterinarian jeared might prove to be contagious pleuro-pneumonia. The herd was visited by the State board on November 18, and the decision arrived at was that the animals were suffering simply from sporadic disease. No *post mortem* examination was made, and they are now reported as doing well.

Some trouble was reported among cattle at Hartland and Milford, but, on examination by the commissioners, the disease was decided to be sporadic.

I visited the herd of Mr. L. B. Mead, of North Greenwich, which I found suffering from contagious pleuro-pneumonia. Although the trouble was of long standing, some of the cows certainly were in a condition to convey the disease to healthy or non-infected animals. There were ten cows, one pair of oxen, one yearling, and six calves in this herd.

The herds of Daniel M. Griffin, Joseph B. Husted, David D. Hawley, and Mr. Porter were visited, but no cases of the plague were found. Reports from Watertown, Waterbury, North Brandford, Hartland, and Milford were of such an assuring character that I did not deem it necessary to visit those points.

NEW YORK.

I am indebted to the New York commission for the following statement made February 12, 1880:

Putnam County.—On the line of the Harlem Railroad there have been lately slaughtered 176 animals. Of these 40 were acute cases. The others, having been exposed to the contagion, were killed to prevent the spread of the disease. The beef was marketed.

In the town of Kent Joseph R. Sprague has an infected herd of 60 head of cows, steers, and calves. They are now in quarantine.

Westchester County.—In Yonkers Mr. Austin had a herd of 27 head, which had been reduced by the ravages of the disease to 8 animals. Mr. Peirpoint had a herd of 11 head, which had been exposed to infection. Two of these had been killed. Mr. Cheever, on Odell's farm, has a herd of 12 head that have been infected. Mr. Coyle has one animal infected.

In Croton Falls, Bedford Township, Mr. Butler, who generally keeps about 50 animals, has lost by death and slaughter his entire herd, with one exception.

New York City.—In the city there are believed to be but five infected stables left. These are in quarantine, and are located as follows:

No. 1. West Seventieth street. Old chronic cases.

No. 2. West Seventy-eighth street. Acute cases.

No. 3. East Ninetieth street and Madison avenue. Acute cases.

No. 4. East One hundred and twentieth street and Fourth avenue. Acute cases.

*No. 5. East One hundred and twenty-first street and Fourth avenue. Acute cases.

Long Island.—The whole western end of this island, as far back as

---

* The last-named stables were infected from One hundred and twentieth street.

Jamaica, is more or less infected. The stables of Gaff, Fleischmann & Co., of Blissville, originally the hot-bed of the disease, are now perfectly free from all contagion. Jamaica is located some 10½ miles back, therefore the infected district includes Brooklyn, New Utrecht, Flatbush, Gravesend, Flatlands, and New Lots, in Kings County, and Long Island City, Newtown, Jamaica, Flushing, and Creedmoor, in Queens County.

Suffolk County.—At the extreme eastern end of the island are extensive unfenced ranges, used as common pastures. The plague prevailed among herds grazing on these ranges, but it is now believed they are thoroughly freed from it, as the last known cases were destroyed at Montauk August 28, 1879, and at Bellport August 11, 1879. This portion of the island has been subjected to numerous examinations, and is now regarded as entirely free from the plague.

Staten Island.—A year ago one case of the plague was discovered on this island. The animal was killed. No case has since appeared, and the island is now regarded as absolutely free from the disease.

On the 12th and 13th days of February, in company with one of the New York inspectors, I visited several stables in Brooklyn. I found several chronic cases in these stables, but no acute ones. At Johnson avenue slaughter-house I was shown a portion of a characteristically-diseased lung, which had been taken from an animal killed a few hours previously.

On February 14 I visited the stables of Mr. Lang, One hundred and ninth street and Fourth avenue, New York, where I found three cows suffering with the plague. One of these was a very acute case, and I was informed had been afflicted but three days. This and one of the others had been condemned to the offal dock. Mr. Froudie, a neighbor of Mr. Lang, lost a cow on the 12th day of February by the disease. A week before he had bought a cow from a dealer named Louis, and the cow that died was taken sick on that day that this cow came to his stable. The nearest stable to Mr. Froudie's is on One hundred and twelfth street and Fourth avenue. Mr. Froudie had owned the cow he lost for eight months. Lang purchased his sickest cow from a dealer named Franke some four or five weeks previous. She was a "two-titter," and on that account Franke knocked off $5 on her price. She never did well. The other two commenced coughing three or four days before my visit.

On the afternoon of the same day I visited the offal dock and witnessed the autopsy of Lang's cows, alluded to above. Both cases revealed well-marked lesions of acute pleuro-pneumonia *contagiosa*. One of the animals, which showed a temperature of 105° Fahr., and 36 respirations per minute, had the whole posterior lobe of the left lung consolidated and strongly adherent to the costal pleura. The right lung was healthy. The pericardium was thickened to half an inch. In both lungs of the second cow were found a number of small isolated spots of the characteristic lesions of the disease, the largest being about the size of a double fist. Their borders were well defined, and the intermediate portions of the lung-tissue appeared perfectly healthy to the naked eye.

On February 16, at One hundred and twentieth street and Fourth avenue, I found three cows which had been exposed to infection, and were in quarantine. They appeared healthy, and one had just been sold to a butcher named McEvoy.

In Tremont, at the stable of Mr. Bohle, I found two cows, one of which had been put into an infected stable on Christmas. Her temperature was 101° Fahr., and she was breathing at the rate of 30 respirations per minute. The other animal was a Jersey cow; both animals had been ordered slaughtered as soon as they could be got ready for the butcher. A Mr. Connors, a neighbor, had had some trouble with his herd, but they were quarantined and seemed to be doing well. The infection to this herd of Mr. Bohle's was communicated by a cow that was pastured with ten others on a common lot. She developed contagious pleuro-pneumonia, and was killed in the month of August. *Three months and nineteen days thereafter* the second animal was attacked and sent to the offal dock, where she was slaughtered. At the end of three weeks a third, and at the end of four weeks a fourth, animal was taken sick, and both were slaughtered. The first one of these animals belonged to Mr. B. Jorkman, the other three to Mr. Bohle, who, as has been before stated, bought a fresh cow on Christmas and put her in with one remaining from his original herd. This was in direct violation of the law and his instructions. She is now diseased and has been ordered to be killed. These ten animals were strictly isolated as soon as the first cow was killed, and no other infection was then possible. Two of them have since been fattened and sent to the butcher in a healthy condition. The remainder, with the exception of those belonging to Mr. Bohle, are still free from disease.

On the 17th day of February, in company with Professor Law and Dr. Hopkins, I visited the farm of Mr. Joseph Sprague, in Kent, Putnam County, whose herd was infected and had been in quarantine for some time. The herd consisted of 53 head, and were sold during the day by the State commission to butchers who had been notified to attend. The animals brought an average of $6 per head, which was regarded as a low price. Three of the animals were considered too badly diseased for beef, and on being killed showed well-marked lesions of the disease in its different stages. This herd was infected by a cow purchased from a dealer named Robinson.

On February 18, in company with the same gentleman, I visited Croton Falls, Westchester County. We found here a gentleman by the name of Butler, who had lost 31 animals out of a herd of 32 by the plague. His remaining cow was in quarantine, with no symptoms of the disease manifest. On the 15th of June last Mr. Butler bought 17 cows of Mr. Robinson, the dealer above referred to, and they were delivered to him on the 17th of the same month. They had been pastured all the summer on "Hyatt's lower farm" *with a cow that had been sick but had recovered.* The first animal on Butler's farm sickened on September 16, and soon died. The remaining 30 head were either slaughtered for beef or killed diseased.

On February 19, in company with the same gentleman, I visited the farm of Mr. Daniel Austin, in Yonkers, Westchester County. Originally this gentleman had a herd of 27 head, 18 of which had either died of the plague or had been killed for beef in the incipient stages of the disease. Five of the animals were killed for beef, and showed no lesions of the disease. Of the four remaining two are well-marked chronic cases, *i. e.*, having portions of encysted lung. This herd was infected by a cow that had pastured on an unfenced range called "Hog Hill," in the town of Yonkers. She wandered into a field near Mr. Austin's place, where she died on the 27th or 28th of July, and was not buried for some days after. The disease appeared among Mr. Austin's cattle on October 21. The herd of Mr. Odell, on whose farm this cow died, was no doubt infected by the same animal. His herd consisted of some valuable Jerseys, among which the plague appeared on August 28. We killed three of his animals, and they all showed well-marked lesions of the disease.

On February 20, visited Mr. Tice, of Newtown, which is a suburb of Brooklyn, Long Island. His herd was infected about the middle of October. Eight of his animals died, and he had continued to fill their places with fresh ones. We found 12 of his animals suffering with the plague. Two cows were killed—one an acute and the other an older case—and both showed well-marked traces of the disease. His herd was infected by a cow sent him about the 20th of September.

A Mr. Crady, whose stables are in Blissville, a portion of the suburbs of Brooklyn, had lost 11 head of cows, out of a herd of 14, since the middle of September.

PENNSYLVANIA.

I arrived in Philadelphia on February 24, and during the evening visited and had a conversation with Dr. J. W. Cadsden, relative to the prevalence of the plague in Pennsylvania. Dr. Gadsden showed me a private telegram giving him the information that the British Government contemplates raising the embargo on *cattle transported from the Western and Southwestern States through Canada and shipped to Great Britain from ports of the Dominion Government.*

On the morning of February 25, in company with Dr. Francis Bridge, I visited the farm of Mr. J. F. Taylor, located near the town of Marple, Delaware County, Pa. We found the gentleman's herd suffering with the disease. Having selected and paid for four acute cases, the animals were slaughtered and examined. The *port mortem* examination revealed all the lesions of the disease in its acute stage. This herd was infected by a cow purchased by Mr. Taylor in the Philadelphia stock-yards. She was in very good condition, and when she arrived on the farm seemed very tired. Next morning she refused to eat and seemed sick. She died in a few days thereafter with all the symptoms exhibited by those that have since died of contagious pleuro-pneumonia.

On February 26 I visited the farm of Mr. Wynne, near Philadelphia. His herd originally consisted of 34 head. Ten of these had already been killed, and two had died of the disease. An examination of those left developed the fact that the disease was still present in both an acute and chronic form. The owner objected to the slaughter of any of the animals. His herd was infected by some cows he purchased in the Philadelphia stock-yards. The disease broke out about the first of June last.

On the 27th day of February I visited Messrs. Martin, Fuller & Co., who have charge of the Philadelphia stock-yards. They offered me every facility for an examination of the premises. During my interview with these gentlemen, Mr. Fuller said that something ought to be done to relieve the dealers in stock from the oppression of the English embargo—that the European trade is now carried on at a positive loss, and that this loss is clearly traceable to the embargo on our live cattle. He further stated that he was in Europe last season, and found the market flooded at Liverpool. His stock was detained fifteen days in quarantine before it could be slaughtered. Besides the expense of feeding all this time, his animals were positively shrinking in weight—that when they were finally slaughtered he was compelled to accept any price offered. He found dealers there who said they could afford to give from $15 to $20 per head more for the animals if they were allowed to drive them back into the country and slaughter them only as needed.

During the day I met, by appointment, Secretary Edge, special agent of the governor. He seemed to appreciate the fact that more thorough and active measures than those heretofore used are necessary for a complete suppression of the plague. He thinks the better plan would be to pay a good price for all exposed animals, and that in the country all exposed and infected animals should be slaughtered as well as those acutely diseased. Under existing circumstances he does not think it would be politic for the State of Pennsylvania to thoroughly eradicate the disease; indeed, he does not think this possible so long as the southern border of the State is unprotected from importations from Maryland. Until quarantine measures are established against this State, or the State itself takes some action for the suppression of the disease within its borders, the State of Pennsylvania cannot hope for success. The farmers of Pennsylvania will go to the Baltimore stock-yard to buy "frames," and in this way new cases are continually being brought into the State. Under the present construction of the law sufficient means to pay a fair indemnity cannot be obtained, and to kill even diseased animals without funds to pay for them, the secretary believes would result disastrously, as it would prejudice the farmers against a better law, which is hoped for in the near future. His policy is simply an effort to keep the disease within its present limits with the destruction of as few animals as possible. Up to January 1, 1886, the secretary had expended but $2,700 in repressive measures.

On February 28, while examining some cows at the stock-yards, I found an acute case of contagious pleuro-pneumonia. The affected animal was in a yard with some twenty other milch cows, and all were being offered for sale. This animal was seen also by Dr. Bridge.

On March 1, while examining lungs of slaughtered animals at the Philadelphia abattoir, I found one showing the well-marked lesions of the plague. The butcher said the animal came from Illinois, but it was afterwards traced to Cecil County, Maryland.

On the 2d day of March I visited Camden and learned some facts relative to the extent of the plague in New Jersey.

On the 3d instant I attended a meeting of the farmers and stock-raisers in the infected district. The meeting was held in Philadelphia, and was called for the purpose of devising means for the extirpation of the plague. During the day, visited Elm Station, Montgomery County, and assisted in selecting six diseased animals from Mr. Wynne's herd for the purpose of *post mortem* examination.

On the 4th and 5th days of March I was engaged in examining lungs of slaughtered animals at the Philadelphia abattoir. I found no traces of the disease, but on the 4th instant, while examining some cows at the stock-yards, I found a second case of the plague in an animal that came from near Gettysburg, Adams County, Pennsylvania.

The following are the sources of infection and locations of diseased herds in Pennsylvania :

Philadelphia County.—The Philadelphia stock-yards are infected. These yards are constantly receiving and sending out to different localities diseased and infected animals.

Chester County.—Mr. M. Corning, of Chester Valley, has a herd of 27 head, among which the disease has appeared. The herd was infected by a cow purchased from a drover, and the infection could not be traced.

Mr. J. Dickinson, of Chester Springs, has a herd of 28 head. These animals were infected by the owner, who brought the contagion from a neighboring farm where he had administered medicine to a diseased animal.

Mr. G. V. Rennard, Chester Valley, had a herd of 18 animals infected by his neighbor's cattle (Mr. Corning's).

Mr. Rennard's cattle had infected a herd of 14 head owned by Mr. J. W. Wilson, his near neighbor.

Mr. C. Holland Frazer, of the same neighborhood, had a herd of 26 head infected by a purchased animal, which he was unable to trace.

Mr. W. Pugh, of Chester Springs, had his herd infected by Mr. Dickinson, alluded to above, who visited this herd for the purpose of administering medicine to a sick animal.

W. J. and H. A. Pollock, Downingtown, had a herd of 30 head infected by a purchased animal.

Mr. W. Reid, West Chester, herd of 5 head, chronic cases; source of infection unknown.

Mrs. Harmaan, West Chester, herd of 12 head; infected from neighboring cattle.

Mr. W. E. Penneypacker, Cambria, herd of 14 head; probably infected from neighboring herd.

Holmes and Bunting, Oxford, herd of 35 head; infected by Mr. Turner's cattle on adjoining farm.

Mr. M. Young, Bradford, herd of 36 head; infected by Mr. Turner's cattle.

Between the herds of Holmes and Bunting and Mr. Turner was a large meadow. The bulls broke down the two intervening fences, and the herds mingled in the meadow. The herds were separated as soon as men on horseback could separate them, but not soon enough to prevent infection.

Montgomery County.—Messrs. J. L. and A. S. Reiff, Worcester, herd of 15 head. Jacob L. Reiff had bought of five different dealers during May and June, and it was impossible to tell from which one the disease came. Two animals had died, and two others had been killed by order of the State inspector. Five others had been slightly affected, but had recovered. A. S. Reiff purchased a cow of his son in July, about the time of the outbreak. One animal died, and a second one was condemned and killed by order of the State inspector. Five other animals were affected, but all had recovered and had been released from quarantine.

Joseph Tyson, Worcester, herd of 13 head. Mr. Tyson purchased a cow of a man who had previously purchased her at the Philadelphia stock-yards. She was killed on September 24, 1879, by order of the State inspector, but as she had been isolated on the appearance of the first symptoms of the disease, only one other was infected.

Charles T. Johnson, Lederachsville. This gentleman's herd was infected by an animal purchased from a dealer. Up to the date of the first inspection in October last, five animals had died. One was afterwards condemned and killed. Five out of the remaining ten were affected, but had recovered.

Peter M. Frederick, Lansdale. Herd quarantined January 29, 1880. The infection was communicated by a cow purchased in the Philadelphia stock-yards. Two animals had been condemned and killed. The remainder—ten animals—were free from disease on March 4.

Jacob D. Wisler, Worcester. Herd quarantined February 6. Three animals had been condemned and killed, and three others were sick.

John C. Blattner, Worcester, herd of 16 head. The plague had prevailed in this herd in a mild form for the past four months. None of his animals died, and he did not suspect the nature of the disease. His cows were greatly reduced and he had been feeding at a loss.

One of his animals had commenced to lay on fat, and all were free from disease except the altered structure of the lungs, the natural result of the disease.   This herd was infected by Mr. A. F. Reiff's cattle, mentioned above.

W. W. Latrobe, Merion, herd of 14 head.

W. Wynne, Elm Station, herd of 28 head.   This gentleman had lost several animals.   The infection came from a cow purchased at the West Philadelphia stock-yards.

Bucks County.—Aaron Yoder, Dublin.   This herd was quarantined September 25.   The first cow to sicken was one that he purchased two weeks previously.   As she had passed through the hands of three different parties, it was impossible to trace her back satisfactorily.   Three out of the four were affected, but had " recovered."

Isaiah Kletzing, Dublin.   This herd received its infection from Yoder's cattle before they were quarantined.   Three animals had recovered.

Lehigh County.—Charles Krauss, East Greenville.   This herd was quarantined December 13.   The infection came through a cow purchased at the Baltimore stock-yards.   Two animals died and eleven were condemned to be killed.   Thirty animals remain, and are thought to be free from disease.

Cumberland County.—Samuel Hess, Eberly's Mills.   Herd quarantined March 20, 1879.   Infected by cattle coming from Baltimore stock-yards. This herd is in York County.

Delaware County.—R. L. Jones, Upper Darby, herd of 49 head.   Infected by purchase from Philadelphia stock-yards.

Thomas Cunningham, Upper Darby, herd of 21 head.

J. G. Haenn, Darby, herd of 14 head.

J. Likens, Ridleyville, herd of 15 head.

J. F. Taylor, Marple, herd of 36 head.   One third of his animals had died, and the disease was still present.

Lancaster County.—J. F. Turner, near Oxford, Chester County, herd of 52 head.   Infected by adjoining herd, into which the disease had been introduced by some calves brought from the State of New York.

David Williams, Coleraine.   This herd had come in contact with the diseased Oxford herd, and was quarantined before any symptoms of the disease appeared.

Lane Gill, Coleraine, herd of 5 head, adjoining above.

Adams County.—J. Redding, Gettysburg, herd of 13 head.   Infected by purchase from Baltimore stock-yards.

NEW JERSEY.

The following are the locations of some of the diseased herds in New Jersey at the time of my investigations in February:

Atlantic County.—Benjamin Gibberson, Port Republic, herd of 11 head. This herd was quarantined October 29, and again on November 28, as chronic cases.   Eight animals had been affected by the disease.

H. A. Johnson and William Ramsay, both of Port Republic.   The herds belonging to these gentlemen were diseased and in quarantine.

Gloucester County.—Charles B. Leonard, of Paulsboro, has two farms, upon one of which he has a herd of 22 animals, 6 of which are suffering with the plague.   He has 28 animals on the homestead farm, only 1 of which has shown symptoms of the disease.   Both herds are in quarantine.

Benjamin G. Lord, Woodbury, herd of 25 head.   On June 13, 6 of these animals were suffering with the plague.   October 27 there were 21 of these animals sick.   On November 25 the same number were suffering with the disease, and were all in quarantine.   Of the first lot of

25 animals, 6 were attacked and 3 died. He then bought four or five fresh animals. These remained in good health for five months and twelve days, but of the original animals 21 had suffered with the contagion.

Camden County.—An occasional case of pleuro-pneumonia had been found here, but no great amount of the disease had ever existed. A most thorough system of inspection of cattle coming from Philadelphia had been established here, and its rigid enforcement had undoubtedly been of great service in preventing the importation and spread of the contagion. From August 28 to December 15, 217 animals, known to have been exposed to infection, were returned to Philadelphia. Forty-one head of these were suffering with plain and unmistakable symptoms of the malady.

Burlington County.—Howard Stokes, West Hampton, herd of 11 head. Quarantined June 20, but did not obey quarantine regulations.

Job Ewan, Mount Holly, herd quarantined July 11. One acute case.

D. Maloney, Recklesstown. Lost 1 animal on January 26.

William Murray, Jacksonville, herd of 14 head. There have been 4 acute cases in this herd, and 2 animals have died. He will probably lose others. The herd was quarantined July 11.

Ocean County.—E. H. Jones, Forked River. herd of 29 head. There have been 27 acute cases in this herd. Six animals were killed on October 2, and on the 15th of the same month the balance were slaughtered. The infection to this herd was brought in some calves purchased in Fortieth street, New York City. From October, 1878, to October, 1879, Mr. Jones lost 32 animals by this disease.

At the same place as the above, Mrs. Strut has 1 animal, Captain Wilson 3, and James Holmes 23, all of which are infected and quarantined.

Mercer County.—G. E. Neunamaker, Pennington. On November 17 3 of his animals were suspected. On the 20th of the same month 2 acute cases had developed, and the herd was quarantined. One animal was slaughtered on January 17. Two animals recovered and are still on the place.

William Walton, Dutch Neck, herd of 32 head. On May 5, 1 acute case appeared. On May 15 the animal was very sick, and, as other cases were developing, the herd was quarantined. The owner did not believe his animals were affected with the plague, and failed to observe the quarantine regulations until one of the animals was killed in order to prove the fact. The herd becoming seriously affected, Mr. Walton sold, on October 29, all his animals to a butcher. This herd was infected by a cow purchased in New York. She calved, and her offspring, at five weeks old, showed well-marked lesions of contagious pleuro-pneumonia.

Monmouth County.—D. C. Robinson, West Freehold. One cow died of the plague on May 13. On the 19th of the same month another cow showed symptoms of the disease, and the herd was quarantined. On June 11 3 more animals were sick, 1 of which has since died and a second one recovered. The quarantine is continued.

A. D. Vorhees, Adams's Station, herd of 5 head. One of the animals was found sick on October 13, and the herd was quarantined. On the 16th of the same month another animal showed symptoms of the disease. One of the afflicted animals was killed. On November 19 a third animal was taken sick. The herd is still in quarantine.

Pliny Parks, who resides on an adjoining farm, had a herd of 8 animals infected. One was killed and the remainder quarantined on October 16.

D. W. Watrous, Perth Amboy, herd of 13 head. His herd was quarantined March 29, 1879. October 13, nearly six months after, he still had 11 head. On February 5, 1880, having added to his herd, he had 13 animals, 3 of which were sick and the others reported as well (?). The 3 sick animals were quarantined and the remainder were set at large.

Isaac Morris, Metuchen, herd of 14 head. The first case of the plague was discovered in this herd on May 22. The animal was taken to the butcher and killed, and the herd quarantined, which is still continued.

Hunterdon County.—Joseph Exton, Clinton, herd of 51 head. On June 9, 18 of the animals were found suffering with the disease and were quarantined. The quarantine is still continued.

Morris County.—D. Frank Corl, Sterling, herd of 13 head. On March 26, 11 head were sick. On February 20, but 5 animals remained, 1 of these showing old lesions. They are in quarantine.

Benjamin Punyon, Millington, herd of 20 head, 12 of which were sick on June 13, when the animals were quarantined. Two animals were killed; and on the 26th of June 8 animals were sick out of the 18 remaining. Two new cases had occurred, but the others were improving. The herd is still quarantined.

Mary Smith, Chambers street, Newark, herd of 5 head. October 24 1 animal was sick. On the 28th of the same month a second 1 was attacked, and 2 were killed. On January 14 the others were reported as recovered, but were still quarantined.

Allice Kennedy, Roseville, had 1 animal affected with the plague, which was killed August 14.

Union County.—C. E. Winans, Salem, herd of 9 head. Had lost 2 animals up to August 5. The remainder were sick and in quarantine.

Louis E. Meeker, Salem, herd of 13 head. Five animals were sick on August 1, when the herd was quarantined. On January 2, having purchased another animal, he had 14 head. Three of these were chronic cases, and were ordered quarantined for 30 days longer.

J. O'Callighan, Salem, on August 26, had a herd of 9 head, with but one animal sick. Up to November 12 he had lost 5 animals, and had but 4 left. On January 20 he was visited by the State inspector, but refused to drive his cattle in from the field for examination. The officer, on threats of personal violence, ordered him to keep up the quarantine, and left without making the examination.

E. A. Bloomfield, Salem, herd of 4 head, 1 sick; quarantined August 26. Had one chronic case on January 1; quarantine continued.

F. Saltzmen, Roselle, herd of 3 head, 2 sick; quarantined September 3. On January 20, 1 animal was sick and the herd was still in quarantine.

Bergen County.—C. McMichael, Leonia, herd of 21 head; 5 sick; quarantined April 1; on July 11 had 2 animals sick, and on January 21 had but 5 animals left, 2 of which were sick; quarantine continued.

Christian Freund, Closter; herd of 10 head; 5 sick; quarantined November 11. The same report of this herd was made on November 19. It is still in quarantine.

Hudson County.—The disease exists in the following localities:
Stables Nos. 133 and 144 Essex street, Jersey City.

Jersey City Heights.—Mary Mullin. 106 Thorn street; J. Lewis, corner Hutton and Sherman streets; J. Platz, 809 Montgomery street; J. Gurrey, Hopkins street; Martin Staunton, Hopkins street; George Reed

87 Germania avenue; J. Leddey, Nelson and Charles streets; J. Ryan, 25 Laidlaw avenue; Jonathan Meyer. 22 Gardner street; John Bosch, Congress and Hancock streets. These localities are all in quarantine.
Greenville.—B. O'Neil, Brittain avenue; William Shaw, opposite cemetery ; Mrs. Corcoran. All quarantined.
Hoboken.—Benjamin Engle, 200 Newark avenue; John Torpey, 172 Grand street; V. Cohen, old small-pox hospital. (Mr. Cohen, having diseased animals, desired a permit to put a fresh cow in his stables; he was refused, but he stated to the officers that he should put her in any way. This he did, and I afterwards saw this cow in his stable suffering with the disease in its acute stage.) Michael Reynolds, 165 Grand street.
West Hoboken.—J. Claude, Cortlandt street; Harris Aaron, Newark street: H. M. Nass, Hollingen; Mrs. Schmidt, Hackensack plank road; B. Benjamin, Cusset street; —— Kuntzle, Blume street; Mrs. Schlooler, Blume street; Ernest Weiss, Demot street, —— Oldmeyer, Boulevard.
Secaucus.— —— Latenstein, county road; H. Fisher, Secaucus road; —— Loeffle, race course ; Bryan Smith, race course; N. Wohlker, race course ; H. Black, North Bergen.

DELAWARE.

The only information I have as to the prevalence of the disease in Delaware I received in the course of a conversation with Mr. George G. Lobdell, president of the Wilmington Car-wheel Company. His farm is located in Newcastle Hundred, about two miles from Wilmington. In 1858 he had a valuable herd of animals. During this year contagious pleuro-pneumonia broke out among some cattle on a farm about three miles from his place. Fearing the infection of his herd, he commenced to sell off his cattle as he could find purchasers, but before this was accomplished, and perhaps within four months, it reached his farm, and by spring he had but one animal left. For two years after this he was without cattle, but at this time he commenced to stock his farm again. About six years ago the disease was introduced into a herd kept on a farm about two miles from his place. His own cattle remained exempt until about two years ago, when they were again infected. Since then he has been using the fumes of burning sulphur, and has had no fatal cases. Mr. Lobdell informed me that some sort of a law had been passed by the State looking to a suppression of the disease, and that three commissioners had been appointed by the governor to superintend and enforce its provisions.

MARYLAND.

Although it has long been known in a general way that contagious pleuro-pneumonia existed among the cattle of this State, no effort on the part of the authorities has ever been made to ascertain with any exactness the localities of the diseased herds.
On the 8th of March I proceeded to Baltimore, where I at once called upon Mr. William B. Sands, editor of the American Farmer, a gentleman who had greatly interested himself in this matter, and who gave me all the information in his possession as to the location and extent of the plague in the State, as well as kindly furnishing me with letters of introduction to the officers of the different agricultural societies throughout the State.
On the 9th of March I visited Hagerstown, the county seat of Wash-

ington County, where, on the next morning, I called upon P. A. Witner, esq., secretary of the county agricultural society. He said that he did not believe there was any disease in the county; that upon the day before there had been a meeting of the board of agriculture, at which there had been a good representation from all the different sections. Those present agreed that they had never known or heard of a case of lung plague in any part of the county.

I was next introduced to Mr. J. B. Bausman, a cattle dealer of this place. In the pursuit of his business he had been all over the county repeatedly, but had never known of a case of the disease. The drift of cattle in this place was entirely from Western Virginia through to Baltimore, never, so far as he knew, from Baltimore here. In his trade he feels very much the evils of the English embargo. It makes a difference to him of at least $10 per head in the price of his cattle. I then saw Dr. H. J. Cozens, an English veterinary surgeon, who had been located here for the past 15 years, and whose practice extends over the entire county. He had had a considerable experience with the lung plague in England, but had never seen but one case in this country, and that was many years ago, in Virginia. He is sure there is none in this county, nor has there ever been. Several other gentlemen from different localities were seen, but always with the same result. One farmer had a cow which he had recently bought that was coughing and not doing well. I visited her and found her suffering from tuberculosis.

In the afternoon I proceeded to Frederick City, the county seat of Frederick County. Here, upon the 11th of March, I called upon Mr. J. W. Baughman, secretary of the local agricultural society. He did not know of any diseased animals, but took me out to the court-house, where we saw and questioned a number of gentlemen from different parts of the county. None of these knew of any cases of this disease; they were very sure that had there been any unusual sickness they would have known of it.

I next saw Dr. P. R. Courtneay, an English veterinary surgeon. He had been here but a comparatively short time, and had heard of nothing that caused him to think there was any of this disease in the county. He kindly offered to bear the matter in mind, and if any cases of the disease came to his knowledge he would let me know at once. Here, as in Washington County, the whole drift of cattle is from west to east.

In the afternoon I went to Westminster, the county seat of Carroll County, and with my letter of introduction called upon Col. W. A. McKillip, president of the county agricultural society. He was sure there was no disease of the kind in the county, but he said that it was quite a common thing at certain seasons of the year for cattle to be brought here from Baltimore. This I regarded as a very suspicious circumstance, and so asked for an introduction to some cattle dealer in town. This was kindly granted, and I proceeded to call upon Mr. Edward Lynch. He said: "Farmers hereabouts generally make milk for the Baltimore market, and procure their cows from among themselves; but from the time that grass comes up until late in the fall of the year some of them are in the habit of feeding cattle; that the cattle for this purpose are generally bought at the 'scales' in Baltimore; that in this way, last fall, Mr. Samuel Cover, of Silver Run, this county, procured some stock which, after having been on his place for a short time, developed disease of some sort; some died, and some that were sick got well. Also, a Mr. Beacham, of Westminster, had had trouble of a similar nature for some time past." In a general way he knew that the farmers hereabouts were somewhat frightened about contagious pleuro-pneumonia.

12 C D

March 12.—Drove to the farm of Mr. Samuel Cover, above referred to, at Silver Run, and found there three cases of chronic contagious pleuro-pneumonia. This gentleman stated that he had got the disease last fall through some steers that came from Southwestern Virginia, but which had stopped at the Baltimore stock-yards for some little time, at which place he had bought them. Some four or five weeks after he got them home the disease broke out among them. He had at that time some 80 head of neat stock. Of these 15 were sick. When the disease first showed itself he put all the sick animals in a building by themselves, and had all his stables thoroughly disinfected. This was kept up all the time, and the places repeatedly whitewashed. In all, 4 animals died, 2 of them the Baltimore steers; the other 2 were cows which he had had for some time. Mr. Cover further says that *now* when he gets cattle he always puts them by themselves in a building entirely away from his regular cow-stables, and hopes in this way to avoid any future out-breaks among his herds.

Returning to Baltimore on March 17, in company with Dr. Daniel Le May, a veterinary surgeon, I visited a herd of milch cows kept at a dairy in Woodbury, near Baltimore. Here we found 1 acute and 2 chronic cases of the plague. The man in charge said that he had got through with the disease, from which he had suffered greatly, some two months ago, by selling out all his sick animals. From here we went to another large dairy in the same neighborhood. The gentlemanly owner informed us that he had had none of the disease for some time; that his plan was to buy often and sell often. In this way he found that he could keep up his milking stock and keep rid of the disease. From here we visited a near neighbor living on the direct road to the city. In an-swer to questions this man said that he did not know if his neighbor (the one from whom we had just come) called it having the disease or not, but that he drove many a sick one past his house on his way to the Baltimore market. He (our present informant) was free to say that he followed this same practice himself, and had done so ever since he lost his first 8 animals. He supposed this was not right, but his neighbors did it; and so he did. Summer was invariably the worst time there-abouts. The next place visited was about 2 miles distant, and on a dif-ferent road. The dairyman here had suffered greatly in the past, but thought that now, by *selling* the sick ones, he had nearly rid himself of the plague.

March 18.—We drove in several directions around the city and found the disease or its effects in all the herds except one that we visited.

March 19.—To-day we examined a number of the cow-stables in the city itself, in which many chronic and a few acute cases were found.

March 22.—Went to Harford County, where the disease was reported as existing in a number of different directions. However, we concluded to visit the farm of Senator George A. Williams, whose herd of fine Alderneys have been suffering more or less from the scourge for the past two years. Here, among several chronic cases, was one that, al-though he had been sick for some time, was making no progress towards a good recovery. This animal the overseer consented to let us kill. The autopsy showed, well marked, the lesions of the disease. The infec-tion here, as with all the other outbreaks hereabouts, came from Balti-more. At this point further investigations were given up for the pres-ent, and it still remains, in order to properly finish this report, to make an examination of the remainder of this State, the District of Columbia, and Virginia, in all of which places it is believed that contagious pleuro-pneumonia of cattle exists to a greater or less extent.

STEUBEN

TIOGA

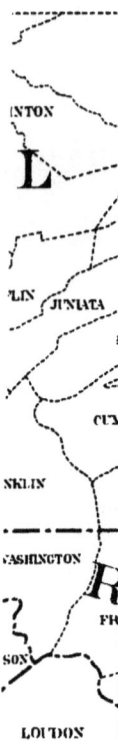

INTON

L

LIN JUNIATA

CUM

SKLIN

WASHINGTON R

FR

SON

LOUDON

CONTAGIOUS DISEASES OF DOMESTICATED ANIMALS. **179**

As a result of my investigations thus far, I find this ruinous foreign plague actually existing among cattle in the following States:

CONNECTICUT.—In Fairfield County.

NEW YORK.—In New York, Westchester, Putnam, Kings, and Queens Counties.

NEW JERSEY.—In Atlantic, Gloucester, Camden, Burlington, Ocean, Mercer, Monmouth, Middlesex, Hunterdon, Morris, Essex, Union, Bergen, and Hudson Counties.

PENNSYLVANIA.—In Philadelphia, Chester, Montgomery, Bucks, Lehigh, Cumberland, York, Delaware, Lancaster, and Adams Counties.

MARYLAND.—In Carroll, Baltimore, Harford, and Cecil Counties. The middle and southwestern portions of this State have not yet been visited.

No examination has yet been made in the District of Columbia or of the infected territory of Virginia, but, as the plague prevailed quite extensively in both of these localities last season, it will no doubt be found still in existence when the investigation takes place.

A map showing the extent of the infected territory accompanies this report.

Respectfully submitted.

CHARLES P. LYMAN, *M. R. C. V. S.*

WASHINGTON, D. C., *April 16*, 1880.

PLEURO-PNEUMONIA OR BOVINE LUNG PLAGUE.

The following article on the subject of the contagious lung plague of cattle is from the pen of Dr. JAMES LAW, and appeared in the National Live Stock Journal for January last. It presents strong arguments in behalf of national interference for the suppression of this destructive malady, and it is inserted here for the purpose of directing general attention to the great importance of the subject, and the necessity for speedy action if the disease is ever to be eradicated in this country. Dr. Law says:

Again and again in these columns we have set forth the need of *national interference* to extirpate foreign animal plagues. For a number of years before the inauguration of the present movement for the suppression of the lung plague of cattle, we continually sounded notes of alarm, and called for a *national*, and not a mere State or local intervention. Now, when at last the matter is likely to secure the attention of Congress, we wish once more to state succinctly the main reasons why this should receive the prompt and vigorous attention of the general government.

In approaching the subject of the protection of animals from native and imported plagues, we are overwhelmed by the vastness of the field, and are fully convinced that the public mind of the country is not prepared for the herculean task of suppressing the whole list. For the present, at least, it would be well that this work should be divided, and that a great portion of it should be left in abeyance until, by a successful experience with the more urgent of these affections, a public confidence in veterinary sanitary measures can be created. A large number of animal contagions, and of dangerous animal parasites, are communicable to the human subject, and these accordingly directly interest the National Board of Health, and demand its intervention on behalf of humanity. * * *

Besides these there are a large number of contagions and parasites that confine their ravages to the *lower* animals; and that must be dealt with from a different standpoint, as impairing our agricultural industry and draining away the national wealth. Some of these already subject us to enormous yearly losses, the so-called hog cholera causing deaths of swine to the value of $20,000,000 to $40,000,000 per annum; the trichina having driven our hams and bacon out of the market in certain countries of Europe; the lung worms, intestinal worms, scab, and foot-rot having depopulated the sheep runs on large tracts which are admirably adapted to wool-growing; and the Texas

MAP
showing the
STATES AND COUNTIES
in which
CONTAGIOUS PLEURO PNEUMONIA
of
CATTLE EXISTS.

THE INFECTED COUNTIES ARE TINTED RED.

fever having caused extensive yearly losses in the Middle and Northern States. But among all these, and many more, one affection stands out as pre-eminently demanding suppressive measures. The lung plague, which has for many centuries desolated the herds of Europe, has invaded our territory, where it has not only maintained itself for thirty-seven years, but has spread three hundred miles from its original center, in a direction opposed to the steady current of cattle traffic. If neglected. it must take but a short time longer to reach the open cattle ranges of the South and West, whence it will be impossible to expel it. It is characteristic of every true plague that the wider its area the greater its power for rapid diffusion. The larger the tract of country infected, the greater is the number of diseased herds and animals, and the greater the danger of shipments from such infected to States and districts as yet free from infection. Again, the wider the infected area, the more extensive is its border, and the greater the number of points at which infection may be conveyed to adjacent herds in territory hitherto uninfected; also, the greater the area, and the more extended the outer line of infections, the greater the likelihood that this shall include a route or channel by which cattle are continually sent out into States as yet uninfected. Once more, the present increasing interest in live stock is determining a constantly increasing draft on the eastern herds for valuable sires to improve the cattle of the West and South. The wider the area of infection, the greater the likelihood that one such sire may be carried out to the plains and infect the fenceless ranges, from which the poison could never be eliminated. No more is wanted. *One animal* landed in Brooklyn has infected seven separate States, and *one infected animal* landed on the plains, or in Texas, will secure the spread of this pestilence over the whole country, and entail its permanent ravages among our herds. The momentous consequences of such an accident cannot be fairly estimated. England, with her 40,000,000 of cattle, lost $500,000,900 in the course of thirty-five years; we with our 100,000,000 should lose over $1,000,000,000 in deaths alone in the same length of time.

This infection of the South or West may happen any day or any hour. The high-class herds of our infected States are not exempt from this disease, and it has been our sad experience to see it in the finest Jerseys and Ayrshires. We have known the owners of such animals (thought to be convalescent, but really carrying large encysted masses of diseased and infecting lung within their chests) anxious to dispose of them and realize some salvage from the wreck of their splendid herds. No ordinary purchaser would have suspected these animals, and yet they carried within them that which, if transplanted to the West, might have proved the starting point for a general infection of the continent.

Like all true plagues, this gains new force with every step made in its advance. As the malady is developed in this country by contagion only it secures an additional advantage with every new animal infected. Every new case of sickness is but another manufactory of the virus, spreading this on the air in countless myriads with every breath expired. No change of latitude or altitude, no modification of season or climate, no alternations of heat or cold, no change geological or atmospheric, no alterations, electrical or telluric, will rob the poison of its virulence or destroy its vitality. The plague once introduced prevails alike on the mountain top and in the valley; on the sea-coast and on the inland plains; in the sweltering heats of Africa and in the frigid winters of St. Petersburg and New York. The one condition of its propagation is the contact of the poison with the body of a susceptible subject.

COMPARISON OF BOVINE LUNG FEVER WITH YELLOW FEVER.

In connection with health laws and their administration, no subject has of late more profoundly exercised the minds of the medical profession and the laity than that of yellow fever. Yet, terrible as this malady is, and fresh as are our recollections of its devastations in Memphis and other Southern cities, it is lacking in the most dangerous features which invest the bovine lung plague with its special pestilential character. Yellow fever cannot prevail in cold weather, and its germs are permanently destroyed by the winter's frosts. No matter how extensive its prevalence in our summer, a severe winter will effectually eradicate it; and on the return of summer it must start anew from another importation, or perhaps from one or two centers where it may be developed in the United States itself. An unusual extension of the poison in one year gives it little or no advantage for the next, and therefore the loss of one or of ten years in applying restrictive measures will not in the least interfere with their final success. But with the lung plague of cattle, as we have seen, it is far otherwise; it advances with insidious but constant steps in summer and winter alike, and every new step achieved is a material and permanent gain in its progress towards a general infection. With this affection the loss of a year is a certain and permanent loss, and will entail a materially increased outlay for its extinction, while in individual cases the loss of a single day or an hour, even, may allow the irreparable infection of the Plains.

Again, yellow fever is generally connected with filth, and will inevitably die out in the presence of absolute cleanliness. But the virus of lung plague is certainly reproduced in the body of the sick animal, and exhaled with every breath, so that it is diffused widely, and in a highly virulent condition on the purest atmosphere. No cleanliness of buildings or drains, no purity of the air, can protect against a contagion like this; the one effectual resort is to stop the production of the virus in the infected system, and to destroy what has been already produced.

COMPARISON OF LUNG PLAGUE AND TEXAS FEVER.

Texas fever is no more dangerous to cattle than yellow fever is to man. It extends from the States on the Gulf in summer only, and in all the Middle and Northern States a severe winter destroys its germs. Moreover, its virus is not diffused on the atmosphere, but is confined to the pastures on which the Southern cattle have grazed. After these Southern cattle have been removed, a simple fence is sufficient to bar its progress to stock in the field adjacent. The loss of a year in dealing with such a disease is therefore of comparatively little importance. The loss is of time only, for next year the fever has to begin anew its progress northward from its permanent home on the borders of the Gulf of Mexico. A neglect for any number of years does not very materially widen the area of its prevalence, and at any time in the future it may be taken up and dealt with as economically and with as good hope of success as it could have been at any period in the past. Yellow fever, therefore, and Texas fever, are to be viewed rather with respect to their present fatality than their future results. In this they are to be viewed as allied in their prejudicial results rather to injurious conditions affecting finance, banking, commerce, railroads, canals, shipping, &c.; the evil is not cumulative, and may be corrected at any time with immediately good results. With lung plague, on the other hand, the peril for the future increases in ratio with the animals and area infected, and the pestilence may at any moment pass beyond our control. If we leave out of view the moral aspect which gives yellow fever its special claim, and consider the three diseases simply as pestilences, there can be no hesitation in pronouncing the bovine lung plague immeasurably more virulent, more pregnant with lasting and irreparable danger, and more imperative in its demand for immediate administrative action than either of the others.

COMPARISON OF LUNG PLAGUE WITH SWINE PLAGUE, GLANDERS, CANINE MADNESS, MALIGNANT ANTHRAX, TUBERCULOSIS, AND MILK SICKNESS.

A comparison of these will furnish abundant reasons why the preference in administrative action should be given to the bovine lung plague. The exotic origin of this plague of cattle, and its recurrence on this continent by contagion only, gives a most perfect assurance that we can permanently extirpate it. But with the other affections named it is far otherwise. There is a strong probability that all of these may be produced in our own land by unhygienic conditions—may be generated *de novo*—a first case may be generated without any antecedent imported one; and thus, although we could to-morrow extinguish every spark of the infections of these diseases, there are reasonable grounds for supposing that before the end of 1880 one or more new cases would have developed of themselves, and would require new efforts at extinction. No measures for the suppression of these diseases can therefore assure a permanent eradication of them, but a perennial vigilance must be maintained to suppress the centers of infection in the already diseased animals, and to remove one by one the mass of unhygienic conditions which are the creative causes of these affections. This is a work which will task the knowledge, wisdom, and energies of many generations yet to come, and cannot be disposed of, like bovine lung plague, by the energetic work of a year or two.

COMPARISON OF LUNG PLAGUE WITH SOME DANGEROUS PARASITIC DISEASES.

Parasitic diseases of animals have an importance that cannot be ignored. It has been shown that ᵃ per cent. of the hogs killed in Chicago are trichinous; that lung worms have killed lambs by the 100,000 in single counties in Iowa, Ohio, Illinois, and elsewhere; that much of the so-called hog cholera in different States is but the ravages of lung worms; that even scab temporarily ruined the wool-growing industry not long ago in Iowa, and still ravages certain other States; and that intestinal worms have been equally destructive in other directions. These mostly spread through the eggs and embryos deposited in moist soil, and in wells, pools, and streams, and may be carried by infected animals from place to place, but they are not wafted far on the air, nor is their preservation and extension so redoubtable as in the case of the lung-plague virus. Moreover, several of the parasites of farm animals infest so many others, wild and tame, that the question as to whether they can be certainly eradicated cannot be answered with the confidence with which we can speak of the extinction of the lung plague.

From these considerations of the most dangerous infections and parasitency of animals may be deduced the conclusion that the bovine lung-plague is that which at present threatens the country with the greatest peril, and which at the same time can be most certainly and effectually expelled from our shores. For the others we can afford to wait; but with this delay is suicidal.

NEED OF EXECUTIVE ACTION BY THE UNITED STATES GOVERNMENT.

The necessity for action being admitted, the question arises, By whom should such action be taken? New York is operating under State authority, and many think this all-sufficient, and point to Massachusetts as an example of the successful extinction of the malady by an individual State. But the same course could never have been successful in Massachusetts had she, like New York, been the *entrepôt* for cattle shipped from infected States. Massachusetts has kept clear of this plague for fifteen years, it is true, but this could never have happened had her weekly accessions of thousands of cattle been drawn from Long Island, New Jersey, Delaware, Virginia, and Pennsylvania, in place of from the great uninfected West. In view of the vital importance of national action in this matter, we submit the following

REASONS WHY THE WORK OF STAMPING OUT THE LUNG PLAGUE OF CATTLE SHOULD BE A NATIONAL MOVEMENT.

1st. The plague is an exotic, and if once suppressed would never re-appear in America unless reimported. It is the duty of the national government to protect the entire nation against a foreign invasion, and against any importation that tends to ruin our native industries. Therefore the national government should extirpate this imported pestilential incubus on our live-stock industry.

2d. The plague is gradually extending to the West and South, and if neglected must lay the whole continent under contribution. It is the duty of the United States Government to regulate and protect inter-State commerce; and if it allows one State to harbor and dispense a disease which is not only ruinous to her own cattle trade, but also to that of her neighbor, it is forgetful or untrue to the sacred trust reposed in it as a national administration.

3d. If the plague were to reach the unfenced cattle-ranges of the South and West it would be impossible to eradicate it, as herd mingling with herd would infect each other, and one herd succeeding another on the same pasture would take in the infection left by its predecessor. Wherever this disease has been introduced into such open common pastures the same result has followed—it has maintained a permanent residence. For examples we need only point to the Steppes of Europe and Asia, to Australia, to South Africa, and to the common pastures around our large Eastern cities.

4th. The State wealth in the South (Texas) and West is made up of live stock to a far greater extent than in the East; and the foreign market for beef is supplied mainly by the West. It follows that the West has a far greater interest in this matter than the East, whether viewed in the light of the preservation of their own herds or the maintenance of their foreign market. The West, therefore, or, rather, the whole country, should contribute to the expense of exterminating this disease.

5th. The disease is more violent and fatal in warm climates than in cold; therefore the relative mortality must be greatly enhanced should the affection be subjected, as it soon must be, to the hot summers of the Gulf States and the Mississippi Valley. This, again, is a cogent reason why the as yet uninfected States should join in stamping out the pest before it shall have invaded their own herds.

6th. The infection of the South and West will inevitably spread the disease over the whole of the Middle and Eastern States. Texas and the plains are at the source of the cattle traffic. Their live stock are sent North and East to finish off, and are shipped to the seaboard in railway cars. If once infected, the Western States must, therefore, rapidly spread the infection into adjacent States and contaminate the railroad cars, loading-banks, and yards, so that these, in their turn, must become active media of the poison. No State, therefore, can be protected if once the West becomes infected. The following may be adduced in further illustration of this:

7th. As the seeds of this disease remain latent in the system for a period varying from ten days to three and a half months, and only become manifest by their effects at the end of this time, infected cattle may be carried from ocean to ocean, or from the Lakes to the Gulf, and remain thereafter for weeks and months in apparent health, and yet spread pestilence and destruction in the end. With such a disease, and with the large cattle-traffic from the West, it is certain that contagion must be quickly carried in the channels of such traffic as soon as the infection shall have polluted its source.

8th. If, to avoid this danger, it is proposed to quarantine all cattle passing from State to State, safety would require a quarantine of at least three months (this was the period of incubation in the cattle that infected Norway and Australia, respectively, and the period was still greater in the case of the bull that contaminated South Africa). Suppose a quarantine of this kind were to be attempted at Buffalo. The regular daily shipment is five trains of thirty cars each, representing 3,000 head of cattle. Detain these for ninety days, and you would accumulate 270,000 head of cattle; so that any such system must speedily break down by its own unwieldiness. It should be stated that the above estimate is still far beneath the real numbers, as the number of cattle-trains passing east is frequently double or treble what has been quoted. Our supposition is based on the regular trains only.

9th. Chronic cases of lung plague, in which a large mass of lung has become encysted, may carry this mass for months, or even a year, before all can be liquefied and absorbed; and at any time during this long period such cases may become actively infecting by the escape of materials from this pent-up virulent mass. Against this no quarantine on a large scale can guard; and as even a most careful professional examination will not always suffice to reveal the presence of the smaller masses, infection by animals in this condition becomes inevitable wherever a large traffic is carried on.

10th. No State can be rendered secure unless all the States are cleared of the pest. In the past year New York has practically cleared six counties, and the disease is now mainly confined to the single county of Kings. But New York, as the great mart for cattle from the South and West, must be again speedily infected unless her neighboring States can be cleared at the same time. If an attempt is made to keep her clear, this State must maintain a most exasperating quarantine on New Jersey and the other infected States south, and must allow no bovine animal to move in the border or river counties, save under police control and by permit; and to make this sufficient, she must maintain a permanent staff of professional inspectors engaged in this business.

11th. Smuggling into one State by residents of another, who can sell and return home with the proceeds before the officers of the invaded State can lay hands on them, is a not unfamiliar experience with us in New York, and must occur to all States adjoining an infected one so long as there is not one executive sanitary administration for both. The power to control equally in both States, and to promptly punish in the one for infractions of the law in the other, would effectually prevent such practices.

12th. With different authorities in adjacent States, it becomes difficult or impossible to trace infected stock smuggled from State to State; and if the second State, having been hitherto uninfected, has no sanitary administration, the smuggled animals may be distributed at will, and the area of infection indefinitely extended.

13th. With different authorities in different States, and with different systems and measures, mistakes and losses are frequent, and much trouble and loss accrue to both State and purchaser, which would be altogether obviated if the system and control were one. As between New York and New Jersey this has been a source of frequent trouble; and cattle have been thrown into the former which the authorities would on no account have admitted, and which the New Jersey authorities would not allow to return.

14th. One method of smuggling, which has caused a good deal of trouble in New York, is the shipment from the Jersey City stock-yards of a few milch cows among a boat load of fat steers, and the running off of such cows into the city and suburban dairies of Brooklyn. This could have been easily prevented had the same power controlled the two States, and had the Jersey City stock-yards been subjected to the same rules as the Union stock-yards in New York. In the latter, cows from safe districts only are allowed in the inspection yards, and all cows entering the yards for fat stock must be slaughtered. In Jersey City milch cows and fat steers occupy the yards indiscriminately.

15th. Uniform action by one authority in all the States would remove the source of much irritation among dealers, who have now to inform themselves as to the restrictions imposed in the different States, and are liable to make mistakes and get into trouble because of the different regulations in force.

16th. It has been decided by the supreme court of Illinois that any State law forbidding the introduction of cattle from a neighboring State, on the ground of suspected infection, is unconstitutional. It would, therefore, appear that no State has the power to exclude the infected cattle of her neighboring State, but must wait for their entry, and quarantine or kill them on her own soil. This would demand State officials and quarantine yards along the whole frontier, so as to keep up a watch day and night, at a ruinous expense. This is fully as preposterous as the proposal to examine and quarantine all animals in transit, and would be practically impossible. If in this matter the individual State cannot lawfully protect herself it devolves on the United States to give the necessary protection, on the principle of controlling and regulating inter-State commerce.

17th. Three of the States now infected have shown no tendency to act for the extinction of this plague; so that, unless the United States comes to the rescue, it will be impossible to eradicate the pest from the country, and the great outlay and effort of the States now engaged in the work of extermination will be money and labor lost. The retention on the continent of a single infected beast, or of one infected place, is sufficient to repeat over and over again the sad results of the landing of the one infected cow in Brooklyn, in 1843.

18th. The United States can alone protect us against a renewed importation. If the quarantine of our live-stock imports is left to the control of the different States, we shall have the most varied and unequal measures. A short quarantine in one State may be looked upon as a premium to her own importers; but this will sooner or later entail the infection of that State, and, through her, of her neighbors; though *they* may have maintained a really protective quarantine on foreign stock. As we anticipated, the attitude of the United States Government towards Canada has led the latter to follow us in imposing a ninety days' quarantine on all European stock. But of what avail is this, if through any of our own States European cattle may be introduced, and with a sham quarantine, or none, may be sent inland wherever they list. A case of this kind has just occurred, and deserves to be exposed in all its shameful recklessness. In the first week in November, a cargo of forty cattle, the property of Messrs. George Brown & Co., of Aurora, Ill., were landed at the port of New York, and at once sent over nearly one thousand miles of rail into the Mississippi Valley, and to within 30 miles of the great centre of the Western live-stock trade. The complaints of New York importers, who are themselves subjected to a three months' quarantine against such discrimination, are just; and in the face of this incident the demand becomes imperative that the Treasury Department shall revoke so much of its present order as permits the duration of the quarantine of European cattle to be determined by the rules of State or municipal authorities. While entirely ignorant of where the blame rests in this case, we cannot too strongly condemn the system that allows cattle to be brought from a country with the record of plague-stricken Holland to be landed at the port of New York; to be conveyed in railroad cars for nearly 1,000 miles; to be unloaded, fed, and watered, at frequent intervals, in the yards used for our native stock on their way East, and to be finally placed among a stock of 100 head, which the owner is ready to sell at any moment, or to any point. It is probably true that Mr. Brown was very careful to see that his animals for importation were selected from districts supposed to be free from infection; but what assurance had he that the disease was not lurking among the herds, and even in the very animals purchased by him? Or, being assured on these points, if such assurance were possible, how could he guard himself against the many chances of infection while his animals were in transit? If such things are to be permitted, we may as well fold our hands and resign our herds to the merciless pestilence, for all our present toil and outlay must finally be lost; and, in spite of all temporary and dear-bought successes, a permanent riddance of the plague can never be secured. If the national government cannot take this matter in hand, and impose an unalterable quarantine of 90 days on all imports of cattle from Europe, then New York, New Jersey, and Pennsylvania may wisely abolish their cattle-disease commissions, and turn their attention to something more promising; for an infection of the West, which, under the present system must occur sooner or later, will more than undo all that they can accomplish, and permanently blight the agriculture of the nation. To-day the national government is on trial. It is not yet too late to avert the crushing load of taxation for the future generations of the American people; but it can much more easily, and by simple inaction, perpetuate a system which must infect our Western stock ranges, and reduce them to the condition of the pestilential Russian Steppes, against the infected cattle of which Austria and Prussia can only protect themselves by patrolling their whole eastern frontiers day and night with *gendarmes* stationed every two hundred paces.

CATTLE PLAGUE OR RINDERPEST.

HISTORY OF THE DISEASE.

The rinderpest (cattle plague, *pestis bovilla*) appears to have been carried from Central Asia to Europe as early as the fourth century, but the first exact description of this disease dates from the year 1711, two years after an extensive epizoötic outbreak of the same in most European countries. It is estimated that in the course of the eighteenth century, not less than two hundred million head of cattle were carried off by the cattle plague. In the beginning of the present century, Prussia, Schleswig Holstein, Saxony, and France, were visited by the plague, which was observed to have followed the movements of armies during the wars of the first Napoleon. In 1828, 1829, and 1830, during the Russo-Turkish and the Russo-Polish wars, the rinderpest was carried from Russia into Poland, Prussia, and Austria. In 1865 the plague appeared in Holland, and was carried thence to England. In both countries the disease carried off one hundred thousand head of cattle in the course of a few months. In 1867, Germany was again visited by the plague, which, however, was prevented by timely measures from spreading beyond the eastern provinces of Prussia. In 1870, soon after the outbreak of the Franco-German war, the rinderpest appeared in Germany in consequence of importations of cattle from Russia, and spread over Germany and France, following the movements of the armies. In the beginning of the year 1877, the disease was again carried into Germany by Russian cattle, and made rapid progress, because the imported animals, apparently healthy, but already infected, were allowed to reach the markets of Breslau, Berlin, and Hamburg, from which cities the infection was gradually communicated to other places. In Dresden the disease spread at once through the whole market. Towards the end of August, 1877, the rinderpest was reported by our consular officers as extinguished in the German Empire; but the danger of its reappearance in consequence of possible movements of cattle from the steppes of Southern Russia to the borders of Germany, though much lessened by the stringent sanitary regulations adopted by the Russian Government, is not regarded as entirely obviated.

Fleming, in his excellent work on Veterinary Sanitary Science, says that, in recent years, several of the most competent veterinarians have endeavored to ascertain the home of the cattle plague, but without much success. Unterberger throws much doubt upon Russia and its steppes being the source of the malady, and he asserts that it is a purely contagious disease in Russia-in-Europe, and also, perhaps, in the whole Russian Empire. It has been seen in Southern Russia, the Asiatic Steppes, in different parts of India; in Mongolia, China (south and west), Cochin China, Burmah, Hindostan, Persia, Thibet, and Ceylon. It is as yet unknown in the United States, Australia, and New Zealand. So far as Europe is concerned the geographical limits of the disease may be given as follows: "Beyond the Russian frontiers, and even in every part of that empire, the steppes excepted, the cattle plague is evidently a purely contagious malady. It is never developed primarily in Europe, either in indigenous cattle or in these originally from the steppes, and it has not yet been positively demonstrated that it may be primarily developed in the Russian Steppes; the most recent observations even tend to prove that in the European portions of these regions the affection is only present through the transmission of a contagium. Consequently, the plague is a malady which is perhaps primarily developed in the Russo-Asiatic Steppes—perhaps elsewhere—but is never seen in Europe except by the importation of its contagious principle."

In Russia the malady is known as *Tchouma, Tchouma reina,* and Fleming regards it as important to note the employment of this term by the Russians to designate the cattle plague. Reynal has pointed out that it proves, philologically, the region in which the disease originates, or rather permanently reigns—in the far east. History demonstrates that the appearance of the disease in early times in Western Europe coincided with the eruptions of the Mongols, and that the contagion accompanied armies; that the route it has followed in more recent years was that of the Huns, and that it remained with these people in the colony they founded on the shores of the Caspian Sea. The word *Tchouma* is used by the Mongols and Nomad Tartars of Central Asia to signify a malevolent deity—something of the nature of a vampire; and it has been adopted, with slight modifications, by all the people who have had any relations with that region.

CAUSES.

Nothing certain or definite as to the causes which develop the cattle plague are known. In Western Europe it relies solely for its introduction and diffusion to the presence of a contagium, carried either by animals suffering with the disease, those which have been in contact with them, or media of different kinds which are contaminated with the virus. Once introduced, it spreads from its point of introduction as from a center; each newly-infected animal becomes a focus whence the disease may radiate in every direction, and it usually attacks those animals which are nearest the foci. It spreads with more or less rapidity as the animals or vehicles charged with the contagium are moved about; even the air may, within a certain distance, be credited as an active agent in the diffusion of the deadly malady. The nature of the contagious matter (contagium), has also so far baffled all the efforts of investigators. Neither microscopic examinations nor chemical analysis of the tissues, blood, and mucous discharges of the infected animals, have led to the discovery of the principle of contagion. It is known, however, that from the very beginning of the disease a contagious matter is formed, which attaches itself to every part of the diseased animal. It is principally contained in the secretions of the mucous membranes, but, being volatile, attaches itself also to the urine, the dung, the blood, the skin, and the breath. It may be communicated to the atmosphere by exhalations from any part of the sick animal, or its carcass. Experience has shown that healthy cattle may be infected by coming near the sick animals, or near anything contaminated by their excrements or exhalations, without actual contact with them. The contagious matter has no effect in open air at a distance of twenty to thirty paces, because the air either dilutes or modifies it so as to deprive it of its power. But in cases where a current of air comes directly from an accumulation of infected matter, and also in inclosed spaces, the contagion may be carried to greater distances. Therefore, the disease may be communicated in a large stable to a healthy animal quite a long distance from the diseased one, or may be carried from one stable to another as far as a hundred feet apart. This happens only when the exhalations are carried over directly from one stable to the other. by a current of air so rapid as not to allow time for the air to dilute or modify the contagious matter. Where one stable is separated from another by a partition which is not air-tight. the contagion is very easily transmitted. Besides these direct means of infection, the disease may be carried to healthy animals indirectly, in many ways. For instance, objects which have come in contact with infected matter, may be carried to a distant place and there spread the disease. Porous substances, such as woolen clothing, wool. hay, straw, &c., are particularly liable to absorb the contagious matter, which may diffuse itself after some time in a distant place. Thus butchers, drovers, and other persons who visit infected stables. may carry the disease from yard to yard, and from village to village. In railroad trucks, the woodwork absorbs a considerable amount of the contagious matter. and, if not thoroughly disinfected, may communicate the disease to animals subsequently placed therein. The dung of diseased animals may spread the contagion to distant places, by being carried away on the wheels of vehicles or the shoes of persons. Dogs and cats may carry it in their fur and birds in their plumage. A small quantity of blood or dung on the sole of a shoe or on the tip of a walking-stick has sometimes been sufficient to carry the disease to a great distance. The modes of possible transmission are, in fact, so numerous as to render it, in many instances, a matter of extreme difficulty to account for the cause of an outbreak of the plague.

The vitality of the contagious matter is variable, according to circumstances. Air is its most potent and reliable destroyer. Hay and straw which have lain above the stables of sick animals have been often used as fodder with impunity after an airing of twenty-four hours. Wool, impregnated with the mucus from the nostrils of sick animals, was found to be innocuous when thoroughly aired for five or six days. Stables and pasture-grounds will be thoroughly disinfected in a few weeks by the action of the atmosphere. In the same way clothing and other porous substances become entirely disinfected by airing. The stronger the current of air the more prompt its disinfecting action. On the contrary, if infected porous substances are not exposed to currents of air, the contagious matter is preserved for a long time. Closely-packed hay and straw, the woodwork and floors of closed stables, manure-heaps, packed-up clothing. &c., may remain infected for several months. A case is recorded of the rinderpest breaking out anew in a stable which had stood empty for four months, but had not been disinfected after a previous outbreak. The flesh and hides of carcasses which had been buried for over three months were found to be capable of infecting healthy animals.

Very high temperature has the same effect in destroying the power of the contagious matter as currents of air, but summer heat is effective only in so far as it promotes the drying up of the contagious particles, and renders them more volatile and more easily diluted by the air.

The contagious matter is not destroyed by cold, not even by frost; on the contrary, its power is preserved, as the drying up of the substances containing it is thereby hindered. Dung frozen through the winter spreads the contagion upon thawing in the spring.

All ruminating animals are liable to the rinderpest, but goats and sheep are less commonly and less severely affected by it than neat cattle. The disease does not affect non-ruminating animals, nor is it in any way dangerous to man.

The rinderpest breaks out generally on the fifth or sixth day from the time of infection, sometimes as early as the fourth, and frequently as late as the eighth or even ninth day. According to some observations, the period of incubation may extend to two or three weeks, but the instances of so protracted an incubation are to be considered as entirely exceptional.

The spread of the disease in a herd of cattle is usually slow in the beginning. Often when the contagion is introduced only a single animal is infected. This one, after the few days required for the incubation, becomes sick and commences to evolve the contagious matter, which infects one or more of the animals in the same stable or herd. Then, again, an interval of time elapses before the disease is developed in the new victims. As soon as several animals are diseased, the contagion spreads more rapidly, and many are attacked at the same time. Want of proper caution on the part of stable-men and other attendants is often the cause of an exceedingly rapid progress of the contagion, which is carried in their clothing from one end of the stable to another.

PHENOMENA OF CATTLE PLAGUE DURING LIFE.

Dr. J. Burdon-Sanderson, one of the commissioners appointed by the English Government to investigate this disease during its last invasion of Western Europe (1865), in speaking of the phenomena of cattle plague and the general character and progress of the malady during the life of the affected animal, says that it is an essential or general fever, and that it can be shown, more clearly than in any human disease of the same class, that the disturbance of the system which is understood by the term fever may exist independently of local changes occurring in particular organs; and in this respect a fact new to pathology has been discovered, i. e., that the increase of the temperature of the body, which is the one and only symptom which all fevers have in common, exists for several days before any other derangement of health can be observed. Although constitutional or general in its origin, the disease is attended with local alterations of structure, some of which are so constant and invariable that no definition of the malady can be complete which fails to recognize and include them. Dr. Sanderson says:

The observations and experiments which have been made, so far as they have been carried out, relating to the phenomena of the disease during life, lead to the conclusion that, with reference to the *constitutional* effects, the disease consists in (1) increase in temperature of the body; (2) increase in the elimination of urea by the kidneys, indicating increased disintegration of tissue; (3) alteration of the physical and chemical qualities of the blood, manifesting itself in impairment of its coagulability and in a marked tendency to capillary hemorrage; and, lastly, (4) a general septic condition of the fluids and tissues, in virtue of which they are unnaturally prone to decomposition even during life.

With reference to its *local* manifestations, the disease appears to be distinguished by an alteration of the superficial structures of the skin and mucous membranes, consisting (1) of minute capillary congestion (*hyperæmia*) of the vascular layer (*corpus papillare, membra propria*); (2) of increased as well as perverted growth of the structural elements, naturally developed at its free surface, this change leading to thickening, softening, disintegration, or detachment of the epidermis or epithelium respectively, but very rarely, if ever, to ulceration or loss of substance in the deeper tissue; and, lastly, (3) of increased and perverted activity of the secreting glands of the skin and mucous membrane, resulting in mucus or sebaceous discharges.

Cattle plague belongs to that class of fevers which is distinguished by marked uniformity in their development and duration. In this respect it resembles small-pox more than any other disease which affects man. The resemblance is, however, generic rather than specific, for in cattle plague the essential phenomenon of small-pox—the eruption—is wanting.

In fatal cases the progress of the disease is divided into three stages. The first stage, comprising the first and second day, is marked by no appreciable change in the condition of the affected animal, excepting increase of temperature. During the second stage, which comprises the third, fourth, and fifth days of the disease, its symptoms develop themselves in quick succession. The appetite fails, rumination ceases, the daily excretion of urea by the kidneys is augmented, while the animal loses strength and weight. The last stage, that which immediately precedes the fatal termination, is characterized by the rapid decline and cessation of the vital functions, and, above all, by sudden sinking of the temperature of the body.

The leading phenomena of the disease may be described as follows, according to the order of time in which they occur:

During the first two days, as has been already stated, there are no symptoms excepting elevation of temperature, so that the time of commencement of the disease can be determined only by the thermometer. But on the third day an eruption, exactly resembling that of thrush, appears on the gums and inside of the lip. The eruption usually commences by the formation of groups of very minute raised points or dots on the surface of the mucous membrane, which are usually first seen a little below the corner tooth on each side. This appearance is in many cases neither preceded nor accompanied by any redness of the surrounding surface, but occasionally a slight blush is perceptible near the elevation. The animal continues to ruminate, and its appetite, pulse, and breathing are unaffected.

On the next day the eruption above described on the mucous membrane of the mouth is found to have extended. The whole of the surface between the lower lip and the gum is studded with raised groups of elevations, while those previously observed below the corner teeth have coalesced, so as to form patches. The animal is listless, takes less food than usual, and ruminates irregularly, but the pulse and respiration are unaltered. On this day alterations may often be observed on the cutaneous surface. In the neighborhood of the vulva and on the inside of the thighs the skin is found to be greasy, as if smeared with an unctuous substance.

On the fifth day the animal is obviously ill. The head hangs down, the ears are thrown back, and the attitude and movements are suggestive of depression. The pulse is sensibly weaker, and often the artery feels hard and thread-like under the finger, expanding scarcely perceptibly with the systolic impulse. The breathing is sometimes almost natural, but more frequently begins to be oppressed and irregular.

With the sixth day the alterations of the mucous membrane of the mouth attain their full development. The under lip is covered with a crust of white opaque material (consisting of epithelium, mixed in most instances with the filaments and spores of a hyphaceous fungus), which is either confluent and continuous or in patches. This crust is usually of the consistency of cream cheese, in which case it adheres so slightly to the surface on which it lies, that the slightest touch is sufficient to detach it. Wherever it is so separated, the bright red vascular surface of the mucous membrane (*membrana propria* of anatomists) is exposed, raw looking, but free from ulceration. Similar appearances are observed on other parts of the mouth, particularly on the upper gum, on the dental pad, on the cheeks and hard palate, and on the lower surface of the tongue near its lateral margins.

During the sixth day the leading symptoms are those which arise from *diminished* contractile power of the heart and voluntary muscles. The affection of the heart is indicated by increased feebleness and frequency of the pulse, and by the extinction of the præcordial impulse; that of the voluntary muscles by the attitude and movements of the animal, which are so indicative of adynamia that many writers have been misled by them into the belief that in rinderpest there is a special paralytic affection of the spinal nervous system. At the same time the mechanism of the respiratory movements is modified in a remarkable and characteristic manner, the modification being dependent partly on the cause above referred to, and partly on pathological changes having their seat in the air passages. The alvine discharges, which during the previous progress of the disease were firmer and harder than natural, now become soft, and eventually liquid and dysenteric. The temperature of the body, which up to the fifth day has gone on increasing, rapidly sinks to below the natural level; this loss of animal heat being attended with a correspondingly rapid diminution in the quantity of urea excreted by the kidneys.

Death usually occurs during the seventh day. It is not preceded by convulsion or any other symptom worthy of special notice.

SYMPTOMS.

One or two days before any other change occurs in the condition of the infected animal there appears an increase of temperature, which is most readily detected by means of a thermometer introduced into the rectum. The temperature is found to have risen by two to four degrees Fahrenheit, from the normal temperature of 102°. At the same time symptoms of fever are observed, such as shivering, muscular twitchings, dryness of the skin, a staring coat of hair, an unequal distribution of temperature throughout the body, and changes of temperature, which are particularly noticeable at the base of the horns.

A very important and characteristic symptom at an early stage of the disease is a peculiar alteration of the mucous membranes. This alteration is very soon noticeable in the vagina of cows, which becomes spotted or striped with red. The next day small yellowish-white or gray specks are clearly seen on the red spots and stripes. These specks are formed by the loosening of the cuticle, which can be rubbed off or detached by the finger, leaving in its place a dark-red depression. The same red spots and stripes and yellowish or gray specks appear in the mouth and nose of the sick animals of either sex.

Fleming, in his work on *Veterinary Sanitary Science and Police,* gives the following description of the peculiar eruptions of the mucous membrane and the skin:

With regard to the mucous membranes and skin, there is much that is not only interesting, but of the greatest practical importance in the way of diagnosis. The development of the symptoms previously enumerated are soon accompanied by anatomical and functional alterations of these membranes, but especially of that lining the vagina and the digestive and respiratory tracts. The vulvo-vaginal membrane is most frequently that which exhibits these changes. It is more or less infiltrated, and of a brown, brick-red, or mahogany color, which is either disposed in streaks, patches, or diffused; and there may be small sanguine extravasations, variable in number. This abnormal color, which is more particularly due to venous injection, has in calves and heifers a greater diagnostic value than in cows, as it is observed, though in a less degree, in animals which are near calving, as well as those which have lately calved.

This symptom, however, is not always observed at the commencement of the disease, but may only appear at an advanced stage. Twenty-four hours after its appearance there are usually seen on the red surfaces small yellow or grayish and slightly salient patches, which might be mistaken for little flakes of mucus, though they are really composed of masses of altered epithelial cells. They adhere but slightly to the dermal surface of the membrane, or are merely lying on it; they are quickly removed by friction, or thrown off by the alterations going on, leaving excoriations corresponding to the situation they occupied. At this period, or a little later, there flows from the vulva a variable quantity of clear or ropy mucus, which, in drying, adheres to the neighboring parts.

While this alteration is going on in the vaginal membrane, or before or after, other analogous changes are observed in the other visible membranes. That of the mouth is more or less hot, and generally, or in patches of variable extent, assumes a deep red, livid, or dark-blue tint, particularly about the gums, though the presence of pigment may conceal this coloration. Ordinarily the derm of this membrane and its epithelium are tumefied at certain points, and the adhesion of these two layers to each other is diminished. In a very brief space there appears at first on the lips and gums, afterwards on the palate and borders and sides of the tongue, little whitish-gray or yellowish elevations the size of a pin-head, due to the proliferation, infiltration, and fatty degeneration of the epithelium in these localities. The number and dimensions of these elevations increase, and sometimes they join each other; their connection with the derm becomes lessened, and soon—frequently within twenty-four hours—the slightest rubbing will remove them in the form of a soft gray mass not unlike bran; they are also thrown off by the morbid process going on. However removed, the derm upon which they were formed is exposed, and in this way are produced these excoriations whose sharply defined bright red color contrasts strikingly with the livid membrane surrounding them. These are the "pestilential erosions" of Kausch, so named from the veterinarian who first described them. These epithelial alterations occur, at times, at the base of the papillæ of the cheeks as early as the appearance of the first morbid symptoms, though, as a rule, it is only towards the second, third, or fourth day that they are most marked.

The secretion of saliva is increased and flows in large viscid streams from the mouth. The nasal mucous membrane is also greatly injected from the commencement of the affection, and becomes infiltrated and swollen; soon after it becomes uniformly pale, or in such a manner as to leave injected streaks or patches; petechiæ also appear in variable number. Towards from the second to the third day, on examining this membrane closely, it will be noticed that there are the same pulpy or caseous epithelial collections observed on the membrane of the mouth and the vulvo-vaginal membrane, and which, when thrown off, leave the derm exposed. In about twenty-four hours after the more evident signs of the disease have appeared a nasal discharge manifests itself; this is at first serous and transparent, but ere long becomes a thick mucus or muco-purulent yellowish matter, which may be mixed with blood, and disagreeably fetid. In drying around the nostrils it forms thick crusts.

The conjunctiva of the eyes are also infiltrated and deeper colored—particularly about the free border of the nictitating membrane—than usual; but this coloration most frequently disappears in the course of the malady, and this membrane is then pale. The secretion of tears is very copious, and, flowing in abundance down the face, by their acridity they may depilate and erode the skin. A thick muco-product fluid collects in the inner canthus of the eye and behind the membrana nictitans, and as the animal becomes emaciated and the eye-ball sinks towards the bottom of the orbit this accumulates.

The skin, which is usually lax shortly after the invasion of the malady, in the majority of cases and in many epizootics, becomes the seat of a diversely characterized eruption, which has been at one time described as squamous, at another papular, vesicular, or pustular, and again as erysipelatous. This cutaneous manifestation more especially appears in those parts where the integument is thin, though it may also invade other regions or even affect the entire surface of the body. The udder and particularly the base of the teats, the scrotum, the margin of the nostrils, the lips, and the vulva, the perineum, and the internal aspect of the thighs, are the localities for which it seems to have a special predilection, but it may likewise be often noted between the jaws, on the shoulders, neck, and withers. The extent and intensity of these exanthemata are very variable. At times they accompany the ordinary symptoms, while at others the eruption is coincident with an intense febrile reaction which lasts for some days and increased temperature of the skin where it is about to appear, with, in certain instances, a more or less abundant transpiration.

This exanthema of cattle-plague consists of (1) a proliferation and abundant disquamation of the epidermis, accompanied by shedding of the hair; (2) the production of small papulæ or nodosities, from which exudes a yellow viscid fluid which, in drying, forms with the hair crusts of variable thickness; (3) the eruption of little vesicles about the muffle, whose contents agglutinate the hairs and gives rise to brownish-yellow crusts; (4) the formation of pustules (the so-called "variola" of Ramozzini) the size of a millet seed or small pea, frequently confluent, and when ruptured and their contents desiccated, producing friable, yellow, or brown crusts, which adhere very slightly to the skin. The duration of the eruption is variable, but in general it does not entirely disappear until from two to four weeks after its manifestation.

In some epizootics erysipelatous tumors have been remarked about the neck, dewlap, or flank. Gas is also developed sometimes in the subcutaneous cellular tissue, ordinarily in the region of the loins, shoulders, sides, or neck, and in rare cases over the entire surface of the body; its presence is recognized by a more or less voluminous tumor, which crepitates on manipulation.

The next day after the appearance of the peculiar eruption upon the mucous membranes, there is a disinclination to eat and ruminate, and with cows a diminution and soon a total absence of milk.

Two days after the manifestation of the above-described symptoms marked changes in the general appearance of the diseased animal are apparent. It lies down very frequently; when standing it draws the hind legs forward as if suffering from colic. The look is distressed, the head drooping, the ears hanging, the breathing oppressed; the pulse becomes rapid and weak, the discharges from the eyes, the nose, and the mouth become thick and purulent, the breath fetid. The iris, which at the commencement of the fever is generally inflamed and cherry red, resumes its natural color with the increase of secretions from the lachrymal duct. Cows far advanced in pregnancy generally calve in this stage of the disease.

On the second or third day diarrhea sets in. The feces, at first thin and watery, then thick and slimy, are filled with detached masses from the mucus surfaces, very fetid and more or less tinged with blood. When the diarrhea has lasted two or three days the disease advances with rapid strides. The animal is so weak as not to be able to rise, the evacuations of excrements are involuntary, the breathing is uneven and rapid, the beatings of the heart are no longer perceptible, the pulse becomes very feeble and the temperature rapidly falls. Death usually occurs on the fifth day from the first visible signs of the disease. Sometimes the course of the disease is so rapid as to reach its culmination within two days.

On the average, 70 to 75 per cent. of the diseased animals die. Those that survive have not had the disease in its most malignant form. Once convalescent the animals recover very fast, but the diarrhea continues for several days after the disappearance of all other symptoms.

In summer, when the cattle are grazing, the disease is less severe than in winter, when they get dry fodder and are kept in close stables.

The symptoms and progress of the disease are the same with goats and sheep as with neat cattle, but the percentage of fatal cases is somewhat less.

Many of the symptoms of rinderpest occur in the lung disease (*pleuropneumonia*), the malignant catarrhal fever, and the mouth-and-foot disease. The lung disease is distinguished from the rinderpest by the absence of the characteristic eruptions upon the mucous membranes; the malignant catarrhal fever, by the dimness of the transparent cornea, which in the rinderpest remains clear; the mouth-and-foot disease, by the ulceration of the foot, the less degree of fever, and its peculiarly rapid spreading from one animal to entire herds.

PATHOLOGY OF THE DISEASE.

Among the lesions observed after death there are several, though no more constant than several of the prominent symptoms, that materially assist in establishing a proper diagnosis. The age and general condition, the state in which the animals were kept before they were affected, their breed, the character and intensity of the disease, all appear to have some influence on the seat and seriousness of the lesions. These vary according to the period at which death takes place.

Fleming says that if the animal is killed at the commencement of the malady, and the symptoms have been comparatively mild, there will nevertheless be found, on examination after death, such alterations in the mucous membranes as congestion and ecchymoses. The latter are more particularly observable on the free border of the mucus folds in the fourth compartment of the stomach (true stomach) and around the pylorus, although they also exist to a less degree in the small intestine, and often in the vagina. When, however, an animal has died from the disease, or been killed when it had attained a certain degree of intensity, the changes are more marked, the body becomes quickly inflated after death, and sometimes even before death occurs. The rectum is elevated and its lining membrane is tumefied and of a deep red color; the tail and hinder extremities are more or less paralyzed during life, and are therefore usually soiled by the feces. The skin exhibits the characteristic eruption, and in those places where there are neither glands nor hairs, as on the teats, it is injected in irregular patches of variable dimensions; the epithelium is thickened, soft or friable, and the integument is often cracked. On removing the skin the vessels which are cut are generally filled with a dark-colored fluid blood, and the flesh is red, blue, or violet-tinted. The peritoneum in some cases may be slightly injected or ecchymosed in patches. The whole of the intestines are generally greatly distended with gas, and in some cases the small intestine may be reddened.

In the interior of the digestive canal are found the most marked evidences of the disease, though they are not always constant and equally intense in every portion of the mucous membrane. In the mouth, pharynx, true stomach, small intestine, and rectum, they are most frequently present. They are least conspicuous and often absent in the œsophagus, the three first compartments of the stomach, and in the cæcum and colon. They may be so trifling as to resemble the lesions of a slight catarrh, while in other instances they are unmistakable and pathognomonic.

In the mouth and pharynx are observed the alterations in the lining membrane and the epithelial changes. It is chiefly where there has been much friction or local irritation that they are most exaggerated, and deep erosions, with loss of texture of the derm of the mucous membrane, may be noted. The œsophagus is rarely affected, though it is not always exempt. In the rumen the quantity of food may be found a little larger than usual. The epithelium on the mucous membrane lining it and the next compartment may be more easily detached than in a healthy state, and a microscopical examination of the cells proves them to have undergone a similar change to those of the mouth. The mucous membrane in t' se compartments is also frequently injected in a general manner, though more deeply in some places than in others. It is not rare to find on this membrane round, oval, or irregular-shaped eschars, disposed separately or in groups, and varying in color from a dark brown to a greenish hue. The elimination of these eschars takes place gradually from around their well-formed borders, and cicatrization afterwards occurs, even in cases which have a fatal termination. Submucous extravasation is probably the cause of these gangrenous patches. Around them the tissues are infiltrated, and more or less injected, while beneath the texture.is injected or ecchymosed, and red or green in color.

The third compartment sometimes contains food, which is hard, dry, and friable; at other times it is soft and pulpy. In the first case, the epithelium of the leaves is readily detached, and adheres to the cakes of aliment removed from between them. This epithelium also exhibits granulo-adipose degeneration. The leaves themselves are injected wholly or partially, and ecchymoses and eschars may be present in them; they are also easily torn. In the fourth compartment and small intestines the contents are at first normal; but they soon change, and there is found a small quantity of thick, yellow, brown, or even blood-colored fluid. The mucous membrane is covered by a viscid, grayish-yellow, or reddish mucus. The cæcum and colon at this period contain a frothy mass of a brownish, sometimes sanguinolent, fluid. The rectum has a thick viscid mucus adhering to its inner surface. If the disease pursues its course, the débris detached from the intestine is mixed with exudations and extravasations to form a viscid, albuminoid, whitish-yellow, brown, or red fluid, in which are shreds and the detritus from the membrane.

When an animal has been killed in the early stages of the disease, and the mucus has been carefully removed from the mucous membrane of the stomach, it is found that the surface of the latter is irregular, and that its tissue is infiltrated and injected to a degree corresponding with the seriousness of the attack and the stage the malady has reached. The abnormal color, varying from a brick-red to a reddish brown, is , generally diffuse, but is most marked at the pyloric portion, attaining its maximum of intensity towards the free borders of the folds. Submucous extravasations are also frequently met with in this part, differing in size from a fine point to a large patch. In the small and large intestines there also exist, at this period, analogous alterations; but,

13 C D

while the redness of the abomasum is usually diffuse, in the small intestine it generally appears in the form of transverse striæ, which are
crossed by lighter colored longitudinal streaks, this intercrossing forming a somewhat regular pattern. These extravasations are common in
the small intestine, but the infiltrations and exudations are not so
frequent in the abomasum. In the duodenum the alterations are usually more intense than in the remainder of the intestine, and it is not
rare to find in it a very marked diffused redness and much sanguine
effusion. The congestion is often greatest around the solitary glands
and Peyer's patches, whose volume is more or less increased. Frequently the areolated aspect of these patches is most conspicuous at the
termination of the first period. The same lesions are found, but in a
less degree, in the large intestines. In these the most salient portions,
such as the borders of the valvulæ, are the parts which are the most
deeply colored and most extensively ecchymosed. The infiltration is
greatest if diarrhea has not been present.

In cases where the disease has made considerable progress, the lesions
are still more characteristic. The mucous membrane of the abomasum
and intestine is deeper colored, often blue or black, and in the duodenum
of animals which have succumbed, it may even be uniformly black,
while the petechæ and ecchymoses are more numerous. In the abomasum, but oftenest in the intestine, towards the fifth day of the disease,
there appears a pigmentation, varying from a bright gray to a slate
color, or even darker, and which takes the place of the abnormal color
due to the blood. This appearance is first noticed in the rectum, and
in the intestines generally its tints seem to be related to the intensity
of the blood coloration, of which these parts have been the seat. It is
therefore in the duodenum, and especially near the pylorus, that it is
deepest-tinted and most extensive. In the duodenum it is diffuse, but
in the remainder of the small intestine it is limited, as a rule, to a double
series of perpendicular zones more or less incomplete, and in the rectum
is usually in the form of longitudinal lines. This coloring matter is deposited in the most superficial layer of the mucous membrane, and is
constituted by minute irregular granules, which, according as they are
disposed separately or in clusters, give rise to the different shades.
Around the orifices of Brunner's glands, and in the texture of the villi,
this deposit appears to be most localized.

The epithelium of the fourth compartment of the stomach rapidly undergoes changes analogous to those observed in the mouth. Their intensity depends upon the part examined, as well as the gravity of the
attack and its stage. In the first and last portions of the small intestine in the cæcum, in the first section of the large colon, and in the
rectum, they are generally more developed than elsewhere. In mild
cases the epithelium, though not yet detached, is always less adherent
to the derm than in health. In more serious cases this layer is found
completely detached over a considerable surface, and especially in the
small intestine. The excoriations thus produced vary both as to extent
and number, and are generally covered by a gray, red, or dark colored
viscid mucus. The matter is tenacious, and adheres firmly in flakes to
the membrane. The extent of these flakes is generally from a quarter
to two inches in length. The color is gray, yellow, red, brown, or black;
their free surface is smooth, and more or less convex; their variable
consistency is less at the border than the center; the membrane beneath
them is injected and spotted with small extravasations, and their margin, in consequence of the retraction of the flake, is separated for a
short space from the border of the erosion.

The mortification which may invade the intestinal wall does not usually go beyond the mucous membrane. In rare and very severe cases it extends to the submucous connective tissue, or even to the muscular layer. The liquefaction of the mortified patches causes a loss of substance in the membrane, and these places are designated "excoriations" or "erosions," according as the derm remains intact or not. Their number is as variable as are the patches. The viscid masses covering the surface of the intestine, as well as the flakes, are produced by the utricular glands of the gastric and intestinal mucous membrane, which are greatly altered and tumefied.

Peyer's glands undergo alterations of a particular character. They lose their epithelial covering, and, in the majority of epizoöties, undergo changes analogous to those of the solitary glands; though in other epizootics they are rarely affected, and when they are the lesions are not always equally marked. Sometimes they are merely covered with a mucous layer, like the other parts of the intestine, and are injected; at other times they are more salient than usual from tumefaction, and they then may contain contents like that of the solitary glands; again, they may be covered by a croupal exudation or false membrane, several lines in thickness, and gray, yellow, red, or blue in color, adhering by its central part to the mucous membrane. The presence of these patches is not a constant feature in the pathological anatomy of the disease; in certain epizoöties it is almost always present, while in others it is exceptional. Among the conditions which appear to have an influence in its production only one is known, and that is the condition of the animal before infection; if it has been well nourished these deposits are most likely to be present.

The prominent alterations in the glands of the mucous membrane appear to consist in an exaggerated proliferation of their cell elements, accompanied by a prompt granulo-adipose destruction of the newly-formed cells. The liver is seldom much altered, but the gall-bladder is very often distended with bile, and its mucous membrane is in somewhat the same condition as that of the intestines. The mucous membrane of the air-passages is greatly altered. That lining the larynx, the trachea, and also the bronchia is injected and marked by extravasations which, particularly in the trachea, appear in the form of longitudinal striæ.

The lungs are frequently emphysematous (interlobular) to a degree corresponding to the intensity of the malady. This condition is chiefly noticed about the borders of the lungs and in the mediastinum, and, passing along the large blood-vessels toward the lumbar region, it may reach the loins. The lungs are also occasionally œdematous. The pleura, like the peritoneum, is occasionally congested in places, and even ecchymosed. The heart is usually flabby, dark or clay colored, and friable, and at times there are subendocardial extravasations towards its base; the blood is darker colored than in health, and coagulates imperfectly or not at all. The kidneys may be tumefied, congested, and more friable than usual. The bladder is rarely empty, but generally contains a quantity of urine, which may be pale, dark colored, or muddy, and have suspended in it shreds of epithelium. Its mucous membrane may also be congested and ecchymosed, and covered with viscid mucus. The vulvo-vaginal mucous membrane presents a very marked redness, which generally extends to the cervix of the uterus. As in the mouth, there are little elevations of altered epithelium on this membrane, with erosions covered by viscid matter. The udder, frequently congested, sometimes contains a small quantity of thick milk.

According to Reynal, the latest observations on the pathological anatomy of cattle plague are those of Damaschino, who has made a complete study of the histological alterations occurring in the disease. This investigator states that the ulceration of the mucous membrane is due to an unique process, which presents a great resemblance to that of pharyngeal diphtheria of man. At the commencement, the lesion consists in an exaggerated production of epithelial cells, which are infiltrated with an amorphous substance, become deformed, throw out multiple prolongations, and acquire an abnormal adhesion, which finally gives them a pseudo-membranous aspect. But beneath these false membranes the young epithelial cells do not submit to the same alterations. Instead of the prolongations adhering to each other and becoming matted together, they are the seat of a purulent transformation, whence results less adhesiveness, and soon the casting off of the pseudo membrane. At this moment ulceration commences, and as these tissues are softened it happens that there is found implanted on this surface fragments of hairs, which are recognized by the microscope. The loss of substance is not always superficial. On the tongue, sometimes, the lesion ceases at a portion only of the thickness of the papillæ, but in other cases it extends throughout their texture. In the stomach it is often deeper, comprising a portion of the substance of the glandulæ, and even the entire thickness of the mucous membrane to such a degree that, without the presence of a thick layer of adipose tissue at these points, the stomach would frequently be found perforated. On the surface of these ulcerations the adipose tissue exhibits all the characters of inflammation proper (nuclear proliferation in the conjunctival parietes). In two cases there was found a lesion of the venal and hepatic parenchyma, consisting in a granular degeneration of the glandular elements. In the liver, the lesion, as is usual, showed a predilection for the periphery of the lobules in the vicinity of the vena portæ; there the cells were found in a very advanced stage of granular degeneration. The epithelium of the kidneys, more especially, showed the peculiar tumefied troubled appearance already indicated, though the granular condition was less marked. The muscular alterations consisted in the presence of numerous elongated bodies, very abundant in the right side of the heart, and incontestably situated in the substance of the muscular fiber. These bodies are blunt at one end, pointed at the other, and are composed of a regular mass of cylindrical cells lying together in such a manner that at the pointed extremity there is only a single cell, at the obtuse end two cells, and in the other part sometimes two, sometimes three cells, clustered on a given segment. It is surmised that these minute bodies are entozoa in their primary stage of development.

MEASURES FOR THE PREVENTION AND EXTINCTION OF RINDERPEST.

There being no remedy known for this disease, human intervention in dealing with it has thus far been necessarily restricted to measures for its prevention and extinction. Most European governments have passed laws and prescribed regulations for the purpose of protecting their repective countries from the invasions of the plague, and for its speedy extirpation on the occurrence of an outbreak. Of all these enactments the regulations now in force in the German Empire, which are given below in translation, are considered as the most complete embodiment of the results of experience and scientific investigation in regard to this subject.

Skin of udder on the sixth or seventh day, showing, in
addition to the usual eruption, patches of redness
on the teats.

Skin of the udder showing eruption in more advanced
stage of the plague.

A.Hoen & Co.Lithocaustic. Baltimore.

RINDERPEST.

Plate III.

Lips and gums, showing apthous condition.

A Hoen & Co Lithocaustic. Baltimore

RINDERPEST.

Plate IV.

Roof of mouth, showing excoriations.

A Hoen & Co.Lithocaustic . Baltimore.

Tongue and throat, showing thickening of epithelium with
excoriation and congestion.

A Hoen & Co Lithocaustic. Baltimore

RINDERPEST.

Plate VI.

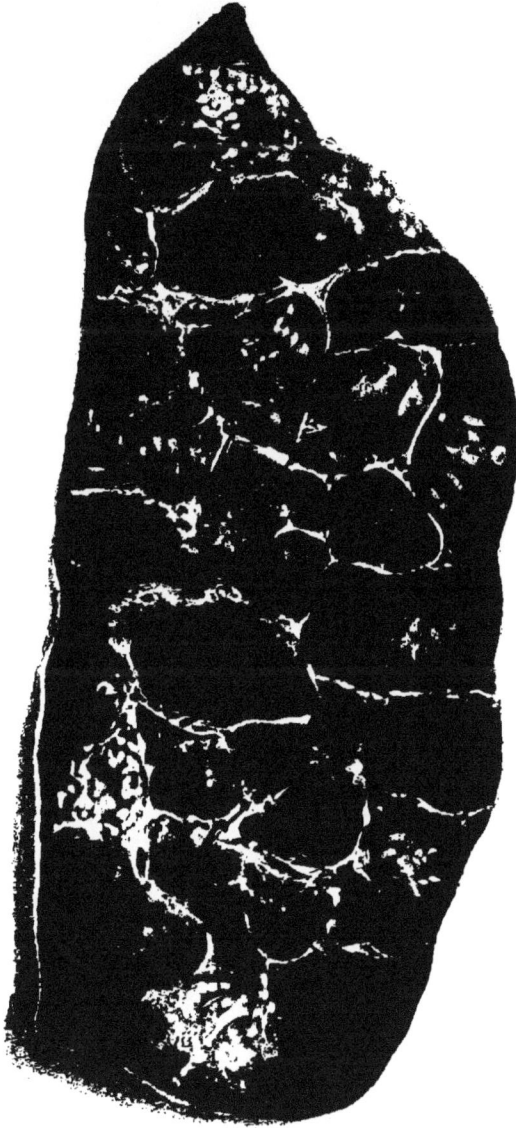

Surface of lungs showing interlobular emphysema, extending
in some places into the sub-pleural tissue.

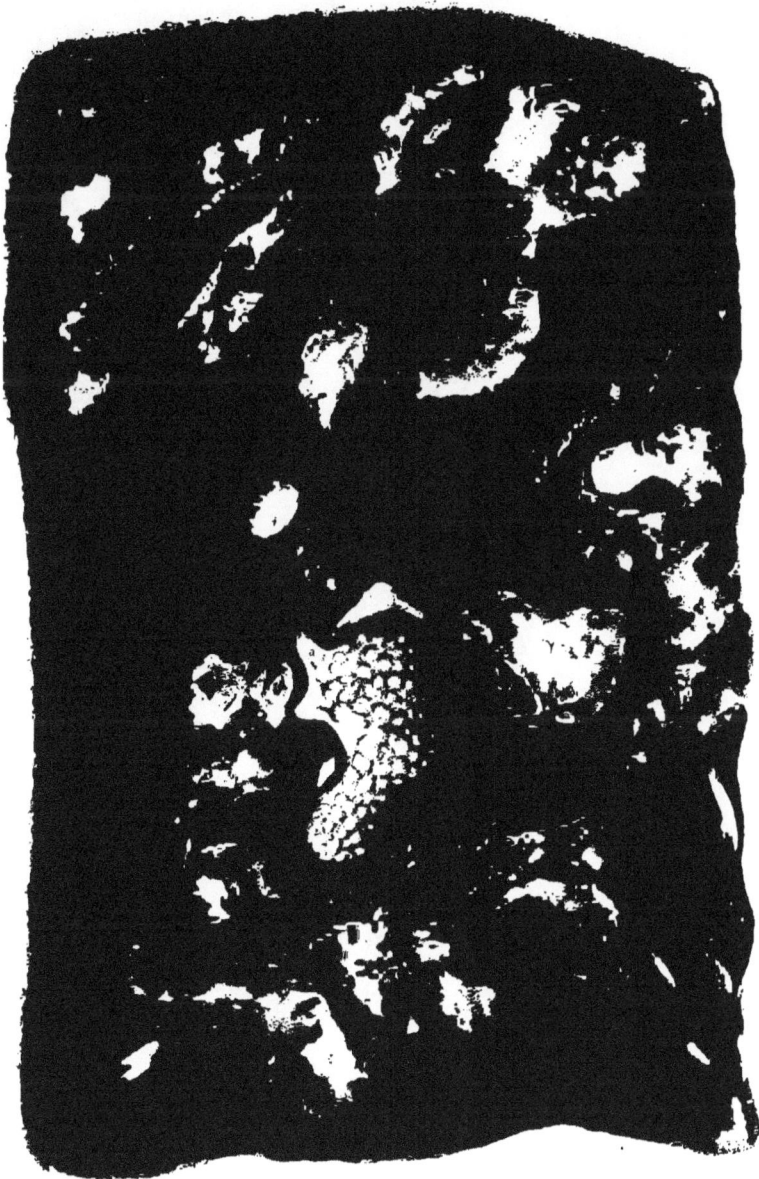

RINDERPEST.

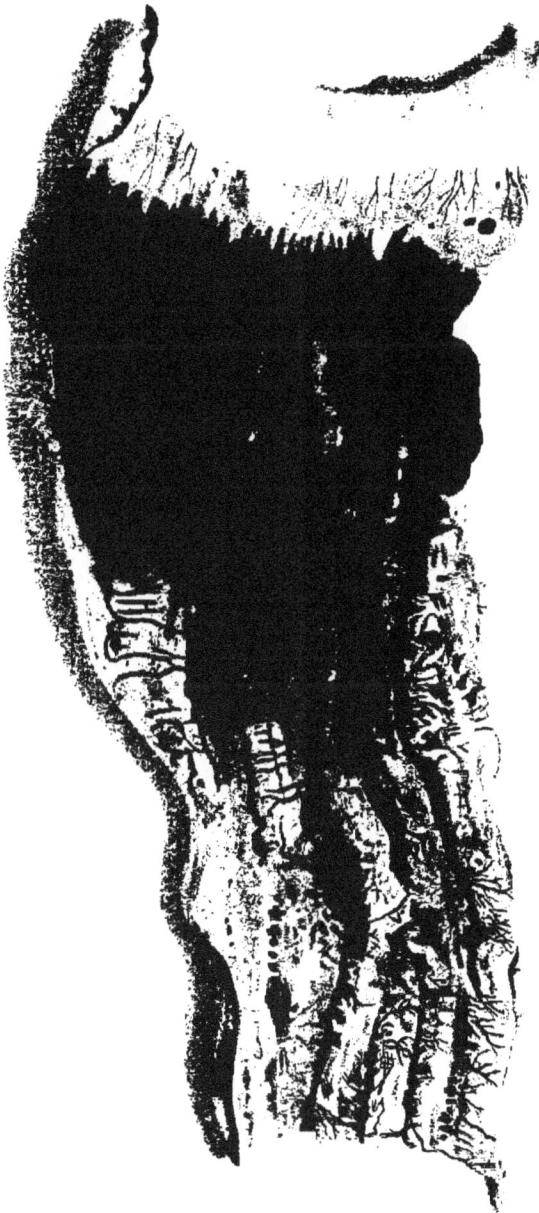

Rectum and Anus, showing deep congestion.

REGULATIONS NOW IN FORCE IN THE GERMAN EMPIRE IN REGARD TO MEASURES AGAINST RINDERPEST.

On the outbreak in distant regions.

ART. I, SEC. 1. If the rinderpest appears in distant foreign parts, which are so connected with the home country by railways or navigation that cattle may be transported in a comparatively short time, then the importation of neat cattle, sheep, goats, and other ruminating animals is to be altogether prohibited from the *infected* regions.

ART. 2. The prohibition of importation is moreover to extend to all articles in a fresh state derived from ruminating animals (except butter, cheese, and milk). On the other hand, the trade in perfectly dry or salted hides and entrails, wool, hair, and bristles, melted tallow in casks and tubs, as well as perfectly air-dried bones, horns, and hoofs, free from soft animal matter, is not to be restricted.

ART. 3. The importation of ruminating animals from uninfected parts of the country in question may be restricted to certain stations, and made dependent upon the following conditions:

a. That it be shown by official certificate that the animals in question have been in an uninfected place for at least thirty days immediately before their departure, and that the disease has not prevailed for thirty kilometers around them.

b. That the transport has been through uninfected places.

c. That the animals in question have been examined and found healthy by an official veterinary surgeon on their passing the frontier.

Easier regulations may, however, be made for the introduction of cattle for slaughter into such towns as have public slaughter-houses connected by branch rails with the railway which brings the cattle.

ART. 4. Further restrictions may be ordered in regard to the introduction of animals, animal products, and articles liable to infection, from countries whence, on account of extensive temporary or permanent infection, there may be great danger of introducing the rinderpest.

ART. 5. What is said about the introduction also applies to the transit.

On the outbreak in neighboring ports.

ART. 6. If the disease breaks out in regions of a neighboring country which are not more than eighty kilometers distant from the frontier, then, for the frontier district, which is to be defined according to circumstances, the prohibition of importation is to be extended to all kinds of cattle, except horses, mules, and asses; to all animal substances derived from ruminating animals, in a fresh or dry state (excepting butter, cheese, and milk); to manure, fodder, straw, and other litter materials, used stable furniture, utensils, and leather work; to raw (or not thoroughly cleansed) wool, hair, and bristles, and to used apparel for trade, and rags.

Persons whose occupation brings them in contact with cattle, for example, butchers, cattle-dealers, and their servants, must pass the frontier only at stated places, and must undergo disinfection there.

Exceptions may be made with the special sanction of the authority, and with such precautionary measures as the special circumstances may require, in regard to the animal products mentioned in Article 2, paragraph 2, as well as in regard to rags packed in sacks, in so far as the importation takes place in closed railway carriages, with an official pass showing that the articles in question come from places entirely free from disease.

Hay and straw, only used for packing, are not liable to the prohibition of importation, but must be destroyed at the place of destination.

ART. 7. If the disease reaches the frontier regions, or continues to extend along the frontier at a distance within reach of the ordinary frontier traffic, then a complete cessation of traffic is to be enforced by a military cordon; but in the neighboring parts of the home country the regulations of Section II come into force.

The transit of railway trains, mails, &c., is to be allowed, even during the cessation of traffic, under the restrictions and with the precautionary measures required by the circumstances.

ART. 8. If the prescribed isolation should be broken through in the cases mentioned in Articles 6 and 7, then the animals liable to the prohibition are to be immediately killed and buried, and all articles subject to infection are to be destroyed or disinfected.

Other articles as well as persons, in case of a violation of the cessation of traffic, directed by Article 7, if it appear that they cannot be disinfected, must be taken back over the frontier by the shortest way, without passing through inhabited places, if possible.

ART. 9. The following measures of supervision are to be introduced in the frontier districts for all localities which are within fifteen kilometers distance from the frontier:

A cattle-inspector is to be appointed in each place, who is to draw up an exact register of the existing stock of cattle, and to make a special record daily of the deductions from and additions to the stock, as well as every alteration therein.

The cattle-register is to be examined at least once a week by the superior authorities.

Notice is to be immediately given in case of disease or death among the cattle.

ART. 10. The directions contained in the present section are also to be applied, with the alterations required by the circumstances, if there is danger of the introduction of the disease by water.

Measures on the outbreak of the rinderpest in Germany.

ART. 11, SEC. II. As soon as any case of disease or death suspected to be from rinderpest occurs among the cattle at a place in our country, the owner must give notice thereof to the local police authorities.

ART. 12. In that case the owner must not slaughter or kill the diseased animals, nor bury or otherwise dispose of any dead animals, until the nature of the disease is ascertained. Till then, the animal must be so kept that men and animals shall not have access to them.

ART. 13. On receiving notice the local police authorities are to send for the competent veterinary surgeon, so that he may verify the disease on the spot. If there be no carcass, an animal is to be killed for the dissection requisite for the purpose. The result of the examination is to be drawn up in a report.

ART. 14. If the disease is ascertained to be rinderpest, the inquiry is to be extended to find out how it was introduced. Further notice is then to be given to the superior authorities, and public announcement is to be made, in which the duty of notification is to be specially pointed out for the neighboring districts as well.

ART. 15. If it be found that there is only a strong suspicion of the rinderpest, a preliminary closing of the farm-yard is to be directed until such time as the disease is undoubtedly proved by further attacks and the requisite dissection or the suspicion is shown to be unfounded. In doubtful cases a superior veterinary surgeon is to be called in.

If the suspicion arises in larger cattle-yards, which are under regular control of the veterinary police, the preliminary closing may be restricted to a separate part of the cattle-yard in question, with the necessary precautionary means.

If the suspicion of rinderpest exists in regard to herds in course of transport, the necessary precautionary measures are to be taken according to the circumstances.

ART. 16. The application, sale, and recommendation of preventive and curative means in the rinderpest are to be forbidden under penalty. Means of disinfection are not to be reckoned among the preventive means

ART. 17. After the outbreak of the rinderpest, the holding of cattle markets, if necessary also other markets, and other large assemblages of men and animals, are to be forbidden within a circuit to be specially fixed, according to the circumstances, but which, as a rule, shall be laid down at not less than twenty kilometers from the infected place. The trade in cattle, and the transport thereof, as well as of manure, fodder, straw, and other litter materials are also to be forbidden unless under special permit. The necessary cattle for the meat consumption must only be bought under the superintendence of the authorities intrusted with the veterinary police.

In the threatened localities the measures of control mentioned in Article 9, paragraphs 2–4, are moreover to be introduced.

For towns of considerable traffic, and for the environs of such towns, special regulations differing from those of this article may be made.

ART. 18. In an infected place, the slaughtering of cattle must be restricted to actual local demand, and take place by direction of the police and under the superintendence of professional persons.

ART. 19. In an infected place the duty of notification extends to every case of ailment in cattle and other ruminants, with the exception of mere external injuries.

ART. 20. The farm-yard in which the rinderpest has broken out is to be, first of all, isolated by watchmen, who must neither enter the farm-yard nor hold communication with its inhabitants, and must not allow the ingress or egress of any persons, except those legally authorized, nor of living or dead animals, or articles of any kind. Only grown-up male persons are to be employed as watchmen, and they must be provided with an easily-distinguishable mark.

Authority to enter the farm-yard can only be given to the person who are engaged in extirpating the disease, and to clergymen, law-officers, doctors, or midwives, for the exercise of their profession; and care must be taken that they are duly identified. At the entrance and round about the farm placards are to be fixed, with the inscription, "Rinderpest."

ART. 21. A partial local isolation is to be imposed on the whole of the circuit to which the infected farms belong, such isolation to consist in the following:

The inhabitants may communicate with each other, but must not leave the locality without special permission, which, as a rule, is only to be granted to such persons as having nothing to do with cattle.

All domestic animals, excepting horses, mules, and asses, must be kept in the stable or shut up. If they are found running about, they are to be seized and slaughtered; dogs and cats to be killed and buried. Conveyances are to be made only with horses, mules, and asses.

The bringing in, sending away, and passing through of all cattle, hay, straw, and other infectable things, are to be forbidden.

At all the entrances to and outlets from the locality, placards are to be fixed, with the inscription, "Rinderpest," and watchmen are to be stationed to enforce the observation of the foregoing regulations.

ART. 22. A local commissioner is to be appointed for each considerable circuit or for several neighboring smaller circuits together, for the time of the continuance of the disease, and he is to have such special inspectors as may be necessary. The notifications prescribed in Article 19 are to be addressed to the local commissioner, who is to see that requisite measures are carried out.

When the outbreak of the disease in any locality is ascertained, the local commissioner is to take measures for ascertaining any new cases of ailment. (Article 13).

ART. 23. If the disease prevails in the majority of the farms in the circuit, the superior authorities may order a complete local isolation.

The locality is then to be entirely surrounded by sentinels (in this case military), and isolated from every kind of intercourse, excepting that of authorized persons, and for the indispensable necessities of the inhabitants of the place, under special precautionary regulations.

The intercourse of the inhabitants is also to be restricted to what is unavoidable. Divine service, schools, and other assemblages are to be suspended. (See article 17.) The beer shops and taverns are to be closed.

The roads leading through the locality are to be barred for the time. If the place is near a railway, no train must stop there, even though it may be a station, unless the station be so situated that it can be entirely isolated and the traffic of the railway station with other places can be carried on without touching the infected place.

ART. 24. According to the extent and the nature of the buildings of any locality visited by the rinderpest, the partial or the complete local isolation may be restricted to separate parts of the locality, and, on the other hand, separate houses and farms in neighboring localities are to be included in the isolation, if necessary.

ART. 25. All cattle sick with the rinderpest, or suspected of being so, are to be immediately killed.

Cattle are to be considered as suspected whenever they have been in the same stable with diseased animals, have had the same attendants, fodder, utensils, or drink, or have otherwise come into direct or indirect contact with diseased animals.

Under what premises other ruminating animals are to be considered as suspected is to be decided according to the special circumstances of each case.

If by killing the suspected animals the stock of cattle on a farm is reduced to a proportionately small remainder, then that also is to be killed.

By authority of the superior power, healthy cattle may be killed for the more rapid extirpation of the disease, though the above conditions may not exist; and this measure may be extended to farms not as yet shown to be infected. (See Article 36, paragraph 1.)

In the larger towns and in slaughter-houses under regular supervision of the veterinary police, the disposal of the flesh and hides of animals which, upon examination, in a living or slaughtered state have been found healthy, may be allowed. But the slaughtering must be done in proper places, under supervision of the veterinary police, and the flesh and the inner parts removed only after cooling, and the hides taken away only after they are thoroughly dry or have lain three days in lime-water (1, 60).

ART. 26. The slaughtered animals, in reference to which the provision in the last paragraph of Article 25 does not apply, are to be buried. For this purpose suitable places are to be used, as far as possible from roads and farms, at such spots as cattle do not frequent. So far as possible waste spots are to be chosen, which are not cultivated at all, or but slightly. The burying-places are also, as a rule, to be hedged in, and to be planted with such plants as grow fast and strike deep roots.

The pits must be dug so deep that the earth may cover the carcasses for at least two meters in height.

ART. 27. The killing and burying is to be done, so far as possible, by the inhabitants of the infected farms, or by such persons of the locality as have no cattle themselves and do not usually come into contact with cattle.

Persons from other localities, especially butchers, are to be employed only when no suitable inhabitants of the locality can be found. To prevent the spreading of the rinderpest by such persons, the proper measures must be adopted. (See Article 42.)

ART. 28. The place where the cattle are to be killed is to be appointed by the local commissioner on consultation with the veterinary officer, taking care to avoid all danger of spreading the disease.

Excrement discharged by the animal in the transport must be removed and buried.

Carcasses must be transported to the pit only by horses or men, on carts, dr... r sleighs, and no parts must touch the ground. The means of transport must be carefully kept separate so long as further transports are expected, and are afterward to be destroyed.

ART. 29. The skinning of carcasses, in reference to which the provision in the last paragraph of Article 25 does not apply, is to be strictly forbidden. Before the burial the hide must be cut in several places and rendered useless. All kinds of refuse, blood, and earth soaked with blood are to be thrown into the pit together with the carcasses. So far as possible, the carcass must be covered with lime before the pit is filled up.

In filling up the pit, layers of stones or brush-wood, if possible, are to be introduced. Watchmen are to be placed at the pit until the isolation is over, or at least for three weeks long.

ART. 30. If a stable in which diseased or suspected cattle have stood is emptied by killing off the stock, then, if the regular disinfection (Article 40 et seq.) cannot be undertaken immediately after the removal of the cattle, any remaining dung is to be burnt or to have disinfected liquid poured over it; the stable, after having all its openings made air-tight, is to be strongly fumigated with chlorine, and then the door of the stable is to be closed and sealed up until the commencement of the regular disinfection. All stable utensils and anything else used for the animals are to remain in the stable.

ART. 31. The foregoing regulations for the isolation of farms and localities may undergo the absolutely necessary modifications for the interests of business, if the disease appears at a time when agricultural labors and pasturage are going on. These modifications are to be decided upon by the superior authorities. The following points (Articles 32 and 33) are to be attended to in such cases.

ART. 32. The isolation of farms cannot, even then, be dispensed with or mitigated; but exertions are to be made to clean the farm as soon as possible. (See Article 25.)

Agricultural works which cannot be delayed are either to be performed by external aid or by the people belonging to the farm, with the necessary precautionary measures.

ART. 33. If the conditions for the isolation of the locality exist, then the isolation of the *Feldmark* (rural district) takes its place; that is, the isolating measures directed in Articles 21 and 23 et seq. are to be transferred to the boundary of the *Feldmark*. The roads leading through the *Feldmark* are to be cut off. The passage and transport of cattle, fodder, &c., on the roads leading along the boundary are to be forbidden.

All inhabitants of the locality who still have farms free from disease and unisolated may carry on their agricultural labors with their own people and their own teams. Cattle-teams in such cases are to be kept as far as possible away from the neighboring estate boundaries, and from the forbidden roads.

ART. 34. In case of necessity, pasturage is likewise to be forbidden in the environs of the infected locality, and the necessary restrictions of intercourse are to be ordered for the immediately adjoining lands, as well as precautionary measures for the management of fields.

ART. 35. In cases of complete isolation, care is to be taken to provide for the most pressing necessities of the inhabitants, victuals, fuel, &c., with the requisite precautionary measures.

ART. 36. In residential and commercial towns, as well as in other towns with brisk traffic, the partial and the complete isolation are not applicable. Other exceptions, required by circumstances, form the provisions of Article 18 et seq., are also allowable. But care must always be taken for the speedy extirpation of the disease by promptly killing all the cattle in infected yards, as well as by proper isolation of the infected localities, and immediate disinfection.

If the rinderpest is found to exist in a public slaughter-house, or in a separately established cattle-market of a large town, then the locality is to be immediately isolated to prevent the disposal of the ruminants and swine which are there. In this case, if the disease has not spread so far as to require the immediate slaughter and destruction of the whole stock of ruminants, the animals not yet diseased may be killed off for the purpose of disposal. But the slaughter, which is also to extend to the swine must take place in the same locality, and within three days, at farthest, under the supervision and direction of veterinary surgeons. For the disposal of the flesh and internal parts, as well as the hides of the slaughtered animals, the provisions of Article 25, paragraph 6, are to be observed.

Measures after the extinction of the disease.

ART. 37, SEC. III. The disease is to be considered as extinct on a farm or in a locality if either all the cattle are dead or have been killed, or if three weeks have elapsed since the last case of disease or death, and if the disinfection has been performed according to the following rules:

ART. 38. The disinfection must be commenced, according to circumstances, as soon as a stable on a farm is cleared of cattle.

It is also to be performed if the stock of cattle has been killed, though no outbreak of rinderpest has been proved. (Article 25, paragraph 5.)

ART. 39. The disinfection must be done on official direction and under professional supervision.

ART. 40. When the stable has been shut up (Article 31), the disinfection begins with its reopening, which is to take place, when possible, within twenty-four hours; care must be taken for sufficient ventilation during the disinfecting operations.

The dung must be taken away and burnt or buried deep in places where no cattle can come within the next three weeks. The liquid manure, collected in trenches, is to be properly disinfected by the application of sulphuric acid and chloride of lime, and to be conducted to pits sufficiently deep.

All walls are to be scraped (the joints cleaned), and then to be well whitewashed; the woodwork is also to be cleaned, washed with strong, hot lye, and after some days to have a coating of chloride of lime solution.

Earthen, gravel, and loam floors are to be dug up, the earth excavated a foot deep, and all treated like the dung. Paved floors of the ordinary kind, that is, with stones laid in sand or earth, are likewise to be taken up, the earth is to be excavated a foot deep, and treated like the dung. The stones may be cleaned. treated with a solution of chloride of lime, and when they have been for four weeks in the air, may be used again.

Wood floors, according to their nature, are either to be burnt or properly disinfected. If the floors are to be taken up, the earth must be excavated and treated as above. Firm, impervious paving of asphalt, cement, or paving laid in cement, is to be cleaned and disinfected.

Instead of chloride of lime, other disinfecting substances known to be effective, such as carbolic acid, boiling water, &c., may be used.

All movable woodwork (racks, mangers, vessels, and other utensils, also the partitions if possible) is to be burnt; ironwork is to be thoroughly heated.

Receptacles for liquid manure and stable-drains are to be treated in the same way as stable-floors, or, if in brickwork, like the walls.

When the disinfection is completed, the stable is to be ventilated for fourteen days.

ART. 41. Only people belonging to infected farms must be employed in the disinfection, or persons who have no cattle themselves; these persons must remain on the farm until the cleansing is over. Only horse-teams must be used for conveyance.

The transport of dung and earth is to be effected according to Articles 28 and 29; the things used for transport, instead of being burnt, may be carefully disinfected, as directed for woodwork.

ART. 42. The clothing of people engaged with dead or diseased animals, and in the cleansing and disinfection, is either to be burnt, or, if it can be washed, is to be soaked in hot lye for twelve to twenty-four hours, then thoroughly washed with soap, and dried in the air; if it cannot be washed, it must be fumigated with chlorine or hung in a dry heat for twelve to twenty-four hours, and then be aired for fourteen days.

Shoes and boots, and other leather articles, must be carefully cleaned, washed with lye, or a week solution of chloride of lime, and freshly greased, then fumigated with chlorine, and aired for fourteen days.

The people themselves must change their clothes and thoroughly wash their bodies.

ART. 43. All fodder which, from its position, may appear liable to have become infected, is to be destroyed by fire at the very beginning of the disinfection.

ART. 44. Dung on the dung-hills, which has been put there during the prevalence of the disease, or within ten days before its appearance, is to be treated like the stable-dung. (Article 40.) The other manure on the dung-hills is to be taken to the fields by horse-conveyance, and, if possible, to be plowed in after three or four weeks.

Until this is done, and for four weeks afterward, no cattle must go into those fields.

If all the dung cannot be taken away immediately, the uppermost layer must be saturated with a disinfecting fluid. But the removal, according to the above directions, must take place as soon as possible.

ART. 45. Even after the complete disinfection of a farm or a locality, and the cessation of the isolation, no fresh purchase or sale of cattle must take place until after an interval of time to be fixed by the competent authorities, which interval must not be less than three weeks from the time when the place was declared free from disease.

Pasture-grounds which have been used by diseased cattle, or cattle suspected of disrease, must not be used again before the lapse of at least two months.

The time when the burying-places may be used again will be fixed by the superior authority, according to the local circumstances in each case.

ART. 46. The holding of cattle-markets is not to be allowed until the lapse of three weeks after the last place in the infected district has been declared free from disease.

When the rinderpest has broken out in residential and commercial towns, or in other towns with a brisk traffic, or in the neighborhood thereof, special regulations may be made, differing from the provisions of Article 45, paragraph 1, and Article 46, paragraph 1.

GLANDERS AND FARCY IN HORSES.

Dr. John R. Page, of Virginia, in connection with Dr. J. J. Terrell, of the same State, made a series of experiments during the spring of 1864, among a large number of horses belonging to the quartermaster's department of the confederate army, for the purpose of ascertaining the identity, nature, and pathology of a fatal disease prevailing among them. These horses were such as had been received by the quartermaster at Lynchburg from the army in the field, and had been forwarded to this point for rest and improvement. The investigations of these gentlemen soon determined the disease to be that of glanders, and they at once took the necessary precautions to prevent its spread and to institute a series of experiments for the amelioration and cure of those animals that were as yet afflicted with but a mild form of the malady. The animals were placed in separate stables, were carefully examined every day for months, and were submitted to various methods of treatment, all of which proved unavailing either to cure or to arrest the disease in its steady progression to a fatal termination. In the course of these observations, extending over a period of six months, a large number of *post-mortem* examinations were made, over one hundred animals having been dissected and the lesions and diseased structures carefully noted. The results of this investigation were afterwards forwarded to the quartermaster-general, who ordered the quartermaster at Lynchburg to have one thousand copies of the report printed for general distribution and use among the quartermasters in the field. This small edition of the work was soon exhausted, and, as many requests have since been made for copies of it, Dr. Page has recently published the monograph with some additions and also with some modifications. The following extracts, detailing the more important results of the investigations of Drs. Page and Terrell, are taken from the last edition of the work:

While the contagious nature of the disease has been admitted from the earliest period, it seems from the recent observations and experience of the surgeons in charge of the Veterinary School and Hospital at Alfort, near Paris, that glanders is not as highly contagious or *infectious* as it has been supposed to be, for of one hundred horses exposed to the contagion it is stated that only eight contracted the disease, and on one occasion, when more than six hundred glandered horses were collected at Alfort, not one of the persons who attended to the horses contracted it. But all medical men will admit that the propagation of any contagious disease depends, in a great measure, upon the predisposing causes that may exist at the time. It is probable, therefore, that the one hundred horses that were exposed to the contagion of glanders at Alfort were just from the pastures, fresh and in good health, possessing, in a great degree, that vital power which enables the system to resist even the contagion of disease. On the other hand, it is reasonable to suppose that the activity of the contagion in Ireland, even in imparting the disease to man, was, in a great measure, dependent upon the predisposition existing on the part of the men and animals from an enfeebled vital power of resistence to disease, caused by bad food and worse hygiene.

The principle of contagion in glanders, whether strong or weak, is admitted by all ancient and modern observers; and we believe it is generally the direct exciting cause. Nevertheless, as a general law, it may be stated that, as is the case with small-pox, measles, and other diseases incident to men in armies, so it is with the first horse that suffers with glanders; it must necessarily have taken it without contagion, so it may occasionally be engendered by causes of which we are ignorant. It may be stated, also. as a general law, "that the contagion once generated multiplies itself without any assignable limit, as the smallest spark may spread the greatest conflagration."

It may be proper to explain what we mean by the term contagion. It is here used in its *specific sense*, meaning the direct transmission of disease, by means of a specific virus, from one animal to another, by artificial or natural inoculation, as in the case of small-pox, cow-pox, syphilis, glanders, and hydrophobia, contradistinguished from the term *common contagion or infection*, as applicable to measles in man and epizoöty and distemper in animals.

* * * * * * *

Horses recognize and communicate with each other by the sense of smell. It is in this way, perhaps, that the mucous membrane of the nose becomes the first apparent seat of glanders. The glandered horse blows the virus from his nose on everything around him, and the disposition of these animals to recognize animal matter wherever found induces the sound horse to investigate the character of the attracting and noxious secretion. He does this by a strong inhalation : the noxious matter is in this way drawn up and retained upon the delicate mucous membrane of the nose, or is perhaps caught by the roughened surface of an abrasion; it finds its proper "nidus," and multiplies its morbific influence; it may be spreading rapidly from point to point, or more slowly manifesting its peculiar elective characteristic in its steady progression. It is possible that a horse may have glanders for several weeks, and even for months, before it becomes an agent of active contagion, from the fact that the disease is often very slow and insidious on the onset, and it is only when the mucous membrane of the nose becomes covered and irritated by chancrous (cup-shaped) ulcers that the smarting out of the virulent matter exists, every surrounding object then becoming a deposit for the virus. Our observation induces us to believe that the chronic form of the disease was that which mainly existed in the animals we have observed, having only seen two acute cases in the whole course of our experience.

* * * * * *

Among the horses under our observation there were several introduced into the stables with the glandered horses without thorough examination, and one, a large bay mare that had a swelled leg, supposed to be due to farcy, but really due to an old injury, attracted particular attention. These animals were very slow in contracting the disease, although in the midst of the most virulent forms that it ever assumes, and we have reason to believe that the disease was contracted in each instance by direct inoculation, inhaling the noxious virus upon the nasal membrane, rather than by common contagion or infection, because, as soon as we discovered the slightest manifestation of the disease we noticed that it proceeded with the regularity of inoculation.

From the experiment already referred to at Alfort, of exposing one hundred horses to the common contagion of glanders, only eight of which became diseased : from the experiments of Gohier, and from our own observation of hundreds of cases, we are confirmed in the opinion that the hypothesis of natural inoculation (or direct inhaling of the virus into the nostril upon the mucous membrane) is the only one that accounts fully for the propagation of the disease. But while we believe that the disease is chiefly propagated by this means, we do not mean to deny the fact of its occasional propagation by common contagion or infection, or even by spontaneous generation, under certain circumstances. All the authors we have consulted agree on this point, and many well-attested instances of the disease being produced in this way are mentioned.

In regard to the spontaneous generation of the disease, we can assert that such generation is rare. During the first year of the war I watched the diseases of the horses in the army in the York Peninsula very closely, with the view to detect the occurrence of glanders or farcy. Early that spring the animals were affected with a very severe epidemic "distemper," from which many died. The deaths were generally produced by the transmission of the disease to the lungs (pneumonia). I did not discover, during this period, any instance of glanders or farcy, and did not hear of any. Late in the month of March I attended a sale of government horses, many of them broken down by fatigue, distemper, bad food, and exposure, but I did not find a glandered or farcied horse among them. The first case I saw during the war was a horse captured from the Federal Army, about the time the confederate army retired from Manassas, and I had reason to believe that this animal spread the disease far and wide among the horses of the confederate army. Personal observation in regard to the food, treatment, and general management of the horses in the army of the York Peninsula satisfied me that there were afforded there as favorable conditions for the spontaneous generation of glanders as could be desired, yet I did not see or hear of a single case of glanders or farcy. As to the degeneration of epidemic distemper into glanders, I have watched closely the course of the disease in animals during several severe epidemics. I have never seen a case of such degeneration, and I believe it rarely occurs. We have conversed on this subject with many persons of close observation and good judgment, who have been dealing in horses for many years, and their experience accords with ours. Still I do not mean to deny emphatically that such occurrences occasionaly happen. Mr. Youatt, in his " Work on the Horse," says "that improper stable management is a far more frequent cause of the disease than contagion. The air charged and deteriorated by the respiration of *many horses kept together in small and illy-ventilated stables* becomes an irritant, or rather a poison, to the mucous membrane of the nose." *   *   *

I have stated before that the disease, as it has occurred under my observation, has been chiefly of the chronic form, having never seen more than two or three cases regarded as acute, and these did not run their course to a fatal termination as rapidly as many authors describe, continuing for three and four weeks before death supervened. As I have stated, also, the disease commences very insidiously, but increases steadily in virulent malignancy to rapid decomposition of the blood, exhaustion, and death.

The first symptom that attracts the attention of the observer to an animal in the onset of the attack of glanders *is a watery discharge from the nostril* (strange to say, most frequently the left), without any apparent fever, loss of appetite, or unusual disturbance of any kind. In every instance of my observation, coincident with the discovery of this watery discharge, I have found the (lymphatic) glands beneath the angle of the lower jaw, on the side corresponding to the affected nostril, engorged or swollen, tumefied, with a tendency to become bound down or adherent to the adjacent branch of the jaw-bone. Several glands may be implicated, but they invariably present to the fingers the impression or sensation of being fastened together in a bundle, and all tied or fastened to the inside of the lower jaw-bone, near the angle, moveable up and down, but confined and restricted by the adhesions to very limited motion, *and they never suppurate or form abscesses.* I have examined over one hundred horses, free from disease, to ascertain the natural condition of these glands under the angle of the lower jaw, and, except in young animals that had not completed the second dentition, I found very few in which the glands could be recognized, and none presenting *the characteristic engorgement.*

In this stage of the disease the mucous membrane within the nostrils will have lost *its natural pink or rose color, and most frequently, but not invariably, presents a pale buff aspect, with purple dots, or patches of congestion* (bloodshot looking spots), as it extends up the cavity of the nose, and these patches or dots will be found, on closer examination, to be an aggregation of small over-distended blood-vessels.

In many instances the mucous follicles (little glands or twisted tubes that secrete mucus) will seem to be elevated and prominent on the pale buff surface of the membrane of the nose, and sometimes they, too, will present a purple hue. Several days, perhaps a week, may elapse, and the animal is still without fever, cough, or loss of appetite ; the coat may stare, or become more rough, and there is an appearance of ill health. The watery secretion from the mucous membrane of the nose now becomes more abundant and *glutinous, and begums the hair on the edge of the nostril, to which the food and dust adhere.* The (lymphatic) glands beneath the angle of the lower jaw become more sensibly engorged or swollen and adherent to the bone. The purple congestion of the nasal mucous membrane increases. The most prominent mucus follicles on the membrane of the nose degenerate into small ulcers, and the discharge soon assumes a thick yellowish purulent character, and becomes still more abundant.

Cough now supervenes, more or less frequent, produced by the general irritation and inflammation of the air passages, and the tendency *to sneeze and snort out the virulent matter begins.* The disease may now progress, with this continuous purulent (fetid) discharge from the nose for weeks, and indeed for months, while the ulceration is extending from follicle to follicle throughout the course of the mucous membrane into the remotest cavities of the nose, in the interior portion of the skull (frontal sinuses). During this time hemorrhage from the nose often occurs, and proves very exhausting to the animal. The constitution is now evidently gravely affected, the animal becomes emaciated, the legs swell, and swellings are apt to occur about the head and lips. The superficial (lymphatic) glands beneath the skin become affected, especially those on the inner side of the hind legs and under the shoulders. Farcy now supervenes, and becomes thoroughly established. Farcy is not, however, a necessary concomitant. I have found it to exist in about one-fifth of all the cases I have had under observation. But whether farcy supervenes or not, the animal steadily grows worse, though in good weather and under favorable circumstances it may rally and improve somewhat. Generally, however, the blood becomes by this time so empoisoned that death soon occurs. In this stage the mucous membrane of the nose often assumes a dark gangrenous (rotten) aspect, and the discharge is bloody, very fetid, and the respiration is labored and difficult, accompanied often with a harrassing and exhausting cough, giving to the poor animal an expression of the greatest anxiety and suffering.

With a few horses under my observation, the disease, although thoroughly characterized, seemed to progress so slowly that I was induced to believe that they might, under favorable circumstances, live for months. Several of these animals were killed, and I found the disease to have progressed even to a carious (ulcerated) and necrosed (decayed) condition of all the small bones (turbinate, ethmoid, and sphenoid bones) of the head, together with an extensive deposit of matter (tubercle) in the lungs and in the (mesenteric) glands of the bowels, showing that they were incurably diseased; yet they would in all probability have lived for months, multiplying victims to the contagious virus. Farcy is said to be curable " when uncomplicated with glanders," but I have never met with such a case. I am of the opinion that cases of cure must be exceedingly rare, as in every case of farcy-glanders I have found extensive deposits of tubercle in the lungs and in the (mesenteric) glands of the bowels.

The anatomical and pathological (diseased) characters of the organs involved will be clearly shown by a summary of the facts taken from accurate notes of the first nineteen cases, made at the time of the examination.

Of the nineteen cases, seven died at the natural termination of the course of the disease. The twelve others were killed (by the injection of air into the external jugular vein) in different stages of the disease, with the view of ascertaining any differences that might exist in the pathological characters in the tissues of the organs. Of the nineteen cases nine had farcy intercurrent in the course of the disease.

The skin of the head and face having been carefully removed, the subcutaneous veins about the face, lips, nose, and lower jaw were found enlarged and their coats much thickened. The lymphatic vessels accompanying the veins elicited particular attention, and in many cases they were found large enough to be easily dissected out and perforated by straws. The walls were evidently thickened, and the lymphatic ganglia (small glands) here and there in their course, especially those beneath the angle of the lower jaw, were found swollen and filled with gelatinous "matter" like glue, causing the glands to impart to the finger the sensation of elasticity before they were cut into.

The skull being now cut through longitudinally with a saw, and the two segments pulled asunder, exposed to view the nasal fossæ (cavities) throughout their entire extent, the vomer or septum of the nose, the bones of the skull, the brain and its membranes. The mucous membrane at its commencement in the nostrils was of a pale buff color, with linear purple streaks of congestion extending up the middle of the septum, or partition of the nose, along the course of the principal vein and lymphatic vessel, traceable directly from the chancrous (cupped) ulcers up to the cavities in the ethmoid bone.

The whole mucous membrane often presented a worm-eaten appearance, so studded was it with chancrous ulcers. The turbinate ethmoid and sphenoid bones (small bones at base of skull) were found carious (ulcerated) and their cells or cavities destroyed by the infiltration of a cheesy, inspissated pus. The lungs were found, in every stage of the disease, congested, with large purple patches beneath the surface of the pleura (or membrane covering the lung), and when these patches were cut into they were found to be made up of veins congested, their coats thickened, and the accompanying lymphatics containing pus, with a gelatinous matter here and there deposited in their course. The lung structure (parenchyma) being cut into, the same gelatinous matter or pus generally exuded in the line of the incision, having evidently been deposited in the course of the small veins and lymphatic vessels. Cavities of all sizes were found in the upper portions of the lungs, filled generally with cheesy pus. The mesenteric glands (in the course of the bowels) were found tumefied, engorged, and much enlarged in many cases throughout their course, with occasional deposits of cheesy pus or gelatinous matter of a dark appearance. In every case I have examined the whole venous and lymphatic system of vessels were diseased, their coats thickened, the lungs filled with cheesy, tuberculous matter, and cavities containing fluid pus. We found tuberculous matter in the spleen in one case only of the nineteen examined at this time, but found it to exist very often afterwards in the spleen and the liver also. A number of the surgeons of the Confederate army on duty at Lynchburg at the time witnessed many of these post-mortem examinations, and they all agreed that they had never seen such similarity of pathological characters in different cases in any disease in the human system.

I have reproduced the disease several times by inoculation of the virus, and in order to show the similarity of its course and effects I will narrate one case from my notes.

On the 7th of March virus of glanders was taken from the nostril of a horse affected with the disease and inoculated in the nostril of a sound horse, just at the point where the skin and mucous membrane become continuous. On the fifth day of the inoculation, in the seat of each point of insertion of the matter, a specific pustule appeared, and the lymphatic glands beneath the angle of the corresponding jaw-bone were felt slightly tumefied. On the eighth day the pustules were ulcerated, and presented a well-defined characteristic chancrous appearance. The mucous membrane of the nose now rapidly assumed the characteristic buff and purple aspect, and the discharge of a thin, watery fluid from the nostril commenced. On the twelfth day the lymphatic glands beneath the angle of the jaw were much more sensibly tumefied or swollen, with a tendency to come together in bundles, and to be fastened down to the adjacent branch of the jaw-bone, yielding to the finger the sensation of decided induration. The lymphatic vessels running along with the veins, about the lips and face, were enlarged and corded. On the fourteenth day of the inoculation all of the symptoms of confirmed glanders were manifested. On the seventeenth day lymphatic vessels (or the small glands in their course) about the lips and face suppurated, and the vessels leading to the space beneath the lower jaw were indurated and corded. There was also considerable swelling of the right hind leg. On the twentieth day there was an increased flow of the watery discharge from the nose, and upon close observation numerous small ulcers could be seen studding the mucous membrane within the nostril. There was also a good deal of constitutional disturbance, as fever with the embarrassment of the respiration. On the twenty-fifth day the abscesses in the lymphatic ganglia, or knots on the face, nose, and lips, having discharged their contents, seemed disposed to heal, but the lymphatic vessels could still be sensibly felt. The appetite of the animal was very good up to the twenty-fifth day of the inoculation, and it was well supplied with nutritious food. At this period, however, rapid emaciation set in, and in a few days it became so weak as not to be able to get up when it lay down. It lingered until the thirty-third day of the inoculation, when it was killed by an injection of air into the external jugular vein.

The *post-mortem* examination revealed the following anatomical and pathological (diseased) characters: The ulcers in the mucous membrane of the nose, which were the points of inoculation, were almost entirely healed, but the mucous membrane, from its commencement in the nostril throughout its entire extent, was covered with numerous small chancrous ulcers, and was of the *pale buff color, streaked with purple congestion*, just as seen in the cases occurring naturally. The blood vessels of the mucous membrane on the septum (partition between the nostrils) of the nose presented an abundant arborescent purple congestion, increasing as it extended higher up. The lymphatics extending upwards, with the blood vessels on the septum of the nose, were very much enlarged, were dissected out carefully, and penetrated with straws. They were much larger than we had ever before seen them, and could be traced up to the ethmoid bone very distinctly, which bone was found carious, its bony structure disintegrated, and its cavities filled with pus, showing conclusively that it had become diseased at an early period after the inoculation, through the medium of the large and engorged lymphatic vessels, traced directly to it from the chancrous ulcers in the nostril. The sphenoid bone was also dark and diseased. The turbinate bones were not absolutely carious, but were fragile, dark, and diseased. The mucous membrane in the sinuses or cavities of the turbinate, ethmoid, sphenoid, and frontal bones were softened, sodden in appearance, and in many places covered with a yellow gelatinous matter. The cavity of the chest being carefully opened, the pleura or investing serous membrane of the lungs was found inflamed, and in many places adherent to the walls of the chest. The right lung presented on its surface and beneath its investing membrane numerous purple patches, which, being cut into, revealed great vascular congestion (blood-shot condition), and a gelatinous deposit which was found to exist, not only in the course of the veins and lymphatics just beneath the pleura or investing membrane, but also in the substance or structure of the lung, which being incised exhibited pus exuding freely in the line of the incision. A great many small, cheesy, miliary deposits of tuberculous matter were found scattered through the lung structure, and in the upper lobe of the right lung, near the apex, a cavity as large as a hen's egg was found filled with fluid pus. The left lung was not as much diseased as the right, but presented very much the same condition. The liver, spleen, and heart were of healthy appearance. The right and left cavities of the heart were found filled with frothy blood, as well as the substance of the liver, but these were the usual effects from death by injection of air into the veins. The entire chain of mesenteric or absorbent glands of the bowels were found enlarged, and a greater number of them filled with pus than we have generally found in cases of the natural disease. This case of the disease, produced by inoculation, is interesting not only because it establishes the fact of the identity of the reproduction, but that it shows the activity of the virus from the day on which it was inoculated and the analogy of its destructive course in the system to that of the natural disease.

The diseases with which glanders may be confounded in its incipiency are: (1) Common catarrh or cold; (2) ozena or indolent ulceration of the mucous membrane of the nose; (3) strangles or colt's distemper; (4) epizootic or epidemic distemper.

1st. *Common cough or cold* is always manifested by some intensity of redness in the mucous membrane of the nose in place of its natural pink or vermilion tint, with a disposition to sneeze and a tendency to cough, which soon passes off without any engorgement of the lymphatic glands beneath the angle of the lower jaw.

2d. *Ozena* is a rare disease, and is very gradual in the development of its ulcerations, which are not chancrous or cup-shaped in appearance; is unaccompanied by early glandular engorgements and constitutional symptoms.

3d. *Strangles or colt's distemper* is peculiar to young horses between the fourth and sixth year; commences usually with fever; cough and severe constitutional irritation, soon followed by a copious yellow mucous discharge from the nose and a ropy fluid from the mouth; inability to extend the head or put it down to the ground; loss of appetite; a considerable swelling under the throat, *posterior to the point at which the glands in glanders are involved, and in the middle line of the space under the jaw, with great tenderness on pressure and tumefaction in the fermenting tissues, followed by early suppuration.* The animal is unable to eat on account of the pain produced by mastication; swallows with difficulty on account of sore throat, and the effort frequently produces violent convulsive cough. The swelling in the space beneath the jaws is uniform and extensive; is exceedingly sensitive to the touch; in a few days becomes soft and prominent, containing fluid pus; it soon "bursts," and the animal is greatly relieved; all the symptoms abate, and convalescence is rapidly established, unless there is some intercurrent disease.

4th. *Epizoötic or epidemic distemper* attacks horses of all ages and conditions, but, as is the case with all epidemics, it possesses peculiarities that will be diagnostic. The attack is sudden, ushered in by shivering fever, cough, heaving at the flanks, redness of the membrane of the nose, constitutional irritation of an intense character, and great debility. The respiratory apparatus sympathizes, and the inflammation extends rapidly throughout the extent of the bronchial mucous membrane. There is sore throat, and the horse "quids his hay and gulps his water." The glands are not greatly enlarged, but the submaxillary and the parotid glands are much swollen and the soreness and tenderness are very great. The discharge from the nose soon follows, is abundant, thick, and becomes rapidly purulent and often fetid. The breathing is accelerated and laborious and the animal refuses to eat. When death occurs in this disease, it is generally due to pneumonia, rarely from general debility and exhaustion. The disease is in great measure attributable to atmospheric causes, of the nature of which we are ignorant.

Glanders may be distinguished from the foregoing diseases, even where not complicated with farcy, by the following characters: It attacks horses of all ages and conditions, and is highly inoculable, furnishing a specific virus, as much as cow-pox. The horse with glanders does not seem to be really sick, *but a watery discharge commences insidiously to drop from the nostril,* most frequently the left, *not accompanied by fever, cough, loss of appetite, or any constitutional disturbance.* The watery discharge is persistent, but gradually becomes glutinous and adhesive, and begums the hairs around the edge of the nostril with particles of food and dust. Coincident with this discharge, *the* (lymphatic) *glands on the inner side of the branch of the lower jaw-bone, on the side corresponding to the affected nostril, become enlarged, either separately or in bundles, and adhere to the bone, and seem to be bound down by adhesions. The glands are movable, but the motion seems to be restricted;* they are not hot and are but slightly tender; they steadily increase in size, and become more indurated as the disease progresses, until they yield to the finger a sensation of specific induration. *These glands never suppurate, but remain specifically indurated,* and become more adherent to the jaw-bone as the disease progresses. The mucous membrane within the nostrils will at this time have lost its *natural pink color; it will be of a pale buff or leaden hue,* with purple streaks increasing as it ascends up into the nostril. The discharge, having become glutinous, is soon streaked with yellow pus and occasionally with blood, the purple streaks soon degenerate into numerous small (follicular) ulcers, and the discharge now becomes entirely purulent, abundant, and often bloody.

The (lymphatic) vessels around the mouth may now become corded, with small knots here and there in their course, and these knots occasionally suppurate. Farcy then supervenes, and the diagnosis is clear. The hind legs (one or both) are apt to swell, and the same corded appearance occurs along the course of the veins, on the inner side of the leg, as described in the lips, and the knots may also suppurate. The animal rapidly emaciates after the discharge becomes purulent, but still manifests a desire to eat and drink until the respiration becomes so oppressed as to render it unable to do either. The course of the disease from the first is steadily progressive, it may be slowly, but very surely, to a fatal termination. Inoculation, if resorted to, will always decide the matter in six or eight days.

Prognosis or probabilities of recovery are exceedingly unfavorable. Youatt says "that in a well settled case of glanders it is not worth while, except by way of experiment at a veterinary school, to attempt any remedies. The chances of a cure are too remote and the danger of spreading the disease is too great to risk remedial measures."

To show how insidious this disease may be in its early stage, how slow it may progress, and how destructive it may be by spreading the virulent contagion to other animals, I will mention the following case that came under my knowledge: A gentleman belonging to the cavalry was killed on picket-duty not long before the Confederate army retired from Manassas, and the thoroughbred mare he rode fell into the hands of the enemy. In a few days, however, the mare was recaptured, recognized, and returned to his family, who sent her to a farm in Bedford County, Virginia, to be taken care of. When she reached the farm the manager observed that she had a slight cough and some watery discharge from the nose, but thought it was due to cold. Her appetite and general condition being good, she was turned into a clover pasture and seemed to improve.

There were seven work-horses and three colts on the farm, all in fine condition. The colts were constantly with the mare, and the work-horses were in the same pasture whenever they were not at work. In a very short time the colts sickened and died with what was considered "a very bad distemper." The mare, meantime, was in quite good condition, but the discharge from the nose had become more abundant and yellowish. The work-horses soon took the disease also, and one or two of them having died, it became necessary to put the mare to work, as her condition was considered better than that of any of the horses on the farm, all having contracted the disease. At the expiration of less than twelve months, all of the horses and the three colts had died of the disease, and the mare, the source of the contagion in the first instance, was still alive, and would have, perhaps, lived (capable of doing some work) for several months longer if she had not been killed.

I have stated that the animals in which this disease is said to arise spontaneously have been generally half-starved, over-worked, and of broken constitution. Confinement in dark, damp, ill-ventilated stables, especially those under ground, may predispose to its occurrence, but, from my observation of a large number of cases, I am of the opinion that it is chiefly propagated by direct contagion, rather than infection. In other words, in a large majority of cases, the disease results from the contact of a particle of glandered virus with the mucous membrane of the nose, rather than from the breathing of the infected air in the immediate vicinity of a diseased animal.

I had notes of several cases in which animals crippled, but free from disease, were put in the same sheds with those with glanders, only separated from them by a partition of boards, which did not prevent the passage of air. These animals were attended also by the same men who attended to the diseased horses. Of the three horses thus exposed, one contracted the disease at the end of three weeks, and the other two escaped for a much longer time. These animals may have contracted the disease by infection, that is, by inhaling the noxious air; but from the fact that the first manifestation of the disease is in the form of an ulcer on the mucous membrane of the nose, I am inclined to believe that it is most often due to the direct implanting of the specific virus upon its surface, just as the particle of vaccine virus is implanted on the skin of the human subject.

The only satisfactory result in the treatment of the disease in this, its first stage, occurred in the case of my own riding-mare, in the month of September, in 1865, while going from Lynchburg to the lower country. I stopped after dark one night at a gentleman's house where I was kindly received, and my horse put up in the stable. The next morning, when I was going to leave, there being no servants, we had to go to the stable to saddle our horses. While doing so, I discovered that the horse in the next stall to mine had a running from the nose, which I soon discovered was due to glanders; and there was nothing to prevent these two from putting their noses in direct contact. I was then informed by the gentleman that he had lost several horses from, what he considered, a very "obstinate and fatal form of distemper," and that he had only recently purchased the animal in question, and was not aware that it had taken the disease until then.

Being thus warned of the danger of the inoculation of my mare, I watched the condition of the mucous membrane of her nose carefully every day, and on the eighth or ninth day after the exposure I discovered some congestion in one nostril, which was soon followed by the formation of a small ulcer, which was immediately cauterized freely with the solid stick of lunar caustic (nitrate of silver). This was followed by injections twice a day of a weak solution of the crystallized nitrate of silver in the affected nostril for ten days or more. The glands beneath the angle of the jaw *did not become affected*, although the ulcer in the nostril did not heal entirely for nearly eight weeks. About two weeks of the time I gave her, in her food, as much sulphate of soda as she would take. I would have preferred to have administered the hyposulphite of soda if I could have obtained it. For a week or ten days after the appearance of the ulcer her appetite was indifferent, and she was evidently feverish; but these symptoms gradually passed off, and her recovery was complete by the middle of November, and ten months afterwards I sold her for $300 to a man who carried her to New York, where she was sold; and a year after that I was written to, by her then owner, for her pedigree, so we may infer that she was cured of the primary disease, and, that all danger of the constitution being involved had passed away. There may be doubts as to whether this ulcer was due to the specific virus of glanders. I can only say that it had exactly the appearance of those I observed so often in the early period of the natural as well as in the inoculated cases of the disease, and it only differed *from the latter in that, that the lymphatic glands did not simultaneously become affected.*

INOCULATION OF CHARBON.

The results of some interesting experiments in inoculations for charbon are given in the *Journal d'Agriculture Pratique* for March, 1879.

During the course of last year, M. Pasteur received from the minister of agriculture of France a commission to examine the causes of spontaneous charbon, that is to say, of that type which breaks out suddenly in a stable or a sheep-fold, and to seek for preventive or curative means for treating it. A report of this *savant*, published in the *Recueil de Médecine Vétérinaire*, gives the results of the experiments made in the department of the Eure and Loire, with the assistance of Messrs. Chamberland, fellow of the university, and Vinsot and Brutet, veterinary surgeons.

Starting from the fact deduced from preceding works, (1) that charbon is the *disease of the bacteremia*, that is to say, of the little microscopic organism which Dr. Davaine was the first to prove existed in the blood of carbuncled animals; that the bacteria obtained in a state of purity, in the midst of inert matters, and inoculated on healthy animals, suffices to produce disease and death, while nothing contained in the charbonnous blood surrounding the bacteria would produce the affection in question, M. Pasteur adopted the following course of research: Is the cause of spontaneous charbon simply bacteria, and, if so, what is the *habitat* of this organism? If this is the case, where is this organism to be found?

The skillful experimenter has endeavored, in the first place, to inoculate healthy animals by means of their food. Sheep were fed with fresh lucern moistened with a liquid impregnated with corpuscle germs of bacteria. The sheep submitted to this treatment did not all succumb; at times there was no mortality, but whenever death did occur it was found to be due to charbon, with this peculiarity, that the disease had had a long period of incubation (from four to six days). M. Pasteur concludes from this that "if in the department of the Eure and the Loire the germs of charbonnous bacteria exist spread over the food, or on the soil, they ought to be probably at once abundant and of different inoculation; in other words, this inoculation to be efficacious requires special conditions." These conditions may be easily realized. It is sufficient to combine with the animal's food anything which may wound the mouth or the principal digestive organs. Thus food the consistency of lucern mixed with different species of thistles, which are found in the fields, and tainted with bacteria, causes a larger mortality than the infected lucern given alone. The examination of anatomical lesions leads to the conclusion that the origin of the disease is in the mouth or in the back part of the throat, as well when the infected lucern alone is concerned as when infected forage mixed with matters which may wound the animal. "From this the inference is that when the disease is communicated by food which appears to be incapable of causing wounds, these must have previously existed."

Thus it is in the primary digestive organs that these lesions principally occur in spontaneous charbon, from which it follows that charbon is communicated spontaneously to animals by food covered with germs of bacteria, but only when the animal has wounds, or wounds itself in eating—wounds which otherwise would be quite insignificant if some accidental circumstance did not render them dangerous.

"How may intense heat and dry weather facilitate the development of charbon?"

Probably, according to M. Pasteur, because they add to the tender-

14 C D

ness of the buccal mucous membranes, which are more easily wounded
when the food is hard and the mineral dust taken in with the food is
more abundant. A very simple preventive against the spontaneous development of charbon is the practical conclusion of these researches.
The means consist in suppressing as far as possible everything which
may wound the animal, particularly in its mouth, to remove from its
food everything which may scratch the primary digestive organs, to remove thistles and all other prickly plants, and to be careful during winter and summer not to give them their food too dry, or mixtures of straw;
to moisten these before giving them, or, better still, to ferment them
with green fodder. The introduction of silicious dust from the road is
also to be feared.

All occasions of the diffusion of the germs of charbon from dead animals should be carefully guarded against. It is probable that the department of the Eure and Loire contains a larger quantity of these germs
than any other department, and as the disease has been so long established there it sustains itself, as the dead animals are not treated in such
a way as to destroy all further germs of contagion.

When spontaneous charbon breaks out on a farm, M. Pasteur ad-
vises, in the first place, to paint the mouths of cows, oxen, bulls, and
sheep every day with a substance or solution which cicatrizes the wounds—
chlorate of potash, perchloride of iron, cinchona powders, &c.—but the
preference may be given perhaps to a saturated solution at fifteen degrees of chlorate of potash, and to wash the eyes and nose of the animal
with the same solution. For sheep, which cannot be treated one by one,
it suffices to dissolve the chlorate in their drink. This may perhaps
answer for the large animals. While M. Pasteur made the experiments
which are thus briefly summed up, M. Toussaint, professor of physiology
at the veterinary school of Toulouse, pursued analogous researches
in the department of the Eure and Loire, and arrived at the same
conclusions. According to M. Toussaint charbon is never spontaneous,
in the literal sense of the word, because it sometimes apparently de-
velops itself. 'There must be inoculation by bacteria or their spores;
it is by the wounds of the mouth that they penetrate the system, and
the food serves as the means of communication. These are important
points acquired by science, but it remains to be ascertained where and
how the germs of bacteria are preserved. To settle this question, M.
Henry Bonley, in the *Recueil de Médécine Vétérinaire*, says: "It will require long and difficult labor; for the spores of the bacteria and the
bacteria themselves have no specific character which may distinguish
them in the midst of the numberless species which pollute the smallest
blade of grass in all decomposing matters, and even upon the dry earth
when the slightest moisture suffices to revive them."

INOCULATION OF MALARIAL FEVER.

Equally interesting and important as the discoveries of Pasteur are
the results of experiments recently made in inoculations for marsh or
malarial fevers. In the spring of 1879 the physical cause of poison to
which intermittent fever is due was made the subject of special investigation by Signor Tommasi, professor of pathological anatomy at Rome,
in conjunction with Professor Klebs, of Prague. According to the statement laid before the Academy of Rome (Transunt. Acad. Science, iii
(1879), p. 216), the investigation was rewarded with complete success.

The two investigators spent several weeks during the spring season
in the *Agro Romano*, a locality notorious for the prevalence of this par-

ticular kind of fever. They examined minutely the lower strata of the atmosphere of the district in question, as well as its soil and stagnant waters; and in the two former they discovered a microscopic fungus consisting of numerous movable, shining spores of a longish, oval shape, and .9 micromill. in diameter. This fungus was artificially generated in various kinds of soil; the fluid matter thus obtained was filtered and repeatedly washed, and the residuum left after filtration was introduced under the skin of healthy dogs. The same thing was done with the firm microscopical particles obtained by washing large quantities of the surface soil. The animals experimented upon all had the fever with the regular typical course, showing free intervals, lasting various lengths of time up to sixty hours, and an increase in the temperature of the blood during the shivering fits up to nearly 42°, the normal temperature in healthy dogs being from 38° to 39° centigrade. The filtered water only caused changes in the temperature of the body, even when five times the original quantity of water was administered, and the trifling fever actually caused was not of an intermittent character. The animals artificially infected with intermittent fever showed precisely the same acute enlargement of the spleen as human patients who have contracted the disease in the usual way, and in other spleens of these animals a large quantity of the characteristic form of fungus was present. The fungus was also found abundantly in the lymphatic vessels of the animals. As the fungus here grows into the shape of small rods, Tommasi and Klebs gave it the name of *bacillus malariæ*.

The strictly scientific methods pursued in this investigation do not seem to admit of a doubt that the learned experimenters really discovered the cause of the disease in question. The discovery may be regarded as another of the series of which those in connection with inflammation of the spleen and diphtheritis were earlier examples. Against the intermittent-fever poison, says the London Journal of the Royal Microscopical Society, which is connected with this newly-discovered microscopic fungus, the medical art was formerly as powerless as it is still against diphtheritis and inflammations of the spleen. For intermittent fever, however, medicine was provided as a remedy when the virtues of quinine were made known, and it may be reasonably expected that, as in the latter case, so against the poison of the diphtheric fungus and that of splenetic inflammation, medical science will sooner or later discover their appropriate antidote.

It has been announced in several European publications that Drs. Marchiafava and Valenti have since detected *bacillus malariæ* in human patients in a more advanced stage than in animals originally dissected.

EXPERIMENTS WITH CHICKEN CHOLERA.

During the session of the French Academy of Sciences, held in February, 1880, an interesting communication was received from M. Pasteur on the subject of virulent diseases, but treated more particularly upon the disease known as "chicken cholera." The bird which is a victim to this disease, observes M. Pasteur, loses its strength, and its wings droop. The feathers on its body rise, and make it look like a ball. An unconquerable sleepiness overwhelms it. If it is compelled to open its eyes, it appears as if awakened from a profound sleep, and soon closes its eyelids again. Frequently it dies in mute agony without having

212 CONTAGIOUS DISEASES OF DOMESTICATED ANIMALS.

changed its position. If it happens to move its wings for a few seconds it is with great difficulty. This disease is caused by a microscopic organism, which M. Pasteur has bred in a suitable manner, and with which he has inoculated Guinea pigs and fowls. The inoculation of the pigs did not always produce death, but did produce an abscess, and fowls inoculated with the contents of this abscess soon died. A few drops of a culture of this microbe placed on a piece of bread or meat fed to the fowls is sufficient to cause the infection to enter the intestinal canal, where the little organism multiplies in such great quantities that the excrement of the fowls thus infected kills others which are inoculated with it. These facts, M. Pasteur says, permit us easily to account for the manner in which the disease is propagated in poultry-yards.

Evidently the excrements of the sick birds are the great cause of contagion. Nothing can be more easy to arrest this than by simply isolating the birds for some days, by washing the yard with an abundance of water, and especially with water acidulated with a little sulphuric acid, which easily destroys the microbe, and by removing all the manure before admitting the birds again. All cause of contagion will have been removed during this period of isolation, because the birds already attacked will have died, so rapid is the disease in its action.

By a certain change in the culture of this microbe its virulence may be diminished, and while the fowls inoculated with the most virulent virus are all killed, those infected with the diluted virus sicken but do not die. If they are allowed to recover, and are again inoculated with the more infectious virus, the injuries produced are local, and do not cause death. Chicken cholera is then of the character of those virulent diseases which do not repeat themselves. Suppose that this microbe of the diluted virus may be fixed in its proper variety, according to M. Pasteur, and that we are not always obliged to have recourse to its original propagation when we wish to use it, it may be made to serve as a veritable vaccine, transmissible from animal to animal as the vaccine of variola is transmissible from man to man.

A number of experiments will alone enable us to judge how well this prediction is founded. In any case the observations of M. Pasteur opens a new road to the study of the virulent diseases, which are for the most part perhaps only parasitic diseases, such as charbon and chicken cholera. They will without doubt give to physiologists the idea (or suggest to them the propriety) of seeking a vaccine virus for contagious epizoötics, such as foot-rot and contagious pleuro-pneumonia, which cause so many losses in agriculture.

The following is a translation of the more important points presented by M. Pasteur for the consideration of the academy:

On the subject of extracts from solution for propagation of the microbe of chicken cholera, a question comes up for consideration. We have shown that these extracts contain no substances capable of preventing the culture of the microbe. But will they contain anything suitable for the vaccination of chickens? I have prepared a solution, the volume of which was not over 120 cubic centimeters. Filtered and evaporated in the cold, and always under conditions incapable of changing its purity, this culture left an extract which had been redissolved in 2 cubic centimeters pure water, which was then all injected under the skin of a healthy chicken. A few days thereafter the chicken inoculated with a very infectious virus took the cholera, and died under the same conditions as chickens which had not been vaccinated. Experiments of this kind lead to an observation which, in a pathological sense, is as new as curious. If a fresh and healthy chicken receive a subcutaneous injection of extract of a filtered culture of the microbe, corresponding to a very abundant development of the parasite, the chicken, after a nervous disorder, which disappears after a quarter of an hour, and is often manifested simply by a slightly increased respiration, and a movement of the beak, which it opens and closes at short intervals, the chicken, I say, takes the form of a ball; remains motionless, refuses to eat, and exhibits a most pronounced tendency to sleep, as in the case of the disease by inoculation with the microbe; the only difference being that the sleep is much lighter than in the real disease, as the chicken awakes at the slightest noise. This sleepiness continues about four hours, after which the chicken becomes alert, and carries his head up, eats and clucks as if nothing had happened. I have repeated this experiment several times, observing the same facts, and in each test I took care to determine that an extract of pure *bouillon*, in which no microbe had been cultivated, gave place to no analogous manifestations. I was led to the conviction that during the life of the parasite it acted as a narcotic, and that it is this narcotic which gives rise to the very pronounced morbid symptoms of sleepiness in the disease of chicken cholera.

By the act of nutrition, the microbe becomes the cause of the severity of the disease and consequent death. This is easily understood. The microbe is *aerobic;* during its life it absorbs large quantities of oxygen, and consumes many of the principles of its medium of growth, which is easily determined by comparing extracts of chicken *bouillon*, before and after the culture of the little organisms. Everything indicates that this oxygen, necessary to its life, is taken from the blood-globules through the different vessels, and this is proven by the fact that during the life, and often long before the approach of death, the comb of the sick animal is seen to become violet, though the microbe either no longer exists in the blood or is there in quantity so small as to escape microscopic observation.

This kind of asphyxia would be one of the most curious characters of the disease under consideration; yet the animal dies from deep-seated disorders caused by the culture of the parasite in its body, by pericarditis, and other serious discharges, by alteration of the internal organs, and by asphyxia. But the act of sleep corresponds with a product formed during the life of the microbe acting upon the nervous centers. The independence of the two effects in the symptoms of the disease is still further established by the circumstance that the extract of a filtered culture of the microbe puts to sleep chickens vaccinated to the maximum extent.*

These facts will undoubtedly provoke considerable thought among pathologists, and notwithstanding the possibly already exaggerated length of this paper, the academy will kindly permit me to briefly point out some other peculiarities of the disease called chicken cholera. This disease is well known; it is terrible and rapidly mortal, especially when caused by direct inoculation of its microbe. It is certainly very remarkable that sometimes, as we shall show, it appears chronic. In fact, in certain cases we see inoculated chickens which, after having been very ill, do not die, but on the contrary experience relative recovery, yet they eat little, often become œnemic as indicated by the discoloration of the comb; they gradually become very thin, and finally succomb after weeks and months of languor. This fact would have only a secondary importance if, under these singular circumstances, it did most frequently happen that the microbe is formed on the body at the time death occurs, a manifest proof that the parasite was preserved in the body of the animal from the time of the last inoculation—always present, always acting, but to a very mild degree, since it but slowly brings on death. It is found located undoubtedly in some vaccinated part, which is therefore unsuited to an easy culture. Vaccinated chickens especially exhibit this form of the disease, which, it is true, is not frequent. It might be considered that under such conditions there should occur a transformation of virulent virus, but this does not take place. In the case to which I refer, the virulence of the microbe is, on the contrary, increased, as may be established by separating them from the blood of the dead chickens, by means of the special cultures, and afterwards inoculating other new chickens with them.

Such facts as these will help us to understand the possibility of long incubations of virus, such as rabies, for example, which, after having existed a long time in the body in a latent state, as it were, suddenly indicates its presence by the most active virulence and death.

Does not human pathology also explain it? Alas! how often do we see diseases of the virulent order, such as measles, scarlatina, and typhoid fever, have serious consequences of long duration, and often incurable. The circumstances I have just mentioned are of the same nature, and we are here able to fix upon the true cause.

In very well vaccinated and well conditioned chickens there sometimes appears upon one part of the body or another an abscess filled with pus, causing no trouble in the health of the animal. It is remarkable that this abscess may even be due to the cholera microbe, preserved as if in a vase or cup, and without doubt powerless to propagate itself because the chicken had been vaccinated.

The disease may be conveyed from the pus of this abscess, by culture, or by inoculation of new fowls, which, after being abundantly developed in the inoculated region, it kills in the ordinary manner. These facts in every particular recall and rationally

---

* NOTE.—However, I must isolate the narcotics and determine whether, by a suitable dose, death cannot be produced, and whether, in such case, there would occur the internal disorders common to the real disease.

explain the abscesses of the Guinea pigs, to which I referred in my former communication. It is very possible that the muscles of the pigs reproducing the microbe more slowly and with more difficulty than those of the chickens, the disease is confined to an abscess, and recovery becomes possible.

I fear I shall abuse the time allowed by the academy if I do not conclude. The subject is so vast and so fertile, that I beg leave to be allowed to present at the next session a report of other observations, to which I shall add further facts which I hope to collect.

Lavoisier has said, one can give nothing to the public if he waits to have covered the entire field presented, and which extends in proportion as the work proceeds. So it is useful to me to have the opinions of competent men to inform or restrain me, either of which cannot fail to strengthen and enlarge these researches.

Regarding the discoveries of M. Pasteur as of great importance, the ministry of agriculture and commerce of France have issued the following circular, a copy of which has been sent to all the leading poultry breeders and dealers of that country:

The contagious affection special to poultry, known under the name of chicken cholera, causes serious losses to agriculture. However little importance it may seem to have when it attacks only an isolated subject, it acquires true gravity when, as is most usually the case, it appears in rather large flocks, which it may decimate, and even sometimes completely depopulate, in a few weeks. The disease may, therefore, occasion considerable loss on farms where the production of poultry and eggs is a matter of lucrative speculation.

The development of the disease may always be arrested, and the object of the present instruction is to make known to agriculturists the means for attaining this end. All farmers know how to distinguish chicken cholera. As soon as the disease invades the flock the birds become languid and sleepy. The temperature of the body rises; the comb becomes violet in consequence of a modification of the circulation; finally death often intervenes in a few hours after the first appearance of the symptoms. Recent scientific researches have proven that this disease is produced by a microscopic organism which is developed in the intestines, passes into the blood, and there multiplies with extraordinary rapidity. This parasite is evacuated with the feces, and may subsequently pass into birds which pick over the manure or eat the seeds which have been soiled by the affected fowls. If a bird dies of supposed chicken cholera, all the poultry should be immediately removed from the yard and the birds kept isolated from each other. The yard and the houses should then be thoroughly cleansed by removing the manure and washing the walls, perches, and soil with a large quantity of water. The water should contain about six fluid drachms of sulphuric acid to one gallon of water, and a coarse broom or brush should be employed for the washing. After the lapse of ten days, if no death has occurred in the mean time, the disease may be considered as having disappeared, and only those birds which show prostration, dullness, and sleepiness, need be further kept in isolation. These means, so simple in their application, are regarded as sufficient to arrest the progress of the contagion and prevent its return. Applied in the beginning of the disease, they limit the losses to an insignificant figure.

CATARRH IN SHEEP.

Dr. C. D. Smead, Logan, Schuyler County, New York, writes as follows, under date of March 26 last, concerning an outbreak of catarrh among sheep in that locality:

I have not as yet examined a sufficient number to give you as extended a report as I wish, but will give you the results of my investigations thus far. I was consulted, on the 20th of February, by a farmer living in the town of Hector, concerning his flock, which, by the way, was composed of 104 sheep, about one-half grade Merinos and one-half grade Cotswold breeding ewes. He told me that the first time he had noticed anything wrong was three days previous, when he saw one of his best ewes standing apart from the flock, with head elevated, snuffling at the nose, and a violent heaving of the flank. The animal grew rapidly worse, and died in about twelve hours. Soon another one was taken with similar symptoms, and died in twenty-four hours. Within four days from the observance of the first case, four animals had died and three were sick. On the fifth day two more were taken, making nine in all. On the 22d I made a post-mortem examination of those that had died, which numbered six of the nine, and found the membranes of the nasal chambers of all of them highly inflamed. Of those that had died within a few hours from the first appearance of disease, I found the chambers of the nose literally swelled full, and death had been caused by strangulation. Of those that had lingered for three or four days, I found that the inflammation had extended to the larynx and the bronchial tubes, and a slight congestion of the lungs, death being caused as in other catarrhal affections. So far I was ready to pronounce the disease as simply of an acute catarrhal affection caused by the preceding wet weather; but my attention was called to some peculiar-looking white specks upon the inflamed and putrid membranes, which, upon close examination, proved to be minute worms, the largest of which were about one-eighth the size of a timothy-seed divided lengthwise. Some were so small as to appear as a mere speck. The membranes were, so to speak, perfectly covered with them, and had I examined with a glass doubtless many more would have been seen. As to the origin of these parasites or their nature I have as yet no positive knowledge, and it may require much time and experiment to ascertain from whence they came. I have examined sufficient to convince me that the disease is caused by these parasites.

The above-mentioned flock were in good, comfortable, well-ventilated buildings, and had been for two weeks; previous to that they had been confined in a yard, with an open shed to run under. For food they had during the forepart of winter oat straw and corn-stalks, and about one-half pint of oats, each, daily. At the time the disease broke out among them they were being fed good, well-cured meadow hay, and one-half pint of oats and corn. Believing that the disease was caused by the worms, and also that the commencement of their ravages was in the nasal chambers, I reasoned as follows:

First, that in order to effect a cure the cause must speedily be removed. Believing that carbolic acid could be used in sufficient strength to destroy the parasite, and also act beneficially on the inflamed membranes, I prepared a solution of crystallized carbolic acid one part, glycerine four parts, water twenty-six parts, and ordered it used by injecting with a syringe into the nose of every sheep as soon as the first symptoms of the disease appeared; also that four drachms of the hyposulphite of soda be given internally, and repeated every twelve hours. Thus far it has proved successful in the flock, except the first-mentioned nine, of which only one is alive. About one-third of the flock have been affected. Other flocks have suffered with the same disease in the same and adjoining towns, with about the same results. I only give the history of this one as a sample. To sum up as briefly as possible, I believe the cause, as before stated, to be worms.

Second. I believe that all sheep attacked will die if they are not promptly treated with some medicinal agent that will destroy the cause.

Third. I believe it to be highly contagious (but in what manner I have no positive knowledge), and all sheep should be immediately removed from the flock as soon as the first symptoms appear.

Fourth. From present experience in the treatment of the disease, I believe that carbolic acid properly diluted and injected into the nostrils within a short time after its appearance will effect a cure in quite a number of cases at least. I have recommended its use to all the owners of the diseased flocks as far as known, and I am waiting results. I propose to continue the investigation, and, if possible, ascertain the cause of the disease.

SANITARY REGULATIONS AND PREVENTIVE MEASURES.

The subjoined correspondence relates to sanitary regulations, preventive measures, and so-called remedies for swine plague and diseases incident to other classes of domesticated animals. In the absence of any positive remedy for swine plague, the sanitary regulations and preventive measures recommended are worthy the careful consideration of every one interested in raising stock. The first and most important province of sanitary science is the prevention of contagious diseases; therefore, in the presence of a highly contagious and fatal disease among domesticated animals curative measures of an experimental kind should not be undertaken, except in regions in which it is indigenous, or in hospitals where the greatest care can be exercised to prevent a dissemination of the malady. Even in contagious maladies that are not very destructive, and which may readily yield to medical treatment, prevention will, in almost every instance, be found more practicable and satisfactory than the trouble and expense of administering drugs and the consequent depreciation in the value of the animal. Moreover, curative measures are not generally successful in a majority of the spreading diseases, hence the great importance of the close observance of proper sanitary regulations and preventive measures. An intelligent system of hygiene, by fortifying the body and preserving the different organs in health, is the most certain safeguard against the development of disease. A suitable and sufficient diet, an abundance of air and light, a regular amount of exercise, no undue exposure to inclement weather, cleanliness, and healthy locality, form the basis of a good hygiene, and in outbreaks of certain classes of destructive contagious diseases is all that can be relied upon for the preservation of dumb animals.

COLORADO.

Mr. Philip Zoeler, Pueblo, Pueblo County, says:

Although no contagious diseases prevail among farm animals in this locality, your report on diseases of swine and other domesticated animals is highly prized by all who have been fortunate enough to secure a copy. There is a weed called " *loco*," found growing abundantly in some localities in this State, which is more destructive to horses and cattle than all diseases combined. If it does not kill an animal outright it makes it worthless forever after. The symptoms are about as follows: Swelling of the head, moping appearance, moves about as if foundered, are afraid of everything, lose the luster of the eye, imagine they see objects in front of them, and jump over a straw as if it were a big log. Animals are mostly affected by it in the winter and early spring, when grass is dry and scarce on the range. A horse once poisoned by the weed can never thereafter be regarded as a safe animal, either in harness or under the saddle. No remedy or antidote, so far as is known, has ever been discovered. As most of our animals roam over a large range, perhaps no remedy could be applied in time if we had one.

GEORGIA.

Dr. H. R. Casey, Appling, Columbia County, says:

While reading your work on diseases of swine, &c., it occurred to me that I would state to you a fact perhaps not generally known. Some twenty-five years ago, while engaged in active practice, I was called to see a case of cutaneous disease. The patient was an old man and lived in the country. He said he had contracted the disease many months before—had had several doctors to attend him, but all were unable to either diagnose correctly or to cure him. I asked him to look back and see if he could connect any event or anything that could have originated such a state of affairs. He said, "I remember that I had a pet pig that had the mange badly, and desiring to cure it I caught him, and after washing him well with warm suds and water I gave him a good scraping." I replied: "The veterinary surgeons say that mange is not communicable from the lower animals to man, but I pronounce your malady a case of mange." I put him on an alterative treatment, and by the use of frequent ablutions of warm water and castile soap, followed by paintings with iodine, he was cured. I wrote an account of this case which was published in the Augusta Medical and Surgical Journal, and this was republished in several foreign medical periodicals—one, I think, in France, and one in Germany. It was held at one time that glanders was not communicable from the horse to man, but this idea has long since been exploded.

Mr. James P. Phillips, Clarksville, Habersham County, says:

I have read with pleasure the report on diseases of swine, &c., and I desire to state that I consider the report of Dr. Detmers as exhaustive, thoroughly covering every point as to swine plague except as to sporadic or spontaneous cases.

I keep inclosed all stock. In 1858 I had brought from East Tennessee sixty-five fat hogs. I lotted them away from my home hogs, fed them one month, and then slaughtered them. There was not a diseased hog in the lot. During the winter I had a herd of sixty head, thirty-two of which I fattened in the orchard. One died of the plague, but no others were attacked. In 1875 I had a lot fattening in close pens, four to a pen. One died of the plague, but none have since died on my farm. My friend C. K. Jarret lost forty fat hogs by the disease in 1875. My neighbors generally have lost more or less every year. None of them keep their hogs inclosed. Mine have some green pasture the year round.

The disease is unmistakable; the symptoms once seen will never be forgotten. My whole aim has been directed toward preventing the malady. I trust a great deal to isolation, to frequent changes of pasturage and locality, and to feeding sound grain. So far, no remedies have been discovered. Tennesseeans used to attribute the disease to Kentucky still-fed hogs, which were driven through the State on their way to North and South Carolina, Georgia, and Alabama. I can show the disease in our mountains, through which there are no roads, and over which hogs have never been driven.

Distemper among our cows is the most serious trouble against which we have to contend. The annual loss is great—during some years devastating. When the poor man loses his cow the loss is irreparable. I have made many *post mortem* examinations, and have invariably found the omasum—the manifold—the diseased organ. If discovered very early, the animal may be saved by turning on green corn. I have given all sorts of purgatives without beneficial results. Several years ago, by accident, I had the fencing around seventy-five acres of pasture burned. My neighbors' cattle flocked to my cultivated grasses, and that season I lost seven fine Ayrshire and Short Horn cows. The next season I lost one Short Horn, but have had no cases since; and yet, on each of the adjoining places there have been losses each year. In one of our mountain districts the farmers had made a law unto themselves, and though most of them used oxen, after a certain time in the spring they neither drove their own oxen down nor allowed others to drive their cattle up into the district; but after the building of the A. and C. Air Line Railroad the advantages of market and trade increased, and they broke loose from their law, hauled off produce at all seasons, and the consequence has been that during the past fall they lost most heavily. I believe you would do a great good if you would keep Dr. Detmers in the field until he masters, or, at any rate, fully examines into, the causes of each and every epidemic, contagious, and infectious disease incident to our farm animals. I have read carefully what each of the examiners have written, and, like all farmers, am glad to hear what scientists have to say; but at the same time we know that preconceived opinions, or the bending of symptoms to suit cases, cure neither flocks nor cattle.

Mr. J. B. Vandeventer, Mount Sterling, Brown County, says:

I have this to say of your special report No. 12, that after a careful perusal of the same I consider it one of the most valuable and instructive works ever issued from your department. The reports of the different veterinarians are the most complete and exhaustive I have ever seen on the subject of so-called hog cholera, and I have read everything that I could lay my hands on that I thought calculated to throw light upon the subject. I would be glad if every farmer and stock-raiser in the country could have a copy placed in his hands, for I am satisfied that if our farmers could become possessed of the information it contains, and would act upon its many valuable suggestions in reference to the management of their swine, they and the country at large would be benefited at the rate of many millions of dollars annually.

Mr. R. W. Hunt, Galesburg, Knox County, says:

Every disease affecting swine is called hog cholera. But whether cholera, swine plague, pneumonia, or what not, it has been very fatal here in many cases, some farmers having lost entire herds consisting of a hundred or more. It seems to me that the Department of Agriculture could render the country no better service than to discover the cause and either a cure or successful preventive for this plague. To accomplish this it would be necessary to send competent persons—those well versed in the diseases of domesticated animals—into the districts infected by the disease. They should enter the field in the fall, say in October, and remain until-mid winter, studying the symptoms of the disease from its incipiency to its determination, making such post-mortem examinations as may be necessary, and conducting experiments looking to the discovery of a cure or preventive. Should such an investigation require all the energy and means of the department for one year and result in the discovery of a preventive or cure for the malady, it would be the best expenditure of money the government ever made. I doubt if anything can ever be accomplished in any other way.

Mr. John S. Armstrong, of La Salle County, in a recent interview with the Commissioner of Agriculture, made the following statement in regard to his experience and treatment of swine plague:

I live on Fox River, twenty miles below Aurora, La Salle County, Illinois. Last year I had some trouble with my hogs, and lost a good many. This year they commenced dying again. They dug up a hog that had not been buried deep enough, and that started the disease again. I examined the hogs and found them lousy. I did not regard this as the cause, but rather the result of the cause. I made a preparation of lard, axle-grease, and sulphur, and applied it with partial success. The under part of their bodies was of a light purple, and raw. I applied this preparation and it seemed to help them very much for a time, but in the course of about three weeks they commenced to cough, and I saw I would have to do my work all over again. I used to have sheep—as many as 1,200 at a time; once they got the itch among them. I cured this with tobacco juice, copperas, and sulphur.

Question. In what proportions?

Answer. I do not recollect. I think I used a little lime also. I thought I could kill the lice with this preparation, but I did not expect to kill the parasites that are said to prevail in this disease. This preparation also proved a failure. On the Monday following I got four pounds of tobacco, boiled it up, and put in about equal parts in two kettles. I boiled it a good while, perhaps half a day. I took the leaves out of one and the juice out of the other, and changed them back and forth to get the strength out of them. The tobacco was the strongest plug tobacco I could get. The next morning I run the liquor off and strained it. By that time one kettle would hold the contents of both. I then took a screen, a sort of net, and put over it; divided it and put in five pounds of copperas. I then took some tallow, I think two or three gallons, part tallow and part lard, put it in the juice and again boiled it down as low as I could. It fermented and boiled over. I then got two gallons of lubricating oil, crude petroleum, and two gallons of coal-tar; I guess at that. I boiled this, and to my surprise a portion of the coal-tar went to the bottom. I then put in some coal-oil and warmed it—did not let it boil but made it pretty hot. I put that on the hog and it killed the parasite as well as the lice, and it restored the hogs. I reported this to the doctor, and he said: "You have in that compound a strong preparation. In the first place tobacco is a most destructive thing. The copperas has iron in it, and carbolic acid is the best destructive in thousands." I did not know that carbolic acid was a production of coal-tar. From what he told me about the carbolic being good to keep the insects from getting in, and recommended its injection into the nostrils, thinks I, if that is the case I will try feeding it. I did not know how much coal-tar I could put in; I did not know how much carbolic acid they could take, but I concluded to try something. I took half a gallon of boiled oil, put twelve ounces of carbolic acid in it, and half a gallon of oil, and warmed that. I mixed salt, sulphur, and copperas, pretty near a pailfull, about one-quarter sulphur, one-quarter copperas, and one-half salt; mixed that in the pail, and put in the acid, and boiled together and stirred it up. I got about a bushel of stone-coal and pounded it and poured that on and mixed it. I have been giving that internally. I have given it three different times to my own hogs and others.

Question. How do you feel it?

Answer. O, they are very fond of the stone-coal. I don't know how much they will eat.

Question. How long have you been doing that?

Answer. The care was effected about a month ago. I think the skin is more affected than any other part of the body. I am inclined now to the opinion that whenever the system becomes weak from any cause that the disease will set in, and the lungs sympathizing with it the cough is brought on. I am also inclined to think that it is possible that this cattle fever may be remedied in the same way, by giving them internally carbolic acid and making an outward application of both. In regard to the proportions, I have been telling others to do just as I have done. The proportions that I first experimented with were four pounds of tobacco, boiled down until the strength was taken from it; two and a half gallons of grease; then put that in, and the least that is necessary to neutralize the coal-tar is sufficient. I may not have got the best proportions. I got the tobacco all in the grease by evaporating the water from it; then about the same quantity of lubricating oil, and about the same quantity of coal-tar. I warmed this and put it on hot. I think the coal-tar has the staying qualities. By putting it on the surface and using it internally as I did, I believe a cure can be effected.

Mr. Ira Rowell, Danvers, McLean County, says:

I have carefully examined the report on diseases of swine, &c., and think the commissioners have given us much valuable information, especially as regards swine plague. I have had considerable experience with this disease myself, and I am of the opinion that the fatality of the plague can be greatly lessened by a proper preparation of the animal. The weakest and poorest animals are always the first to contract the disease. Since I have been giving my hogs proper care, and regularly using preventive preparations, I have had much better success than formerly. I keep them on an orchard of bluegrass pasture until about the middle of July, where they get plenty of apples and an occasional feed of oats. Every day I take a bucket of water and add to it a handful of sulphur, a little salt, ashes, and copperas, and with this I sprinkle their beds thoroughly. I also add at the rate of one teaspoonful of pure carbolic acid to the bucketful of pure water for drinking purposes.

Mr. A. L. Miner, Momence, says:

I have carefully read your report on diseases of swine and other animals, and was glad to find that the commissioners who had charge of the investigation arrived at conclusions I had previously formed as regards swine plague. As there are a great many facts of special value to stock-raisers, and as I have been engaged in raising hogs for breeding purposes for over twenty years, my experience should be worth something. The examiners were right in saying that an ounce of preventive is worth a pound of cure. While I have never had the disease termed cholera among my hogs, I have never been able to make any headway in doctoring one that was sick. While the cholera has not visited this locality, hundreds of hogs died of it last year in the west end of the county. My practice is to have good pastures for my hogs, and good shade during hot weather. I feed to keep in good flesh, and give salt, and salt and wood-ashes regularly once a week. I give them pure water to drink and what swill we make about the house. As I keep about one hundred hogs on an average the swill does not go very far. I keep in a dry place, of easy access to the animals, a trough about 14 feet long, in which the following ingredients are constantly kept: One peck of wood-ashes, one quart of salt, two pounds of pulverized copperas, one-half pound of black antimony, one-half pound of sulphur, one peck of charcoal, and one gallon of bone-dust. A little observation of hogs in trying to eat old bones will show the necessity of giving them bone-dust; salt is a necessary ingredient of the blood—it gives tone to digestion : ashes neutralize the acids, and charcoal assists in the same way and also absorbs the fetid matter; salt and ashes act as a destroyer of worms; sulphur purifies the blood; copperas is also a blood purifier and tonic : bone-dust furnishes the lime and bone forming material for a perfect animal. With this treatment and management I can only say that I have lost but three animals by disease in the past ten years.

Mr. E. Hemsinger, Burnt Prairie, White County, says:

We agree with the conclusions of Dr. Detmers as to the contagious and infectious nature of the so-called hog cholera, and heartily indorse his mode of stamping it out. If a compensation is made for diseased animals, all right; but in order to rid ourselves of the plague we would gladly submit to it at our own loss of sick animals. Ours is a pork-producing district, and has suffered as much from swine plague as any other in the United States. We have three rivers, all close together, the Big and Little Wabash and Skillet Fork, and it is a well-established fact that the disease prevails much worse along water-courses than elsewhere. It is a mistake to suppose that the services of a veterinary surgeon are necessary to properly diagnose this disease, at least in localities where the disease has prevailed to any great extent. A swine-breeder of our county readily recognizes it—would not often fail to do so while riding at a gallop. If, when the plague first made its appearance, every hog had been slaughtered at the owner's expense, and by that means the disease had been extirpated, our people would now be hundreds of thousands of dollars better off. Stamp it out is our motto. Pay if you can, but if you cannot, stamp it out anyhow.

IOWA.

Mr. I. P. Winterstein, Fox Point, Tama County, says:

I have read your report on diseases of swine with great interest. For the last ten years I have been trying to discover the cause of hog cholera, and, according to my observation, the following are predisposing causes of the disease: First, keeping too many hogs together; second, the drinking of bad and impure water from mud-holes and hog-wallows; third, neglecting to give salt at regular intervals, or not at all; fourth, feeding too much strong food, such as corn, and too little slop, or slop with too small a quantity of milk in it. I do not think any one can be a successful breeder of hogs in large numbers without keeping cows. Hogs will drink almost any kind of slop if composed of part milk. A clover pasture is a necessity to the well being of pigs. Small hogs should have separate quarters, and be kept warm and clean. Salt should be mixed with wood-ashes—all the feeder can command. I have never yet known a hog to get sick that was given all the wood-ashes, charcoal, and salt it would consume. In my opinion, charcoal is the very best preventive for diseases of swine that can be given.

Mr. John Stuart, Traer, Tama County, says:

I have read and studied special report No. 12 of your department carefully, and I am happy to find that Drs. Detmers and Law have gone down to the very foundation of the matter, and give the disease, of which ninety-nine one-hundredths of the losses result, the very appropriate name of swine plague, by which it is to be hoped it will continue to be known in the future. Hog cholera has come to mean anything that kills the hog, and no one disease in particular. When over four years ago I undertook the self-assumed task of fathoming this heretofore mysterious disease, I had no suspicion that it would cost me more than a tithe of the time and money that it has. My experiments extend over a period of eighteen months, during which time I have traveled from the Missouri River to the Atlantic Ocean. I have had an average of fifty experimental hogs at one time, bought from herds that were dying off. I lost none in all that time. The reason that they did not die on my hands I now know was because I bought not those that were in a dying condition, but those that were already recovering from the plague.

I will here call your attention to one way in which the farmers are being imposed upon by hog-medicine men. The impostor takes care that by pure accident he drops around fifteen or twenty days after a herd begins to sicken, and is exceedingly sorry that he did not happen around in time to save the whole lot; at that late date he will only be able to save those that are not too far gone. The truth is, the animals are already convalescent, and have passed through without any serious injury. He goes off with say 25 cents for each hog that would have recovered without his aid as well as with it, and the farmer is happy over the result, never suspecting that he has been swindled. It is frequently the case that hog raisers, after having tried many remedies, and finding no good results while the disease is in the period of incubation, finally give some of the nostrums (such as are referred to on page 48 of the report as the "specific remedies"), and finding the deaths less frequent, think they have at last found the true remedy, and are afterward astonished to find that in future attacks these same remedies are of no use in arresting either the spread of the plague or curing the animals affected.

I am more than ever convinced that to the absence of iron on the surface of this alluvial country is traceable by far the greater number of deaths from swine plague. In all my travels, which have been extensive, and in all my investigations, which have been conducted with care, I have not been able to find a single instance in which the animals have had access to iron in the clay or water, or been supplied with it artificially, where there has been serious trouble. I differ in opinion with those commissioners as to the best way of treating the diseased animals, and especially in their suggestions of stamping the disease out by the slaughter of all those affected. Should every hog in the United States be destroyed, the disease would certainly be stamped out for the time being, but by the very first shipment from Europe we would be liable to have the plague reintroduced. From the investigations of Dr. Kline, of London, it is evident that it existed there prior to its appearance in this country. There is but little doubt of its having been brought from there in the first place.

I have investigated as many as twenty different cases of new outbreaks of the plague this winter, and traced the introduction of these contaminated centers to animals that no doubt left their former homes in good health but caught the plague in transportation. Every person should be prohibited, under heavy penalty, from allowing their sick hogs to run at large, and also from leaving the dead so that dogs and other animals can be the agents of spreading the contagion. Were a stop put to these things, I am fully persuaded that as many as one-half of the losses would be saved annually.

I have just returned from a tour of investigation in this State and in Missouri. In Washington County it is now believed that a hog-medicine man, who went through the county, by some means spread swine plague in the herds of those who would not buy his medicines. At any rate, since he was there the disease has swept through the county and carried off probably $100,000 worth of animals. I do not know that that man did spread the disease, but I do know that were I disposed so to do I could spread it and the owners would know nothing about how their animals became affected. That being the case, it would be safe to keep strangers away from the herds that are in good health. At Des Moines I went before the committee of the house on agriculture for the purpose of having the legislature pass a law to prevent the introduction and spreading of swine plague, and had that body adopted my suggestions it would have reduced the deaths in the State one-half annually. They divided on the question of whether they would allow people to sell or compel them to burn or bury dead hogs, as if it made any difference what we do with dead hogs; it is the live ones we want to save. They don't seem to know how to save people of the State $1,000,000 annually, which might be done by a little of the right kind of legislation. My experience, as well as my investigations, force me to the belief that swine plague of itself does not cause death, but that in cases where even slight inflammation of any of the vital organs exists, that and the plague is almost certain to cause death. But the most general trouble is the deficiency of limpid blood, which any person may prove to his own satisfaction by comparing the amount of blood of two hogs of the same size, one kept exclusively on the black loam, and the other where it can obtain iron in such form that it can assimilate it. The appearance will be found to be about as one is to two—the one dark and thick, the other red and limpid. The problem as to what the disease is having been ascertained beyond peradventure, the next question of importance would seem to me to be what is best to be done about it. To which I would answer, get that knowledge as widely diffused and as quickly as possible. There has been so much nonsense written that farmers are not inclined to read, and if they do the language is generally beyond their comprehension. Another trouble is, so many men think they know more about the matter than any one else that they will not read; and still another trouble is, the country is overrun with impostors, in the shape of hog-medicine men, who have imposed on so many that people have become sick of the whole subject and thrust it out of mind as far as possible. But my experience is, that when I succeed in getting a number of hog raisers together and discuss the diseases of swine, and especially swine plague, in the light of recent discoveries, proving my position from what they themselves know, every one believes as I do.

KANSAS.

Mr. J. Hodgins, Centralia, Nemaha County, says:

I have perused with much interest the copy of report of investigation of diseases incident to domesticated animals, and must cordially thank you for the able manner in which you have brought forward and presented to the public a subject of such vast importance to our people. For over thirty years I have been engaged in raising stock, and have owned all classes of farm animals, but have been so fortunate as to lose but few by disease. In 1866, during an extreme cold spell of weather, it was reported to me that sixty hogs had died of disease in the mill-yard at Centralia. I went there to investigate the matter, and asked the owner if he knew what caused the disease or death of the animals. He replied that it was the "cholera"; but I found that one hundred and fifty hogs had slept in one mass or pile, and that the under ones, in coming out in the morning, had cooled off too suddenly. This brought on the disease of which they died. At the Seneca feeding-yards the same occurrence was noticed three or four times under similar circumstances. At another time I went to investigate cases of reported cholera. I found the hogs in a deep, filthy mire, with no dry sleeping or feeding place. They were fed exclusively on corn, and had no salt or clean water. My conclusions were that hogs should have clean pens, an occasional change of food, salt, and coal, with but few in a herd.

In the early history of our agriculture such diseases were unknown, because, perhaps, the treatment of the animals was then different. We must go back to first principles. I doubt not but we will continue to have contagious diseases, but the best remedy will be found in preventives.

I have had some experience and loss among young cattle by a disease known as black-leg. My remedy is a good one, and almost infallible if made in the early stages of the disease. It is to bleed the animal in the hind feet, by cutting a perpendicular gash between the hoof and the dew claws; then rub down the animal's legs until the blood travels freely. Many cattle are annually lost here by pasturing and feeding them on corn-stalks. They are left in too long at first. Having recently been taken from the grass, they glut themselves on the dry food. At first they should not be left in over an hour at a time, and then turned to water and salt. About two hundred head were lost in this county last fall by this carelessness in feeding.

My children raise from 400 to 700 fowls every year, and have for the past ten years. We have never lost a fowl by disease, while many of our neighbors have lost all they had. We give them clean quarters, plenty of food, and clean, fresh, cool water, charcoal, and burned bones. When a fowl is set on eggs for hatching we sprinkle one teaspoonful of sulphur in the nest.

Mr. T. G. Hunt, Chetopa, Labette County, says:

During the prevalence of the so-called hog cholera here one of my neighbors tried an experiment with the disease in the way of a remedy. In a herd of about one hundred and twenty head there were about thirty sick. Several animals died before he made any attempt to relieve them. Afterward, when my hogs were attacked, he told me the remedy he was using, and we both succeeded in arresting the disease with all that would eat. The remedy was as follows: Corn scorched until about half charred, and then ground into a coarse meal. Three and one half bushels of this was put into a common coal-oil barrel and mixed with water. To this was added one-half pound of niter, one pound each of flour of sulphur, copperas, and rosin, and four pounds of soft soap. Of those that would eat this preparation not one died, and the disease soon disappeared.

KENTUCKY.

Mr. J. W. Darrow, Vanceburg, Lewis County, says:

Through the politeness of Colonel Rand I have been furnished a copy of your valuable and exhaustive report on diseases of swine, &c. I do not so much approve the course of the examiners in hunting down this disease as I do their efforts to find a remedy. Still I am aware that a majority of people would indorse the course followed and contend that it was necessary to define the disease before attempting a cure. This swine plague, as the commissioners have seen proper to name it, is more like consumption than any other disease I am acquainted with. I have been experimenting with it for the past four years, and have come to the conclusion that I have at last found a preventive for the disease, that is, I can take a herd of swine and prevent the plague from attacking them with only a fence between. Before giving my remedy I wish to go back some forty years, when we had no swine plague. Then every farmer had a few hogs which he allowed to run in the woods until late in the fall. They were then put up and fattened in the course of two or three months and no disease made its appearance among them. Now we confine them to close quarters, and fatten them artificially. They are not allowed to run at large, and if they were we have so many of them that there are not enough such things as burdock, poke-root, &c., for the large number of hogs raised, for you know that every good farmer exterminates every such weed he finds growing on his place. When a boy, some forty or fifty years ago, I have seen swine dig down from one to two feet after burdock and poke-roots—not one hog alone, but from two to a dozen at a time. Each would no doubt get a small quantity of the root, perhaps a sufficient quantity to satisfy its desire. In the absence of these roots my plan is to raise parsnips, which seem to contain like preventive qualities with the above-named roots. Two bushels of these roots are perhaps enough for twenty-five head of hogs. I cut them into small pieces and feed them in the spring of the year. I have been keeping this remedy, or rather preventive, to myself, having communicated it to but one man, the Rev. Mr. Martin, now a resident of Ripley, Ohio. He came to me one day and said that his hogs were all dying of cholera. I gave him some parsnips and directed him how to use them. A few days afterward I asked him how his hogs were getting along. He replied that the parsnips acted like a charm. In the town of Vanceburg, during last winter, three-fourths of all the hogs in the town died of the disease. Out of a herd of a dozen I lost one. This occurred before I knew that the disease had invaded my herd. You will doubtless think strange that I did not help my neighbors. My answer is that I had but a few roots on hand. I planted parsnips three different times, but the seeds failed to germinate. There was still another consideration—I was waiting for a reward to be offered for a preventive of the disease. I do not claim that I can cure the so-called hog cholera, but by feeding these roots occasionally the disease will be prevented. As there is no use hiding our light under a bushel, I have concluded to give this information to the public. You are at liberty to use it in any way you please. Should I happen to prove a benefactor, I hope to hear from you again.

MARYLAND.

Mr. Wm. Douglas, Baltimore, says:

Accept my sincere thanks for the report received from your department, in which I find an article treating of pleuro-pneumonia in cattle. I am farming and dairying near this city, where this disease has been a great scourge. Since reading your work on the subject I have vaccinated with the best results. If our laws were as stringent as they are in England, of which I am a native, the disease would soon be eradicated, as it is not so terrible here as I have seen it in the old country.

I have just lost one hundred head of hogs by the disease known as cholera.

MICHIGAN.

Mr. John Barber, Edwardsburg, Cass County, says:

I have read the work on diseases of swine and other domestic animals, kindly sent me by you, and I desire to say that I regard it as a very valuable book. I am very glad the government has at last come to our assistance. Farmers as a class do not expect appropriations in their behalf, and this move comes to us unexpectedly though thankfully. The book is highly prized by farmers, and many good lessons can be learned from it, as it contains the experience of many persons. I am especially interested in the work, as I have for many years past handled a large number of hogs. I commenced raising hogs nineteen years ago, and have continued in the business ever since. My books will show that I have handled on an average two hundred and fifty head annually of my own raising, besides a large number that I usually buy for the purpose of fattening. I never lost any worthy of mention until last summer. As long as I took charge of them myself I never lost any. For ten years past I have kept a large trough or box under an open shed where my hogs could get at it at all times. In this box I always kept a mixture of ashes, sulphur, salt, and copperas, say one bushel of ashes, four quarts of salt, one quart of sulphur, and one-half pound of copperas, all thoroughly mixed and renewed as often as eaten up by the animals. The amount of this preparation that they will consume in a year is surprising. I keep clean fresh water where the hogs can get at it as often as they want to drink. In addition to the above I also feed about one ton of charcoal per year. I pay 12 cents per bushel for it, and call it cheap at this price, as it is worth more to me than the same amount of corn. I have a place for the charcoal near where the hogs get their water. I keep it here constantly, and often sprinkle salt and lime over it. They seem to eat it with a good relish. I am of the opinion that this treatment has kept my hogs healthy. My animals are of the Poland China breed, and I have found them more profitable than any crop I have ever raised. They bring me more clear money for the amount invested than any other stock.

I am satisfied that this swine disease is contagious, but mine took it without any possible chance of getting it from other hogs. The germs may have been carried through the atmosphere as was the epizootic epidemic among horses a few years ago. But, as I said before, I never had the disease among my hogs as long as I attended to them myself. Two years ago I moved off the farm and left my hogs in charge of a foreman or hired help. I had about two hundred head of as fine hogs as any one would wish to see. They were of the Poland-China breed, and were smooth, healthy, and thrifty. My man did not keep up the practice of salt, ashes, and sulphur, and charcoal and lime, as I had done, but dropped suddenly from the practice. In the fall, after they had been fed all the corn on twenty acres of first-class corn land they began to die. I procured the best remedies I could get, and also employed a hog doctor, as he termed himself. Together we devoted all our time trying to save the animals. We did about everything we ever heard of any one doing, but all to no purpose. We lost all except nine head, and those that were left never afterward amounted to much. I cleaned out the pens and yards and used disinfectants plentifully. The following spring I purchased sixty head of hogs which I knew to be healthy, but they, too, soon commenced to die. I lost only a part of this lot, as some lived and did well. I am of the opinion that pigs bred on the farm will live and do well, but that those brought from other farms will take the disease.

I am a firm believer in preventives, but have no faith in cures except by bleeding. If that could be done, I think it would result in saving some of the animals. My reason for saying this is, after my hogs had nearly all died, and I was completely discouraged, I took a hog that, to all appearances, was beyond the possibility of recovery, and cut off its ears and tail. It bled profusely, got well, and afterwards had a fine litter of pigs, all of which lived and did well.

I have a brother who lived with me for some years and helped to take care of the hogs. He afterwards removed to and located on a farm in Iowa, where he commenced to raise hogs extensively. He followed my plan by always keeping before his hogs the preventives I have mentioned. He also sprinkled slaked lime and a little salt on the ground just before feeding. On this he dropped the corn and compelled the hogs to eat and snuffle it up their nostrils. And now I will tell you what I know to be the fact, for I was there myself and witnessed what I am going to tell you. Mahaska County, where he resides, is a large hog-producing county. During the time alluded to the farmers of the county lost nearly all the hogs they had. On every side of my brother's place the hogs were dying. We could scarcely travel through the neighborhood for the insufferable stench from the dead animals. They died so rapidly that the owners did not attempt to bury them. One man, within a half mile of my brother's place, lost seventy large, fat hogs, and in the immediate neighborhood about all the hogs died of the disease. But my brother did not lose a single animal. He was the only one in the neighborhood that used the preventives I have mentioned. People came to look at his hogs and declared that they could not escape, but they did, and I believe it was all owing to the preventives he used. I also believe that my hogs were saved in a like manner, and that I never would have lost any had they been kept supplied with the preventives.

MISSOURI.

Mr. E. G. Halsey, Thompson's Store, Audrain County, says:

I have carefully read your report on diseases of domesticated animals, and must say that while those who conducted the investigation did apparently all in their power to discover the nature and cause of the malady known as swine plague, their work so far amounts to but little, inasmuch as no specific course of treatment or remedy within the reach of all, or that can be applied by ordinary farmers, is given. I hope the investigation will be continued. We have too much at stake in the hog to give the matter up without expending every possible effort to discover some remedy by which the plague may be stopped. It may, as some of the examiners intimate, be expensive, but let it be made. I have experimented to some extent with the plague myself; have tried almost everything, and have found but one remedy that seemed to do any good. For it I can say this: That I never lost a hog that lived thirty-six hours after taking the mixture, and I never had one to get sick after beginning to doctor the herd. But, then, as one of your examiners says: "They might have got well anyway, and the well ones may not have got sick." The remedy is simply this: Three pounds of common dog-fennel to one gallon of water; boil for one hour, and then take the sirup thus obtained and give one-half pint twice a day to the hog. The best way to administer it is to mix with mush or slop. It does not cost much to try it, and nothing is lost if it fails. I do not know the chemical properties of the plant, or how it operates on the system of the hog; but I do know that my hogs got well when I gave it to them and died before I tried it. At the risk of being denounced as an old fogy I send it to you, believing that all who try it will be satisfied.

Mary D. Inloes, Baker's Grove, Barton County, says:

I am very thankful to you for a copy of your special report No. 12. I have given it a thorough perusal, and was very much pleased with the information it gives concerning the treatment and management of farm stock. You will perhaps think it strange for me to address you on such a subject, but as I desire you to know what I know I will make no apology for so doing. I live on the border of Barton and Vernon Counties. We have a good country and fine outlet for stock, but are not exempt from the so-called hog cholera. When we find our hogs getting sick, we put them in a pen where they are well sheltered from wet and inclement weather. We then give them clean, pure water to drink, and make a swill of wheat-bran, to which we add one tablespoonful of copperas and three of sulphur, a handful of salt, one teaspoonful of blood-root, and a small quantity of charcoal. This preparation has invariably cured our hogs. We continue it, giving it twice a day, until the animals have recovered. My neighbors have used the preparation with like satisfactory results.

Some time ago I had a very sick mare. I had been keeping her in a very warm stable, but she got exposed to a cold rain and contracted a severe cold. This brought on lung fever. When taken she had a severe chill, which was followed by high fever. She lost appetite, her head drooped, and her chest in the region of the lungs sunk in. I blistered her over the lungs with a preparation of alcohol and turpentine, rubbed her legs from her knees down to her feet, and then gave her one ounce of aloes. Then I boiled one-half peck of oats, put them in a sack and put the sack over her head, drawing it as far down as I could. As soon as she commenced to tremble and stagger I took it off and wiped her face dry. I gave her nothing but boiled oats and wheat-bran mashes to eat, and tepid water to drink. She is now well and hearty, and ready for service whenever wanted.

I have used the following recipe for bots with good results: To a bran-mash add a teaspoonful of blood-root and about one tablespoonful of pulverized alum, and mix well. This cured one of my horses that was badly affected with bots.

Chicken cholera frequently prevails here. All we do for them is to give them slaked lime and put copperas in their drinking-water. This is a good preventive, if not a cure.

Mr. W. J. Beasley, Four Mile, Dunklin County, says:

After a careful perusal of your report on the diseases of swine and other domesticated animals, I am fully satisfied that it is a most valuable work, and if the suggestions it contains are carried out the result will be a saving of a large per cent. of the animals which are now annually lost by the various diseases common to them. This will be more especially the case with swine. I have myself experimented somewhat with diseases of swine, and generally with good success. At first I tried many different remedies, but without any success whatever. My last effort, which was with a preventive rather than a remedy, was very successful. It was with sour slop. I took a barrel that held about sixty gallons, which I filled with wheat-bran, and then added water and let it stand until it had soured and thoroughly worked. I then added one bushel of corn-meal. I let this stand two or three days, and then added sufficient copperas to give the liquid a pale yellow color. Then I added from three-fourths to one pound of pulverized sulphur; and, while stirring this thoroughly, added sufficient lime to make the solution feel slick when tried between the fingers and thumb. I fed my hogs regularly with this solution about three times a week, and continued it as long as any of the animals appeared sick or droopy. I have had nothing to complain of since using this preventive. It causes the hogs to shed off nicely, and their skin and hair keep smooth while those that are deprived of it become rough and mangy. While I am satisfied no cure can be found for the disease, I am of the opinion that there are a number of preventives. So far as the contagiousness of swine plague is concerned, I have for years been satisfied that it can be carried from neighborhood to neighborhood, and from one section of the country to another, on the wing as well as on foot. It prevails in this county more or less every year, and some seasons its ravages cause very heavy losses.

Dr. Fred. J. Ballard, Independence, says:

I have had an opportunity of investigating a case of the disease which is so prevalent among the hogs in this section of the country, misnamed hog cholera.

The investigation was necessarily limited, but I will try and give a plain and simple statement of the case. The hog examined was of Berkshire breed, and weighed about sixty pounds. It was brought to the farmer's house, together with two others, before it was weaned. These hogs have had the liberty of the yard connected with the house, been fed on kitchen slops, and have had free access to corn, and were otherwise well taken care of; had good water and a dry shelter.

The most prominent symptoms noticed were a disinclination to stir or move around, loss of appetite, with a continued tendency to sleep or drowsiness. This condition lasted for about a week or ten days, when death ensued in a very easy manner. The hog I examined died October 10, and on the next morning the following facts were developed:

On opening the abdomen, the first of a morbid character that was noticed was a very extensive inflammation of the bowels, showing a highly vascular action of the circulation of the serous surface of the smaller intestines or jejunum. The liver was somewhat enlarged and almost black, while here and there tuberculous deposits, varying from the size of a bird-shot to that of a buck-shot, filled with a clear watery substance. The kidneys were normal, and nothing could be found in either the cortical substance or the medullary. In the pelvis of the kidney and the ureter some albuminous urine was found.

On closer examination of the inflamed patch of intestines was revealed the cause of the hog's death, viz: A large globular mass of organic growth, about six inches in its anteroposterior diameter and a little less in its lateral. This mass had the appearance of a cancerous growth, *schirous* in appearance, enveloping the mesenteric glands, and almost obliterating the bowels which passed through the mass. It also embraced several convolutions, and in their track or mucus surface was found a dark-colored substance that looked like coagulated blood. Posterior to this globular mass the bowels were nearly empty; nothing was found in them but ordinary mucone, and one worm of the long, round variety. The colon or large intestine was apparently in a normal condition, but the rectum ruptured very easily. Some of the muscular tissue was examined and found to be healthy.

On opening the thorax the lungs and heart were next examined. The pericardium was in a very tender state, so that it ruptured very easily. The lungs, also, were in a bad condition: all of the left and half of the right were in a condition approaching gangrene, and with decided crepitation, while the pleura, both pulmonalis and costalis, shared in the same general inflammatory condition. The brain was not examined.

The microscope revealed in the lung, after a maceration in acetic acid of three days' duration, and about the same time in alcohol, numbers of long worms that, under a magnifying power of two hundred diameters, looked to be about one-eighth of an inch thick and four inches long, and could be plainly seen moving about in the field of the microscope. Other varieties existed in the same lung, but owing to the absence of a higher magnifying power they could not be determined.

The above is the result of my investigation.

About the same time Prof. J. Q. Egelstou, of Olathe, published a letter in the Globe-Democrat, of Saint Louis, in regard to a parasite in the lungs, and spoke of them as long, hair-like worms. I claim that I discovered the microscopic entozoa first, and date it October 16, 1876.

To further pursue my investigations, I made it a business to go to slaughter-houses, and, as the animals were disemboweled, made notes of the condition of the various parts, and finally in the lower lobe of the left lung of a fine large hog I was rewarded by finding in a bronchial tube a nest of about six large parasites three-fourths of an inch long, which in all probability belonged to *Cysticirci*.

It was subsequently denied by Professor Riley that a parasite could live in the lung tissue of either a hog or a man. With all due respect for him as an entomologist, I do not think he is much of a morbid anatomist, or he would not differ so entirely from standard authorities.

I sent one of the worms found in the bronchial tube of the hog to Mr. J. Monteith, secretary of the State board of agriculture, with the request that he pass it over to Professor Riley, which he did, and Mr. M. stated that the professor spoke of writing to me, but he never did.

These parasites must certainly produce irritation, the irritation creates the cough so much spoken of by others, and the whole has a tendency to produce the state of the lungs mentioned in my above report.

NEBRASKA.

Mr. H. T. Vose, Syracuse, Otoe County, says:

Within a radius of a mile and a half of my place there has been in one season a loss of at least 1,000 head of swine. One man lost $2,000 worth, others $25, $80, and $100 each. The first fact demonstrated was that corn feeding was not the only cause of the disease, as Mr. George McKee's hogs, the first lot referred to above, were not and had not been fed on corn, but were given mill feed of barley, oats, and wheat, he being a miller. The second point, which seemed to be well established, was that the disease was infectious or contagious, as a small lot put with his herd took the disease. The third point demonstrated was that drug medication was not of much avail. Fourth, that the germs of the disease were carried by the water-courses from place to place. Fifth, that safety from contagion was to be found only in the isolation of each lot. Sixth, that the abnormal appetite for excrementitious matter indicated a need, and that this need seemed to be satisfied by feeding charcoal and animal food. Concluding from the trichinia characters, as well as from the vermicular presence in swine, that worms of some sort might be the cause or a chief incident of the disease, I have fed meat to engorgement and as a cathartic, thinking that these parasites would detach themselves from the organism of the hog and attach themselves to the animal food, and then, by the purging action, be expelled from the system. I gave charcoal as an absorbent and neutralizer, and lastly, when I could not get the animals to take any food, I gave them ice, which they would readily eat; and thus I gave both drink and food, the first of which proved an active agent in the reduction of fever and amelioration of its effects. As meat food, I have given rats, rabbits, pigs (not diseased), hens, the offal of healthy animals, &c., and I place more reliance on this food, with charcoal or dead, fine coals and ice, than any or all kinds of drugs so called. I have now on hand five animals that were seriously affected with the disease. They bled from the nose, vomited, ate excrements, shivered, nosed in the ground, arched the back and shrank the sides, heaved the flanks, &c. They were treated as above indicated, and now seem to be in a fair way of recovery—are eating heartily and gaining rapidly in every respect.

Mr. William D. Davis, Xenia, Sarpy County, says:

I learn from your valuable report on diseases of domesticated animals that black-leg prevails to a great extent in many localities. Having been engaged in stock-raising in this county for the past twenty years, during which time I have lost a great many young cattle by this disease, I wish to communicate to you the result of an experiment tried by myself and others. In January last I had three calves attacked with every symptom of black-leg. They were lame in the fore-leg, the hair stood straight out, and there was great swelling of the body. Having tried every known remedy without effect, and as a last resort, I administered to each animal one tablespoonful each of saltpeter and carbonate of soda. This gave them relief in three or four hours, and they recovered. While I do not claim this as a certain remedy, it is certainly worthy of further experiment. I shall treat further with it, as occasion may offer, until I am satisfied of its virtue. The calves referred to were strong, robust animals, about nine months old. I wean my calves as soon as grass begins to dry, afterwards feeding them corn and hay.

NEW YORK.

Mr. Thomas H. Morris, 94 Reade street, New York, says:

I send you the following recipe for glanders in horses, which, in my experience, has proved very efficient as a blood purifier in cases of inflammatory diseases: One spoonful each of sulphate of zinc and yellow rosin, two spoonfuls of salt, and four spoonfuls of hickory ashes. Pulverize, sift, and mix thoroughly, form into balls with molasses, and give once in three days, or give dry in feed if readily eaten. I have steamed animals with this preparation when they were unable to swallow, and have used it with good effect in heaves. It is also a good preventive and frequently cures distempers. I have used the above preparation so often that I have no hesitation in recommending it to others. When given on the first appearance of disease it will be found to have an excellent effect.

NORTH CAROLINA.

Dr. C. Wells, Concord, Cabarrus County, says:

As a preventive of hog cholera mix a tablespoonful of turpentine with a handful of cornmeal for each hog, and give once every two weeks. After the turpentine has been mixed with the meal, sprinkle a little dry meal over it, and the hogs will then eat it readily.

OHIO.

Dr. J. S. R. Hazzard, Springfield, Clarke County, says:

The so-called hog cholera has devastated many herds in this part of the county. My hogs were attacked for the first time this winter. Several died before we were aware that the fell monster was upon us. His strokes were rapid and deadly, but we are persuaded we checked the ravages of the fatal disease by daily administering bromo-chlorallum not only to the sick but to the whole herd. I gave about a teaspoonful two or three times per day to each hog. During the incipient and fever stages they would drink seemingly to slake their thirst. Several well-developed cases recovered, and several escaped the disease entirely. Two animals are still lingering with prospects of recovery. Poland Chinas succumbed to the disease, but crosses between Poland Chinas and Jersey Reds more readily. None of the pure-blood Jersey Reds have yet died, but two are sick with prospects of recovery, as above stated. Bromo-chlorallum is a good disinfectant, antiseptic, and alterative. My hogs were all taken with a chill, and the first ones attacked had black, tar-like discharges, but none suffered in this way after taking the medicine. Whether this remedy will be effectual in staying the ravages of this terrible pestilence, experience alone can determine.

PENNSYLVANIA.

Mr. N. B. Willierd, Loyalton, says:

During last April and May our horses suffered with a disease unknown here previous to that time. The symptoms were a loss of appetite and flesh, followed by swelling of the throat. In my opinion, the disease was brought on by exposure to inclement weather during the past winter (1878-'79). It resembles somewhat diphtheria in the human family. I used linseed oil mixed with spirits of turpentine as an outward application. In addition to this, I took live coals and sprinkled rosin over them, and compelled the animals to inhale the vapor. The first case I had was with a horse that was being used to harrow in oats. When I stopped, the animal commenced to shake as with an ague. I immediately bled him on the tops of the hoofs and in the mouth, after which I rubbed him all over with a bunch of straw, and then covered him with blankets for half an hour; then I removed the blankets and rubbed him again, and continued in this way until he recovered. All my horses were treated in this way. They were soon able to eat and drink, but they were left in such a weak condition that they were unfit for use for several weeks.

Mr. S. Jenkins, Wyoming, Luzerne County, says:

I am in receipt of your report on diseases of swine and other domesticated animals. There has been considerable "hog cholera" and kindred diseases in this county during the past year. Since receiving your report I have made inquiries of farmers, butchers, and veterinarians, and I find but one opinion prevailing among them. They agree that the only course worthy of attention is that of prevention, by changing place of keeping, cleansing pens and folds, and giving more attention to diet; *that a diseased hog is not worth curing.* Such an animal will never do well afterwards. The attack is likely to return and spread the disease among the healthy; and upon the whole, after the disease has once fairly taken hold of an animal, death by the knife is the best result for all concerned. That a temporary cure in some cases can be effected there is no doubt, but there is no profit in keeping the animal, particularly in the case of the hog, even under the most favorable circumstances, while it is possible that great evil and serious loss may result. The plates showing effects of disease on vital parts, and the recommendations of the investigators, are highly commended by those who have had experience in the maladies and their treatment, and exactly correspond in most respects with their observations. The knife should be added to your remedies.

TENNESSEE.

Mr. R. Douglass, Gallatin, Sumner County, says:

A question of great importance to the farmers is to prevent the loss of sheep by grub. A very simple remedy is to make a strong decoction of tobacco and apply about a half ounce in each nostril with a syringe, while the sheep is on its back. This will cure nine times out of ten. In regard to so-called hog cholera, I think a preventive can be found in crude petroleum. A small quantity of this should be put on the corn or in the slop about once or twice a week. In addition to this, a small quantity sprinkled on them will be found beneficial. I bought a barrel, and by using it in this way I have never suffered the loss of any hogs.

Mr. J. T. Richardson, says:

If you desire both a preventive and a cure for the so-called hog cholera, I will give it to you. This remedy, if simultaneously used, and should prove as effective with others as it has with us, would exterminate the disease from this country. Strong language, you will say. All right. I will tell you why I use such strong language, and let you judge for yourself. Previous to 1868 I became so discouraged in trying to raise hogs that I seriously contemplated raising mutton and beef as a substitute for pork. I had tried every remedy and preventive to no purpose, and had discarded everything of the kind. But a neighbor of mine, who was a large hog-raiser, lost no hogs, and he finally told me that I need lose no more if I would follow his directions and use a certain remedy. I did so, and for ten years neither I nor others who have used the remedy have lost a single hog by the disease, unless in cases where the remedy was not used. The remedy is as follows: When you find that the disease has invaded your herd, take a half-bushel basket and go out and dig it full of poke-root. Chop the roots up and put them in a kettle of water. Put into this, also, three or four hands (some twenty or twenty-five leaves) of tobacco. Add a peck or more of red-oak bark, and boil until the poke-root becomes soft and mushy, like boiled potatoes. Then put in shelled corn, and after it has become completely soaked feed it to the hogs. The quantity thus prepared will be sufficient for about forty head of hogs. Every sick or well animal that eats the corn thus prepared is safe. Feed every day until, from the increased appetite of the sick animals, you can pronounce them out of danger. We care nothing about the so-called hog cholera except the trouble of fixing and administering this preparation. The fear of losing hogs by the disease does not disturb us in the least.

Mr. E. Link, Greeneville, Greene County, says:

I have given the report on diseases of swine a careful perusal, and have gained much valuable information thereby. I have had the disease therein described among my own hogs four or five different times. Generally it has more seriously and fatally affected pigs and shoats than older animals. I was assiduous in trying remedies recommended by those who had had more to do with the disease than myself, but generally with most indifferent success. I am now decided in the opinion that specifics which favorably affect some cases will prove wholly powerless in others. Once, in a lot of over twenty shoats, several of which had already died at the rate of one a day, and several others were ailing, I think all would have died had I not fed them chlorate of potash. I think I fed them a fourth of a pound in meal at two feeds, and they all recovered rapidly. If in this case the blood needed oxygenizing, I accidentally hit upon the remedy. At another time the same remedy, about the efficacy of which I was quite sanguine, proved to be no remedy whatever. On another occasion I found a speedy remedy in calomel—an ordinary dose for a man. Afterwards, in other attacks on my swine, I found this of no avail. I regret to say that, outside of the belief that remedies affect variously or not at all the same disease under different conditions, my conclusion is one in which nothing is concluded.

TEXAS.

Mr. C. A. Beeman, Hamilton, Hamilton County, says:

We are not troubled with the swine disease here, but our cattle, especially the female portion, and more particularly cows that have had calves, are affected with a disease for which we have neither name nor intelligent treatment. The animal is first affected in one or more of its legs, which become stiff at the joints. This gradually increases until the cow becomes very lame and poor. The animal finally dies or wears out the disease and recovers. But few, however, recover. I have a nice heifer, with her first calf, now suffering with the disease. She has had it for three months or more. She does not seem to be getting any worse, but is nothing but a pack of skin and bones. She has a good appetite, but drinks very little water. The department would do a good thing if it would make an investigation of this disease, in order that a remedy might be found for it. The disease is very common in this section of the State.

Mr. J. C. McCaskill, Clinton, DeWitt County, says:

I wish to report what I find to be a certain specific for diphtheria in hogs, as I recently lost some of these animals by this disease. A short time ago I discovered one of my hogs in almost a dying condition. It was scarcely able to get its breath, and upon being handled or turned over would have terrible spasms. I took pulverized sulphur and sweet milk and poured it into its mouth, but believing it too far gone to swallow, we made a sulphur match, lit it and held it to its nose and made it breathe the smoke. In a few minutes its breathing became quite easy, and after the second application it got up and walked off, and is now apparently in as good health as any hog I have.

Louis Brandt, V. S., Mayersville, DeWitt County, says:

I learn from a Pennsylvania paper that in the herd of Mr. John Kraus, of Krausville, pleuro-pneumonia has made its appearance, and that as many as fifty persons attended the *post-mortem* examination of one of the cows that died of the disease. This disease is so contagious that if any one of these neighbors who attended the examination should go into their own stables wearing the same clothes, without having previously been disinfected, they would most certainly communicate the disease to their own cattle. This disease, however, is just as easy to cure as any other if attended to in time. Killing the infected animal is an unnecessary loss, both to the owner and to the State. The disease is so contagious that the strictest quarantine should be maintained against an infected stable or premises. The least infected animals should be kept in a separate stable, and both stables should be daily disinfected and the sick animals properly treated. My father lost 150 head of cattle by this disease, and he sent me to Hanover (Germany) to the veterinary school there. The disease prevailed in that city and surrounding country. I treated many cases, and whenever I commenced in time I was successful.

The disease known as murrain in this locality is regarded as infectious; but it is not. In this disease the kidneys are affected. Instead of secreting urine only, the animal loses a great deal of blood; and without removal of the cause, death is certain. If taken in time, and the proper remedy used, the disease can be cured in a few hours.

VIRGINIA.

Mr. R. Saunders, Dumasville, Essex County, says:

My first experiment in treating hog cholera was with an apparently very healthy lot of six months' old shoats, some of which had been taken suddenly with spasms. I took one and bathed it well with spirits of turpentine, which appeared to relieve it, and it finally recovered. But it never became thrifty, or of much account. My hogs were kept on pasture, and I did not see them except at feeding time in the morning and evening. Whenever I found one drooping and apparently sick, I would administer the same treatment, and also give the turpentine in slops. But I had poor success with this treatment, as four-fifths of all attacked died. The animals had good pasture, and an abundance of pure running water. I have found corn roasted on the cob very good in correcting acidity of the stomach. I also occasionally gave concentrated lie diluted with water, about a tablespoonful to each hog. Since using these articles I have not lost a single hog. As these preventives are within the reach of every farmer, I would advise their general use.

Many persons are of the opinion that the disease is brought on by want of proper care and attention. This I doubt, as it seems to be equally if not more destructive among those that have careful attention than among those that have no attention at all. On many farms adjacent to others seriously infected, and with about the same care and attention, the disease never makes its appearance.

Mr. R. P. Buckam, Salem, Fauquier County, says:

I have found that a liquid from leaf-tobacco mixed with swill is not only a preventive but a cure for hog cholera. I have been using this for eleven years, and have lost but one hog during that time. The use of the tobacco causes the animal to break out on the skin with sores or pustules.

Mr. J. W. Alexander, McDowell, Highland County, says:

The following will be found an effective prescription for rot: Eight ounces of common salt, 2 ounces of powdered gentian, 1 ounce powdered ginger, and 4 ounces of colombo. Put the whole in a quart bottle, and fill with water. Give one tablespoonful morning and night for one week; then add to one quart of the above 3 ounces of spirits of turpentine, and give of this two tablespoonfuls at night, continuing the other each morning.

Dr. R. A. Belfield, Richmond County, says:

I must say that your report on diseases of swine and other domestic animals is one of the best prepared and most valuable works ever issued from your department. The information it contains will be worth millions of dollars to the people of this country, as it covers subjects of such vast interest. We have had but very little cholera among hogs in this county. Where it has occurred has been among hogs that were allowed to range on low marshes. The disease is rarely met with on elevated lands.

WEST VIRGINIA.

Mr. W. H. Euess, Glen Easton, says:

Allow me to say that your report on diseases of swine, and ailments incident to other classes of farm animals, is a very valuable and carefully-prepared work, and will be of great value to the farmer if the advice contained therein is heeded. My experience is that the farmers as a class do not read such works and profit as they should by the experience of others; the fact is we do not read enough. If stock raisers would pay more attention to preventive measures, to keeping their pens and stables clean and well ventilated, there would not be so many diseased animals. If we desire to preserve this great industry the matter must be grappled with vigor and prosecuted with energy. I hope, now that a beginning has been made, that our legislators will take such steps as will effectually put a stop to the spread of swine plague in this country, and also stop the sale of diseased hogs in the markets by unprincipled men.

WISCONSIN.

Mr. Edwin Reynolds, Metomen, Fond du Lac County, says:

Accept my thanks for a copy of your valuable report on diseases of swine, &c. While our farm animals are measurably free from disease, there are a few which require attention. First, goiter in lambs, contracted before birth, is the most prevalent and most destructive of any of the diseases which afflict our animals. The farmers of the Northwest have, during the past fifteen years, put forth many efforts to discover the cause and a remedy for this malady. As high as $5,000 has been offered as a reward for a remedy or preventive of the disease. Many flock masters have lost nearly all their crop of lambs by it. I have concluded, as have many of my neighbors, that locality and decayed matter in sheds and yards, with too close confinement, has much to do in producing the disease. A neighbor has a large bank barn facing the south. West of the barn is a shed 24 by 60 feet, running south on rather low ground. There is decomposing matter under the barn, and the ewes seldom have exercise during the winter. The result is that he loses nearly all of his lambs. Two years ago he sold off all his stock and purchased anew from other flocks, hoping by so doing to get rid of the goiter. But this proved of no avail, for his first crop of lambs were infected as badly, if not worse than his former crop. This is but one case out of a number. I have high and dry yards, well-ventilated barns, and good water of easy access. I feed sulphur and salt regularly, and am but little troubled with the disease.

Garget in cows prevails to some extent. This trouble is induced by allowing the blood to become impure. As there is no skunk-root growing in this locality, stock cannot help themselves to any cleansing substance. I procure skunk root of the druggist, and feed it to my cows and other horned cattle. As soon as I find a cow giving bloody milk, I wound and insert a piece of this root into the brisket deep enough to make a running sore. This soon cleanses the system, and in a few days the cow will be all right again.

For milk fever, saltpetre is a sure cure if administered in time and in sufficient doses. This fever is induced by too great a flow of milk before and after calving. Too much care and attention cannot be used in keeping cows in good health, as a sick cow gives impure milk, impure milk makes impure butter, and in all cases of impure food disease is induced in the human family.

PREVALENCE OF DISEASE AMONG FARM ANIMALS.

Answers to the usual circular letter of the department indicate a partial abatement of some of the more destructive diseases which annually prevail among farm animals. Swine plague, which has been more destructive than any other disease to which domesticated animals are subject, has not prevailed so extensively as in former years, nor has it been so fatal in those localities in which it has prevailed. The sanitary and hygienic measures recommended by the examiners in special report No. 12 seem to have been closely observed by many farmers, and the result is a greatly diminished mortality among swine. Still, in a few isolated localities, the losses from this disease have been enormous, one county alone in Missouri (Platte) reporting a loss of $200,000 for the year. Several other counties in the Northwestern States report losses ranging from $10,000 to $100,000.

There has been but little, if any, abatement of the disease known as chicken cholera. The result of recent experiments has satisfied M. Pasteur, of France, that this disease is charbon; that it is communicated by a microphyte, and when prevailing as an epidemic is highly contagious. M. Pasteur's experiments prove that the microphytes, which are the seeds of the disease, are taken into the stomach on food defiled by the excrements of afflicted fowls, in drinking water, &c., and are then taken up by the blood and carried to and lodged in all parts of the system. As no remedy has been discovered for the disease, a liberal use of disinfectants, a separation of the sick from the healthy fowls, and the observance of proper sanitary regulations are recommended as the only means for preventing a spread of the malady.

Three counties in Missouri and five in Kansas report a visitation during the year of Texas or splenic fever of cattle. For a full history of this disease, so far as it relates to its appearance and progress in this country, the reader is referred to the exhaustive paper of Dr. D. E. Salman, which will be found elsewhere in this volume

A disease variously known as black leg, black quarter, bloody murrain, &c., prevailed extensively in some States during the year. It is reported as having been very destructive in seven counties in Arkansas, nine in Georgia, two in Illinois, one in Indiana, seven in Iowa, nine in Kansas, one in Kentucky, five in Michigan, two in Minnesota, five in Nebraska, two in North Carolina, one in Pennsylvania, four in Tennessee, five in Texas, five in Virginia, two in West Virginia, and one in Wisconsin. This disease is malignant anthrax, and generally proves fatal, as it runs its course so quickly that its victims are usually found dead before the first indications of anything amiss are observed. Its symptoms are an extensive engorgement of a shoulder, the neck, breast, or side. Young and thrifty cattle are those most frequently attacked. Afflicted animals show plethora, fever, halting on one limb, stiffness, and excessive tenderness of some parts of the skin, followed quickly by swelling of such parts, with yellow or bloody oozing discharges from the surface, emitting a crackling sound when pressed. "These swellings become firm, tense, insensible, and even cold, and, if the subject survives, may finally slough open, and leave large, unsightly, and inactive sores." Recoveries, however, seldom occur.

In a few Northwestern and many Southwestern States distempers prevailed quite extensively among horses and cattle during different seasons of the year. These ailments were generally caused by exposure to sudden and severe changes of temperature. In those States in which proper provision is made for housing farm animals these distempers are of rare occurrence.

Fifty-five counties report the prevalence of scab and foot rot among sheep during the year. In one or two counties each in Michigan and Minnesota a disease resembling goiter destroyed a good many lambs.

ALABAMA.

Calhoun County.—But few diseases of any kind prevail among farm animals in this county except among hogs and fowls. Every disease that attacks the hogs is called cholera, and the same might be said as regards fowls.

Clarke.—In this immediate vicinity we have had several horses and mules afflicted with farcy, an almost incurable disease. The animals mostly were brought here from Kentucky and Missouri, and were nearly all killed, in order to prevent the spreading of the disease. Diseases also exist among imported breeds of fowls.

Crenshaw.—Hogs are dying in this county at an alarming rate of cholera. A great many fowls are also dying of the same or a similar contagious malady.

Elmore.—Horses and cattle are remarkably free from infectious diseases, while hogs and fowls are especially subject to a very destructive and fatal disease known as cholera.

Escambia.—We have no disease among farm animals here except among hogs and cattle. The disease affecting hogs is known as cholera; that affecting cattle is not known. The animals become stiff and lame in their limbs. About one-fourth of those attacked die.

Franklin.—Horses in this county have been affected with distemper, from which cause several have died. Hog cholera has prevailed, but not with its usual virulence. The value of the losses among this class of animals will perhaps not exceed $1,500.

Henry.—The disease known as cholera has caused the loss of a good many hogs in this county during the past season. A large number of fowls have also died of a disease generally known as sore head.

Lee.—I think 15 per cent. of all the hogs in the county are annually attacked by cholera, and at least one-half of these die. Many fowls also annually die of this disease.

Monroe.—The only disease prevalent among our farm stock seems to be cholera among hogs. But few hogs are bred and raised in this county, and, the disease not being general, the loss has been small.

Pickens.—We have no infectious or contagious diseases among cattle or sheep, though a good many of each die every year for want of proper care and attention. Diseases of swine prevail to a limited extent.

Walker.—Sheep and cattle have been very healthy during the current year, but we have lost by disease a great many horses, hogs, and fowls. I think $20,000 would hardly cover the losses among these three classes of animals.

ARKANSAS.

Baxter County.—About every other year the cholera makes its appearance among hogs in this county, and generally proves very disastrous to this interest. Every year we lose a few horses by blind staggers, and cattle by murrain. A good many sheep die of rot, and fowls of the disease known as cholera.

Carroll.—Distemper and murrain have prevailed to some extent among our cattle, causing a loss of perhaps $2,000. Hog-cholera has also prevailed, but in rather a mild form.

Crawford.—With the exception of hogs, farm stock has been comparatively healthy during the past year. The disease affecting them is generally known as cholera. It is not very general at present.

Fulton.—Since the early part of last fall and the first of winter hogs have been healthy and in fine condition. Previous to this time the cholera, in its most fatal form, prevailed extensively among them. This disease prevails to a limited extent at this time. In some localities cholera prevails among fowls. Horses, cattle, and sheep seem to be free from disease.

Grant.—A good many hogs have died in this county during the past season of cholera, but at present the disease is prevailing to a very limited extent. Fowls are still very seriously affected by a disease of like character.

Independence.—Horses are affected with big head and big shoulder. These diseases can hardly be regarded as contagious. Cattle suffer with murrain and a slow fever, of which many of them die. A great many hogs have died of cholera. Tar water is thought to be good for this disease. A good many fowls are also lost by cholera.

Izard.—A disease is prevailing at this time among horses called blind staggers, which is very fatal; 50 per cent. of those attacked die. It is caused, in almost if not every case, by feeding decayed and worm-eaten corn, and is not considered contagious. The only contagious disease we ever have among these animals is distemper, and this appears only at intervals of several years. It is severe, and, in a few instances, fatal to colts. The losses among swine have been quite heavy, as usual.

Jackson.—There seems to be no infectious or contagious diseases among horses, cattle, or sheep in this county. Quite a large number of hogs have been lost by diseases incident to swine, and chickens have suffered to a considerable extent with a disease known as cholera.

Marion.—The cattle of this county are annually subject to murrain and black leg, of which a great many died. Hogs die at an alarming rate of cholera. At least two thousand five hundred head have been lost in the county during the past season. The disease rages most fearfully during those seasons when there is most sickness among the people. In the healthy pine regions the animals are almost exempt from the malady, but along White River, when there is much sickness among the people, they nearly all sicken and die.

Mississippi.—A considerable number of cattle have been lost by black tongue, and a large number of hogs and fowls by cholera. The greatest mortality among live stock of all descriptions is caused by buffalo gnats and colic.

Montgomery.—The only disease that has prevailed to any considerable extent among our farm animals has been cholera among swine. Chicken cholera prevails to some extent, and generally kills all the fowls attacked.

Polk.—We seldom lose horses here by disease, but cattle frequently die off rapidly with murrain. Cholera is very destructive among hogs. I should think that at least one-half of these animals are annually lost by this disease. Great losses frequently occur also among our fowls. The disease affecting them is also called cholera. The losses are sometimes estimated as high as two-thirds of the entire crop.

Perry.—Swine are the only farm animals in this county that can really be said to be affected by either contagious or infectious diseases.

Pope.—The principal disease affecting horses in this county is known as blind staggers. Almost every animal attacked dies. Those afflicted are well fed and generally in good condition. Murrain prevails among cattle, and about one-half of those attacked die. Swine plague prevails extensively, and about all die that are attacked by the disease. No remedies seem to prove of any value. Drs. Detmers and Law describe the disease very accurately.

Prairie.—No contagious disease prevails among horses. Every few years the distemper passes through the county, but the loss is never very great. Cattle are more or less affected with bloody murrain every year. About three-fourths of those attacked by the disease die. It is supposed to be caused by bad water. We have had no hog cholera of any consequence for two years. Chickens are subject to cholera here as elsewhere.

Sevier.—We have had but little hog cholera in this locality during the past season. The remedy I sent you three years ago, composed of copperas, sulphur, salt, and soap, has proved very successful here.

Saint Francis.—Some horses have been lost in this county during the past year by blind staggers. Cattle have been almost entirely exempt from disease. A few have died of a swelling at the root of the tongue and an affection of the throat, but the disease was not regarded as contagious. Hogs have been badly affected with cholera and quinsy, but I do not think the losses have been as heavy as usual. The condition of all classes of farm animals is about an average.

Sharp.—No kinds of domestic animals in this county are affected with disease except hogs and fowls. The disease which affects these classes is generally termed cholera. It prevails once in about every three years.

CALIFORNIA.

El Dorado County.—Pneumonia among horses and quinsy among hogs are about the only diseases that seriously affect farm stock in this county.

Plumas.—The only disease I have to report among any class of farm animals in this county is an epizootic distemper among horses. Several animals have died.

San Bernardino.—Sheep are almost all troubled with scab, which destroys probably 5 per cent of the young ones. Fowls are sometimes troubled with swelled head, which frequently proves fatal.

San Diego.—In the fall, when food is scarce, a good many horses are killed by eating poison weed. During dry seasons cattle are generally affected with murrain. Sheep are not much subject to disease, neither are hogs, but many of both classes are often lost by starvation. A disease called "swelled head" prevails among fowls during very dry seasons.

San Joaquin.—The only disease that afflicts our sheep is scab, which prevails throughout the State. It yields readily to proper treatment, hence the losses are not very heavy. Fowls are subject to roupe and gapes, diseases contracted by being confined in filthy places, drinking impure water, &c. Sulphate of copper added to their drinking water twice a week, and sulphur to their food an equal number of times, with clean quarters and not too many confined to one place, are sure preventives of disease.

Yuba.—I cannot learn that any animals have died in the county from infectious or contagious diseases. Some few cattle have died from eating poison weed that grows in some localities. A few head of horses have died from different causes, but from common diseases peculiar to this climate. But little disease has prevailed among sheep during the season. A considerable number of fowls died during the past spring from roupe, which is quite common among chickens especially.

CONNECTICUT.

Windham County.—Out of a hundred head of hogs owned by a gentleman of this county about 50 per cent. died. 1 think about 70 per cent. of the young pigs in the county died last spring. Other farm animals seem to be in a healthy condition.

DAKOTA TERRITORY.

Hutchinson County.—Black leg is the only contagious disease that has prevailed among cattle in this locality during the past year. It prevails only on the river bottoms, and is uniformly fatal. Sheep are troubled with scab.

Lake.—Horses and hogs are free from disease. In the summer and fall cattle are affected with "foot evil," but none of them die. The only loss seems to be in hindering their growth. A simple and almost sure remedy is to wash the foot with warm water, and then put pine tar on it. This should be done as soon as they are affected. Fowls are affected with rough, scabby legs, and perhaps 8 per cent. of them die while yet in good condition. Stock generally is in good condition.

Yankton.—Horses, when young, are subject to a very fatal kind of distemper. Many others die of unknown diseases. Young cattle die of black leg, and older animals of the big jaw. We have no remedy for either of these diseases. We have lost some hogs by a disease which affects their legs, of which they lose the entire use.

DELAWARE.

New Castle County.—During the past year about one hundred cows have been lost in this county by infectious and contagious diseases. Horses to the value of $6,000 or $7,000 have also been lost, and a large number of fowls have died of diseases to which they are incident.

FLORIDA.

Clay County.—Hollow-horn prevails to some extent among our cattle. As a remedy we bore the horn and insert salt water. The sand disease prevails as usual among horses. This is a collection of sand in the stomach and bowels. As a remedy, one pint of hog's lard is used as a drench. A disease resembling warts prevails extensively among fowls.

Hillsborough.—There is a heavy annual loss of cattle in this county from two causes, viz., old cattle seem to contract disease from feeding near salt water, or die in the winter season from poverty and the effects of ticks. Calves under one year old die in great numbers from some unknown cause. They are never weaned; penned from March to August or September, and are then turned out on the range to live as best they can with the mother. Probably one-third of these young animals die during the winter months. The owner of a herd of 1,500 head told me that his annual loss was not less than two-thirds. His stock is running out very fast, as much, I think, from in-and-in breeding as from any other cause. Very few bulls are imported.

Lafayette.—Horses are generally healthy. A few die of colic and grubs. Cattle are afflicted with lice and ticks to such an extent that many of them die. They have to depend almost entirely on the wild range, and consequently they suffer greatly during the winter. Hogs do well after they are one year old. Young hogs are frequently attacked by mange, a disease from which they rarely recover.

Lancaster.—A disease known as blind staggers has prevailed among horses to some extent in this county. Many hogs die in the inflammatory stage of the specific swine plague. Horses occasionally die of colic. Sheep and fowls rarely die, or are affected with contagious or infectious diseases

Liberty.—The disease most prevalent among horses in this county is blind staggers, which is caused by bad management, poor feed. &c. Cattle die more from neglect than from any other cause. A good many hogs are lost by cholera.

Madison.—Horses are frequently attacked with staggers, which generally proves fatal. Cattle are healthy. Hogs are often attacked with the disease known as cholera, and large numbers of them die. They are also sometimes affected with thumps, from which they never recover.

Manatee.—Horses are troubled with epizootic, distemper, leeches, and sand; cattle with hollow-horn ; hogs with staggers ; and fowls with gapes, warts, sore eyes, and lice. The losses in this county, from the various diseases which afflict farm animals, will aggregate from $8,000 to $10,000 for the past year.

Marion.—With the exception of cholera among hogs, all classes of farm animals in this county have been exceptionally free from disease during the past season.

Orange.—Nineteen-twentieths of all the horses in this county have been imported from Kentucky and Tennessee. Of those that die a large majority are killed by "sanding," as it is called, the difficulty being a large collection of sand in the stomach. A large number of cattle annually die of salt sickness, and many hogs and fowls are lost by diseases common to them.

Taylor.—Lobelia has been found to be a cure for what is regarded here as hog cholera. It has been discovered in an old English work published in the seventeenth century that wood-betena (perhaps *betonica officianalis* is meant) was then raised and used as a cure for the hog cholera.

Volusia.—During the past year about sixty head of horses have died of staggers. A great many cattle annually die from starvation. Hogs die from various diseases, and in addition a good many are killed and devoured by bears.

Washington.—Horses that are attacked with staggers almost invariably die, as we have no specific for the disease. Cattle frequently die of hollow-horn, and a great many hogs are annually lost by cholera. A fatal type of distemper sometimes prevails among sheep, as does also cholera among chickens.

GEORGIA.

Baker County.—Our losses of farm animals by disease have not been very heavy during the past year. The usual diseases have prevailed, but not in a very malignant form.

Berrien.—There has been considerable disease and loss of cattle and hogs in this county during the past year. Whether the diseases are contagious or not I am unable to say.

Campbell.—A few cattle have died during the past season of murrain, and a good many hogs and fowls have been lost by a disease known as cholera.

Carroll.—Large numbers of hogs are produced here, but a great many annually die of cholera. But few sheep, horses, and cattle are raised in the county. Fowls die in great numbers of diseases peculiar to them. Dough made with blue-stone water is a good remedy. It seems to be too simple for ignorant people to have confidence in.

Chatham.—The only disease of a serious character that has prevailed among farm animals in this county during the past year has been cholera among hogs.

Cherokee.—Horses are occasionally subject to an epizootic distemper, but more die from staggers than all other diseases combined. This disease seems to be brought on by feeding unsound corn. Cattle are also subject to distemper, but the most fatal disease that affects them is murrain. Hogs and fowls die in large numbers from the disease known as cholera.

Clayton.—Farm animals in this county have been comparatively healthy during the past year, but fowls have suffered terribly. Three thousand dollars would hardly cover the losses among chickens alone.

Clinch.—We have lost a good many horses and mules in this county during the past year from blind, mad, and sleepy staggers. Both animals also die quite frequently from eating sand. A large number of hogs have died from cholera. At least one-half of these attacked die. Not more than one-fifth of the horses attacked with staggers recover.

Cobb.—We have been measurably free from contagious diseases among horses and cattle, but the so-called cholera plays sad havoc among our hogs and chickens.

Dade.—There has been quite a heavy loss of farm animals in this county the past year from the various diseases to which they are incident.

De Kalb.—Some few hogs have died the past season of a disease generally known as cholera. Fowls die in larger or smaller numbers every year of a similar disease. Some persons will lose nearly all their fowls during certain seasons, while a neighbor, living not over a mile distant, will not lose one.

Dodge.—We have had less than the average amount of disease among our farm animals during the past season. The losses are therefore light in comparison with some seasons.

Early.—The only losses worthy of mention among our farm animals have occurred among horses and swine. Four or five thousand dollars will cover these losses. The money value of sheep and cattle lost by disease will not exceed one thousand dollars. Our stock has not been visited by any serious epidemic during the past year.

Fannin.—Cholera has killed about 75 per cent. of the hogs in the northwestern section of the county during the past season. Other classes of stock seem to be healthy.

Fayette.—Losses of cattle in this county have been greater during the past year than of all other classes of farm animals combined. A few hogs have died of cholera.

Forsyth.—Staggers prevails more or less every year among horses and cattle in this locality. Hogs and fowls suffer greatly from contagious diseases, mostly, I believe, from a disease generally called cholera.

Gwinnett.—We have no infectious or contagious diseases among either sheep, horses, or cattle. Hogs are occasionally attacked by cholera, as are also fowls, and those that are affected rarely ever recover.

Haralson.—The losses among all classes of farm animals in this county from diseases incident to them will not amount to over $3,000 or $4,000 during the past season.

Heard.—The diseases general prevailing among horses are distempers and epizootic maladies, gravel, grubs, &c. Cattle are not doing very well, and hogs are afflicted with cholera, but not so extensively as in previous years. Sheep are affected with rot, and fowls with cholera.

Houston.—I estimate the losses by disease among all classes of farm animals in this county for the past year at about $10,000. The losses among swine will amount to half of this sum, although diseases have not been as prevalent as in former years among this class of animals.

Jackson.—Hogs, sheep, and fowls in this county are very unhealthy. A great many hogs annually die from a disease supposed to be cholera, but I fear it is the result of improper feeding, and a lack of attention in salting and the use of preventives. Fowls, like hogs, are subject to cholera.

Johnson.—We have no disease of special importance among horses or mules. There have been twenty or thirty cases of black tongue among cattle, of which two or three died. I do not know whether the disease is contagious or not. The cholera among hogs so far this season has been confined to about three herds. Every year we hear of chicken cholera, but it never occurs in flocks that are properly cared for, at least where their quarters are cleaned out often and fumigated twice a month with burning sulphur and tar. As regards diseases among swine, when a hog is sick, let the symptoms be what they may, it is called cholera; hence the great difficulty in ascertaining whether it can be cured or not. I have had one case which I believe from the symptoms was the cholera. He became very poor, after which the symptoms seemed to cease. I put him in a pen with others to fatten, but he became poorer; I gave him to a freedman and told him if he could make anything of him to do it; he fed the invalid until he became tired, and then turned him out; he ran at large with the other hogs, and though we fed him when he came about the lot, he became poorer and weaker until he finally became so weak that he could not get up without help after he had laid down; he finally rambled off into the woods, where he no doubt perished. This was the only case among our hogs, and he lingered five or six months.

Jones.—Hog and chicken cholera has been very destructive here for two years past; other classes of stock have been but little affected. If a remedy for hog cholera could be discovered, Middle Georgia could raise more meat than would be necessary to feed the people. Both hogs and cattle thrive well here. The annual losses among horses will reach from $4,000 to $5,000.

Laurens.—The swine and fowl plagues are extensively prevailing here, proving very fatal and causing much perplexity and loss. But little disease prevails among other classes of farm stock.

Lincoln.—Cholera has proved unusually fatal to hogs in some localities in this county during the past year. I have cured several cases of this disease in the early stages of the attack by drenching with a tablespoonful of melted lard and spirits of turpentine in equal quantities. This is regarded as the best remedy known here.

Lumpkin.—Hogs have been unusually healthy the past season. Some little cholera has prevailed, but the losses have been slight.

McDuffie.—We never lose any cattle except from neglect, want of shelter and food during the winter season. Cholera gets among our hogs frequently. Cause—neglect. Corn steeped in tar and fed to hogs will be found a good preventive of this disease. We lose a great many fowls every year by cholera. As yet we have found no preventive for this disease.

Marion.—Murrain prevails quite often among cattle in this county, and cholera among hogs and fowls. We have no remedies for these diseases, which generally prove fatal. Murrain seems to be brought on by neglect, and much of it might be prevented by a free use of tar, sulphur, and salt given in the early spring of the year.

Miller.—Hog cholera has been very bad here, and the most of the animals attacked died. A good many horses die every year of colic and staggers, blind staggers, or brain fever. Our cattle are subject to a great many diseases, one of which we have no name for. The symptoms are as follows: They lose their appetite and refuse to eat but little. Then they become very sore all over, and their feet grow tender. No remedy has been found, and a great many animals die of this remarkable disease. We have very poor winter pasture, and a good many cattle die of starvation.

Monroe.—Horses and mules are generally healthy, losses occurring more frequently from bad treatment than from disease. We have cholera both among our hogs and fowls. Sheep and cattle generally do well.

Muscogee.—Occasional cases of cholera among hogs (every disease of late is called cholera) are reported from different sections of the country.

Murray.—There has been no contagious diseases among the horses, cattle, or hogs in this county during the past year. The disease most fatal to sheep is rot, and that to fowls is known as cholera.

Oconee.—Cholera prevails to a considerable extent among chickens, and the loss among fowls in this county has consequently been quite heavy. Horses, cattle, sheep, and hogs are healthy.

Paulding.—Cholera has been quite destructive to hogs and chickens in this county. Other classes of stock are healthy.

Polk.—A few cattle have died here during the past year of distemper and murrain. Hogs are often affected with and die of hog cholera. This year they have been healthier than usual. Three years ago they nearly all died. I had a herd of forty-four head, and they all died but seven. During many years over half the fowls in the county are lost by the so-called cholera.

Putnam.—There are no contagious diseases among horses here, but there is a large loss by colic and disease caused by feeding wholly upon Indian corn and Indian corn fodder with no change. The loss from such causes is probably 2 per cent. of all the mules and horses owned and worked in this county. The loss among hogs is sometimes quite heavy, and is caused by cholera.

Severn.—Hog cholera, the most prevalent and destructive disease incident to farm animals in this county, is being treated with better results than heretofore. As good a remedy as we have tried is furnished by the native pine. Cold fresh tar, when made from the light-wood knot, if given in time, is a sure preventive. Some use the turpentine gum run fresh from the green pine, while others use the spirits of turpentine. All are good preventives, and in some cases have been known to render relief after attack.

Sumter.—About the usual number of deaths have occurred among farm animals in this locality during the past year. Our greatest loss has been among horses and cattle. Hogs are quite often afflicted with a disease known as cholera.

Tatnall.—This county affords splendid natural ranges for sheep, but the depredations by dogs are so great that but few are raised. There have been some diseases among this class of farm animals, and also among hogs, but the losses during the past season have not been very heavy.

Terrell.—The general condition of farm animals in this county is good. There is some complaint of hog cholera in the western portion of the county.

Towns.—Bots and distemper have affected horses to some extent in this county during the past year. Milk sickness and hollow horn have also prevailed among cattle, in many cases proving quite fatal.

Thomas.—Cholera prevails among swine, and generally proves very fatal. At least two-thirds of all animals attacked die. Fowls have cholera as well as hogs, and a goodly number die annually of the disease.

Washington.—There has been some distemper among horses, which has proved fatal in several cases. The cholera, as it affects hogs, proves fatal in almost every case. Nothing seems to check its ravages. I estimate our loss in swine for the year at $12,000.

Wayne.—I estimate the losses of cattle in this county during the past year at 5,000 head, valued at $30,000. I cannot say, however, that any contagious disease has prevailed among them. They generally die from starvation and exposure during the winter months. A good many horses and hogs have been lost by the usual ailments.

Whitfield.—Except among fowls no contagious diseases have prevailed among farm stock during the past year. A good many hogs have been lost, but whether the disease of which they died was contagious or not I am unable to say. The same might be said in regard to horses.

Worth.—Hogs would do well in this county but for cholera. Sheep husbandry is the business of the county. It is profitable, and is rapidly advancing. Some rot prevails, but it is not generally fatal. Distemper is the only disease that prevails among our horses. Black tongue is often quite fatal to cattle, and charbon, mange, and thumps to hogs.

ILLINOIS.

Adams County.—The only diseases among horses in this county that have entailed losses upon the farmers are distemper and lung fever. Hogs have been in better health than for several years past. In a few neighborhoods swine plague has done considerable mischief. I believe the improved condition of swine is owing to dry weather.

Carroll.—The only contagious disease prevailing among horses is a rather mild type of distemper. Hog cholera has been less prevalent and fatal than for many years past. A few hog raisers in the northern part of the county have suffered to a limited extent. There has been some cholera among fowls.

Cass.—With the exception of cholera among hogs, we have had no contagious diseases among farm animals in this county during the past year.

Clinton.—During a residence here of upwards of forty years I have never known our domestic animals so free from disease as during the past year. The only diseases of an infectious or contagious character are those common to swine and fowls, and they have not prevailed very extensively during the past season.

Crawford.—Cholera prevails among hogs in this county, as it does more or less every year. Fowls are also afflicted with a similar disease. All other classes of animals seem to be in a healthy condition.

Du Page.—The assessors return the value of hogs lost by disease in this county for the year 1878 at $10,335.90. I presume the losses have been about the same for the current year. I have no data as regards other farm animals.

Edgar.—Horses have been afflicted with distemper and hogs and chickens with cholera. I estimate the losses of hogs in this county at $45,000 by this disease during the past year.

Elkhart.—With the exception of hogs, all kinds of domestic animals are quite healthy. A large per cent. of these die each year of hog cholera. The losses generally reach from 25 to 40 per cent. of those attacked.

Effingham.—Horses that have been attacked by disease and lost in this county are generally of the poorer class. Farm stock generally is in good condition.

Ford.—Quite a fatal disease to horses has prevailed in some localities in this county during the past year. Some 240 head, valued at $80 per head, have died. By some it is supposed to be an affection of the kidneys. The animals have a high fever, and are weak and tender about the region of the kidneys. In some cases the animal eats heartily up to the time of its death. Hog cholera prevails to a considerable extent, and is more general and fatal among pigs and shoats than among older hogs. Our losses in this class of animals for the past year will reach from $15,000 to $20,000.

Grundy.—In speaking of diseases incident to domesticated animals I would say that the total value of annual losses considerably exceed those that occur from purely infectious and contagious diseases, as, for instance, disease of the kidneys, pneumonia, &c. In fact, it is not entirely settled as to what are and what are not contagious diseases. I am pleased to see that our veterinary surgeons have at last concluded to use the microscope in their researches. This will lead to definite knowledge, and it is hoped to at least a partial application of remedies.

Hamilton.—Distemper seems to be the only contagious or infectious disease prevalent among horses and cattle in this locality. I do not know what else to call the disease. Large numbers of hogs and chickens die of cholera.

Hancock.—Cholera among hogs and fowls has not been so general or so fatal the past season as in former years. Disease among cattle has been confined principally to bloody murrain, although there have been a few cases of dry murrain.

Hardin.—Since the year 1868 we have had no infectious or contagious disease among horses, cattle, or sheep. Swine and fowls become much more healthy as light reaches the farmers on the subject of diseases to which they are incident. My opinion is that nine-tenths of the diseases among swine originate in filth and from lice.

Jasper.—The only diseases at all troublesome to domesticated animals in this county are cholera among hogs and fowls and a disease peculiar to horses, which seems to affect the animal in the back and loins. It generally proves fatal, but is not consid-ered contagious.

Kankakee.—A few cases of distemper and lung diseases prevail among horses. Hogs are healthier this year than usual, though a number have died in localities where the cholera prevailed last year. Large numbers of fowls die every year from cholera and the effects of vermin.

Lee.—With the exception of sheep, which have been remarkably healthy for the past five years, there has been much ailment among all classes of farm animals. I estimate the losses for the year as follows: Horses, $12,000; cattle, $3,000; hogs, $5,000; sheep, $1,672; total, $21,672.

Lincoln.—Live stock, with the exception of hogs, seem to be free from any con-tagious disease at this time. I learn that hogs in some localities in the county are dying rapidly of cholera. These animals are generally confined, and it is to be hoped that the disease will not greatly spread.

Livingston.—I give the loss of hogs by disease this season at $2,584, and that of horses at $2,800. Many of our citizens think I have estimated the loss of hogs too low, but I think I have placed the figures full high. There is always a tendency to exaggerate such losses.

Macoupin.—The swine of this county have been but very little affected by disease the past year. All other classes of animals have been entirely free from disease.

McDonough.—There have been more catarrhal diseases prevalent among hogs the past season and less cholera than usual. The diseases are all equally fatal, however, and I think our losses for the year will reach $16,000. There has also been a heavy loss in horses and cattle, and these two classes will aggregate a sum equal to that given in the loss of hogs. The losses among sheep and fowls have not been very great.

McHenry.—The past season seems to have been a remarkably healthy one for stock. A few hogs have been lost in different localities in the county, but the owners do not claim that they were afflicted with contagious diseases.

McLean.—With the exception of hogs, all classes of farm animals have been excep-tionally free from all contagious diseases during the past year. They have been af-fected with the so-called cholera, but the disease has not prevailed so generally as in previous years.

Madison.—Our hogs have not been so generally or fatally affected with cholera during the past year as in former seasons. Other classes of farm animals seem to be free from contagious diseases.

Montgomery.—The value of hogs lost by cholera in this county during the current year will reach $12,000. No other class of stock has been seriously affected.

Ogle.—After a careful perusal of your report on the "Diseases of swine and other domesticated animals," I would say that the descriptions of the swine plague accord with our knowledge of the disease as we have found it in this vicinity. When a herd is affected (or infected) those in large numbers and closely confined suffer the most. I know of no effectual remedy, and think the better plan is to isolate the sick from the well animals, and then let the disease take its course, using only such sanitary treat-ment and disinfectants as may seem best. A few animals will generally survive. Let the pens and surroundings be thoroughly disinfected or purified by lime before any attempt is again made at hog raising on the premises. The scientific and thorough investigation by the commissioners of the various diseases of domesticated animals, with attention to preventives and treatment marked out by them, will, I believe, do much good; especially would this be the case if farmers and stock raisers could only be induced to read this report thoroughly and with a desire to profit by its instruction.

Perry.—Horses and cattle are absolutely free from infectious and contagious diseases in this county. Hogs are less subject this year than heretofore to the so-called hog cholera. Most of these animals die in the spring, when they are young and of little value.

Piatt.—Our stock this year has been remarkably healthy, with the exception of hogs. Our aggregate loss in these animals will reach from $20,000 to $25,000.

Pope.—The only disease incidental to farm animals that is at all destructive is that known as cholera among hogs. It prevails here to a greater or less extent every year, and the losses occasioned by its ravages impose a heavy tax upon the farmers of the county.

Rock Island.—Hogs are the only domesticated animals that are raised to any great extent in this county. The cholera, whether contagious or not, has been quite preva-lent among them this year, and has probably carried off as high as one-eighth of all the animals in the county.

Richland.—I give the aggregate of the losses of the various classes of farm animals in this county for the current year as follows: Horses, $6,060; cattle, $7,940; hogs, $13,845; sheep, $1,431; chickens, $429.

Shelby.—The diseases prevalent among horses are colic and epizootic distemper, and among cattle that of dry murrain is about the only serious ailment. The condition of farm animals is much better than an average.

Stark.—The only positive infectious and contagious disease prevailing among domesticated animals in this county is the so-called hog cholera. While the disease has not been so general and fatal as in previous years, the loss occasioned by it has imposed a heavy tax on our farmers. Cholera still prevails among fowls, but in a mild form.

Stephenson —We have 27,000 hogs in this county, and of that number 20,000 have been affected during the past year by the various diseases to which this class of animals are subject. About nine-tenths of those attacked have died, entailing a loss upon our farming community of some $45,000. In addition to this we have perhaps lost other farm animals amounting in value to $10,000.

Scott.—A large number of horses, cattle, and sheep have been lost in this county during the past year, but whether by contagious diseases or not I am unable to say.

Tazewell.—No contagious diseases prevail among any class of farm animals in this county except cholera among hogs. Chickens are also affected by a disease similar in character. About 10 per cent. of the former and 15 per cent. of the latter are annually lost in this county by this disease.

Vermillion.—Hog cholera has appeared in some localities in this county, but comparatively small losses have as yet occurred. All kinds of stock seem to be in a thriving condition.

Winnebago.—Swine plague made its appearance here about three years since, but prevailed south of this many years previous to that. It prevailed among hogs in this place last year, and is now affecting hogs brought from a distance. Those raised on the farm have not taken the disease. I estimate the losses in the county for the year at $29,000.

INDIANA.

Dubois County.—The only disease among farm animals in this county is that which is generally known as cholera among hogs. So far this year it has not been very general or very destructive.

Hamilton.—Hogs at times are carried off, at an alarming rate and great extent, by the disease generally known as cholera. The money value of the losses in this one class of farm animals will reach $45,000 for the current year. A great many fowls are also annually lost by cholera. I hear of no disease among other classes of animals.

Hancock.—We have been more fortunate with our hogs than usual, as they have suffered much less this than in previous years with the disease known as cholera. Our corn crop is short, and we will not have enough to fatten them.

Harrison.—A cerebral disease prevailed among horses in this county during the past season. It proved very fatal. The only remedy that seemed to give any relief whatever was diluted alcohol and morphine; but the disease run its course so rapidly that there was not much time for treatment. The same old disease has seriously affected swine and poultry during the past season. There are no remedies, and prevention seems to be the only measure of safety.

Jay.—No other classes of farm animals in this county are affected with disease except swine. They have been afflicted with cholera, a disease which has proved a terrible scourge to the farmers of our county. Horses are sometimes afflicted with distemper, but seldom die from its effects. A good many fowls are annually lost by chicken cholera.

Lawrence.—I give the aggregate losses among swine in this county, by diseases incident to them, at $7,713 for the year. This, I think, is rather below than above the actual loss. Other classes of farm animals seem to be healthy.

Marshall.—Hogs and chickens are troubled to some extent with cholera. Probably about 5 per cent. of the entire number annually die of these diseases. Horses, cattle, and sheep seem to be healthy.

Miami.—The only loss of farm animals we have sustained during the past year has been occasioned by cholera among hogs. Chickens also occasionally die of cholera. The value of the hogs that have died will reach $5,000 or $6,000.

Morgan.—There has been no infectious disease among farm animals in this county during the past year except among hogs, and I think the mortality among this class of animals has been 50 per cent. less than in former years. The disease prevalent among these animals is generally known as cholera. Other classes of domesticated animals seem to be in an average condition of health.

Perry.—With the exception of swine and fowls, all classes of farm animals have been remarkably free from disease the past year. Hogs and chickens have both been affected with cholera. I think fully 25 per cent. of all the chickens in this county annually die of this disease. Hog cholera has not largely prevailed this season.

16 C D

Randolph.—There have been but few hogs lost in this section during the present year, except by those diseases which seem to prevail at all seasons and in all sections. Horses, cattle, and sheep have so far escaped all contagious diseases. We have an abundance of pasture, and all classes of stock are doing well.

Ripley.—The only contagious disease prevailing among i .m animals in this county is that among hogs, and they have been healthier than heretofore. Fowls are most subject to disease. Very few of those that are taken sick recover.

Shelby.—Cholera is still making sad havoc among our hogs and chickens. The loss among the former has been very heavy. Horses are occasionally affected with distemper. Cattle and sheep are healthy.

Stark.—There have been no infectious or contagious diseases prevalent among farm animals in this county during the past year, except among hogs. This disease is called cholera, and it is too well known to require a description. Some horses, cattle, and sheep have died from diseases common to them.

Steuben.—The value of horses lost in this county last year by disease was $1,140. Hog cholera prevails to a limited extent. We have no remedy for this disease. All other classes of animals are in good health.

Tipton.—The farmers of this county have lost during the year at least 8,000 hogs by the so-called cholera and other diseases. These, at $3 per head, would make the value of our losses $24,000 for the year. There have been losses in other classes of farm animals, but not by contagious diseases.

Tippecanoe.—With the exception of hogs and fowls, all classes of farm animals have been remarkably exempt from all kinds of diseases during the past season.

Warren.—We have suffered a loss of at least fifteen head of horses this summer from a disease which is here called farcy of the heart. A few cases of lung fever and glanders have also been reported among horses. Cattle have been afflicted with black leg and murrain; hogs with cholera and thumps; and sheep with scab and foot-rot.

IOWA.

Appanoose County.—Black leg among calves and yearlings prevails to a greater or less extent every year, and is the only disease known in this county as affecting cattle. We also have more or less hog cholera every year. About all those attacked die. We have no remedy, and know very little about its cause.

Benton.—The hog cholera has prevailed in some parts of this county to an alarming extent, nearly all the animals in some herds having died with it. I think the loss in this county will amount this season to $30,000. Chicken cholera also prevails, and I should think that at least 15 per cent. of the fowls in the county annually die of the disease.

Calhoun.—Hog cholera prevails in our county to some extent. The animals commence with a very bad dry cough, and swelled neck. Some will eat and seem very hearty up to the time of their death. A good many hogs that weigh from 300 to 400 pounds are lost. All remedial measures have proved abortive. During my long residence here, I have found that pure, fresh water is the best preventive of this disease. Many cattle are dying from overloaded stomachs and indigestion. Perhaps the difficulty may be caused by smut, which is found in great quantities in all corn-stalk pastures. The husks which the cattle feed on are not only very dry and filled with smut, but the animals are frequently two or three days at a time deprived of water. The naturally dry condition of the food, and a failure to secure sufficient water to assist digestion, no doubt produces the disease which causes them to die so rapidly.

Cherokee.—Diseases among farm stock are quite rare in this county. Occasionally we lose a few hogs by a disease known here as cholera.

Crawford.—The cattle of this county are subject to black leg, a disease which seems to be superinduced by eating corn-husks in the late fall. Hogs, as usual, have suffered terribly with the disease still generally known as cholera. The value of the losses during the past year, among this one class of farm animals, will reach $45,000. Cholera, also, affects chickens, and sometimes plays sad havoc with all kinds of fowls.

Decatur.—The only contagious disease affecting horses in this locality is distemper, which prevails in all its various forms. Its fatality is mostly confined to old horses and colts. But little disease prevails among swine, and the general condition of all farm animals is better than an average.

Franklin.—There are no diseases prevailing among any other class of farm animals than the usual diseases among hogs. I think the losses during the past year will reach one-fourth of all the hogs in the county. They are still dying in large numbers. Large numbers of fowls also die every year.

Hardin.—Cholera has prevailed more or less among swine and fowls throughout this county during the entire year, but during the last few weeks it has increased rapidly. Our losses up to this time of swine alone will reach $35,000. The next heaviest loss has been in fowls, which I place at $7,500. To this may be added a loss of horses amounting to $2,500, and of cattle $6,000, making a total aggregate of losses of over

$60,000 for the year. This report is as near correct as a report can be made without a thorough canvass of the county.

Harrison.—Distemper has prevailed to some extent among cattle in this county, but the losses have not been very heavy. Hog colera prevails, and a good many animals have been lost. The fatality has been greatest among pigs.

Henry.—There have been no infectious or contagious diseases among farm animals in this county except among swine, and this has been confined almost entirely to one locality, including one township and parts of three others adjoining it on the east, south, and west. The infected section has been almost depopulated of hogs, many farmers not having a solitary animal remaining.

Iowa.—Distemper has killed a large number of horses and colts in this county during the past year. I estimate our losses among these animals at $20,000. . Cholera is the great trouble among swine, and has carried off at least $20,000 worth of hogs during the year. Cattle are healthy at this time, although for the year I think the value of our losses will reach $15,000. The value of losses among all classes of farm stock will reach from $55,000 to $60,000. As regards swine plague, the observations of Dr. Detmers exactly agree with what I observed in my own herd three years ago. If I had known the facts set forth in his report before the disease made its appearance among my hogs, I believe I would have profited at least $500 by the information.

Jefferson.—A few calves have died of a disease known as black leg and a few hogs of cholera during the past year. I am happy to say that all other animals have continued in good health.

Johnson.—Farm animals generally are in good health and condition in this county. In the early spring a number of shoats and pigs were lost by cholera, say from 500 to 800 in all. They were generally small, and not worth over $1 per head.

Marion.—With the exception of strangles, which has killed a good many colts in this county, horses are in a healthy condition. Scab prevails to a limited extent among sheep. In some localities in the county hogs are dying rapidly of some malignant disease which seems to be contagious. Of course it is called cholera. The lungs and intestines seem to be the principal parts affected.

Marshall.—The old disease common to hogs, and known generally as cholera, has prevailed with its usual virulence and malignancy among swine in this county during the past year Our farmers have lost upwards of 30,000 head by the malady, which were worth at a moderate estimate $126,000. A like disease has also prevailed among fowls, entailing a loss upon our farmers of from $8,000 to $10,000, in addition to the above heavy loss of hogs. Ducks and geese have so far been exempt from the disease which has destroyed so many chickens and turkeys.

Monona.—The only disease prevailing among horses is distemper. Nearly all our losses among cattle occur from dry murrain, a disease supposed to be caused by eating dry corn-stalks. Great losses continue to occur from hog cholera. These losses are largely among pigs and shoats and breeding sows. Cholera is also prevalent among fowls.

O'Brien.—Farm animals in this county are reported as generally healthy. A few horses and cattle have been lost by well-known diseases, and a large number of fowls have died of the disease known as cholera.

Palo Alto.—Glanders or nasal gleet is the only contagious disease known among our horses. Hog cholera is not known in this county. Black leg prevails to a considerable extent among calves and yearlings. Sheep are in a healthy condition.

Shelby.—Our hogs are seriously affected with some kind of fever, for which there seems to be no remedy. The losses for the year will reach from $9,000 to $10,000. Chickens also suffer from cholera, and the losses have been quite heavy.

Taylor.—Fifty-odd horses have been lost in this county during the past year by distemper. A good many cattle have also died of a disease known as black leg. Hogs, as usual, have been badly afflicted with cholera, and the losses have been heavy. A disease similar in character prevails among fowls. We have no remedy for this disease.

Washington.—I think nearly one-fourth of all the hogs bred in this county die each year of cholera. One of my neighbors has lost 190 hogs, and another one nearly as many more. Several others have lost their entire herds. So far my hogs have escaped disease. I attribute this immunity to feeding artichokes freely. I feed usually about 200 head of hogs. I have never known of a diseased herd where artichokes are used freely.

Warren.—All classes of farm animals in this county are reported healthy this season, with the exception of hogs, and disease has not been very extensive among them. A great many fowls have died of cholera.

Woodbury.—The so-called cholera has prevailed to a considerable extent among hogs in this county the past year. I am of the opinion that not one hog in ten that is attacked is affected with cholera. Nearly all the diseased animals that I have examined were afflicted with a lung trouble. The first symptom was a dry cough, and in a few days thereafter the animals would cough up small microscopic worms.

A change of feed and plenty of sulphur will effect a cure. At least such treatment has never failed with me if commenced as soon as the first symptoms were observed. A disease similar to glanders prevails among horses which run on bottom lands. No remedy has been found. A good many cattle have been lost by a disease similar to black leg.

KANSAS.

Anderson County.—Our horses during the past season have been afflicted with a most malignant type of distemper, from which many have died. Another disease, similar in its symptoms to that produced by overfeeding, has destroyed a good many animals. Hogs have the cholera and die in great numbers, especially where large numbers are kept together.

Chautauqua.—Some two hundred head of horses and upwards of four hundred head of cattle have died of disease in this county during the past year. The disease which has so fatally affected horses does not seem to be contagious. The deaths among cattle have been occasioned by Spanish fever and black leg.

Edwards.—A fatal disease, known as Texas fever, prevails among cattle here. No remedy is known. Other farm animals are in very good health and condition.

Elk.—Spanish fever has been very fatal to horses and cattle in localities where exposed to animals recently brought in from Texas.

Ellis.—The principal disease affecting horses is what is known here as blind staggers. In some localities in the county Texas fever has prevailed among cattle, and in others black leg, the latter prevailing mostly among calves. No remedies have proved of any benefit in these diseases.

Ford.—Our cattle are affected by a disease known as "Texas fever," which is brought into this county by cattle driven from Texas for sale and shipment from here. The disease does not prove fatal to Texas cattle, but is very disastrous to native stock. No remedy can be found for the malady.

Franklin.—We have distemper among or horses, and occasionally one dies of colic, bots, or overdriving. Cattle are generally healthy, though now and then a death is reported by black leg. These cases occur more frequently among calves than among grown cattle. Swine plague does not prevail very extensively this season, though it seems to be as fatal as in former years. It is reported as existing in but one township in the county.

Jackson.—The prevailing disease among horses is distemper, and that among cattle black leg. Cholera prevails among swine, but the disease is not very general this season. Have no data to estimate number of fowls. They are bred in large numbers, and at least 10 per cent. of the whole number die every year of cholera.

Johnson.—With the exception of swine, farm animals have not suffered to any considerable extent with either infectious or contagious diseases this year. Hogs continue to suffer from a disease known as cholera, but even this disease has not been so prevalent as in former years. The condition of farm stock is now better than for some years past.

Labette.—During the past year the loss of domestic animals by disease has been unusually heavy in this county. I estimate the value of the losses as follows: Horses, $15,000; cattle, $20,000; hogs, $13,500; chickens, $1,080. A peculiar disease prevails among horses in Southern Kansas. It resembles somewhat a decline or consumption in the human family, except that they have no cough. The animal continues to lose flesh until it becomes so weak that it seems impossible for it to rally. Those suffering from the disease do not lose their appetite, but the food they consume seems to do them no good. Horses brought in by emigrants seem to be more liable to the disease than those raised here. The mule seems to be more hardy, and they are now being more extensively raised than horses.

Leavenworth.—Hog cholera is an annually recurring epidemic in some parts of the county, apparently defying all sanitary and remedial measures.

Lincoln.—A good many cattle have died recently in this country of a disease we call black leg. The animal's tongue becomes so greatly swollen that it cannot eat. The disease then extends to the legs, and the animal soon becomes utterly helpless. It generally attacks young stock. There seems to be no remedy for the malady. Sheep die of a cold, which seems to have its seat in the head, and causes a discharge from the nostrils. They linger but a short time. A good many fowls have been lost by gapes.

Mitchell.—The usual losses have occurred from diseases among farm animals during the past year. Swine are more subject to a disease than any other class of stock.

Montgomery.—The only infectious or contagious disease that prevails here is cholera among chickens. I think the losses, at 10 cents per head, will amount to between two and three thousand dollars for the past season.

Nemaha.—About 50 horses are annually lost in this county from snake-bites and lightning. No diseases prevail among this class of animals. About two hundred head of cattle have been lost, but I do not think any of them have died of contagious diseases. This is a superior stock county, and the general condition of farm animals is good.

Osage.—The most noticeable disease that has recently prevailed among our horses has been nasal catarrh. Nearly one hundred head have been lost by this and other ailments. Black leg prevails among cattle, for which there seem; to be no remedy. I recently lost one-fourth of my own herd of calves by this malady. Hogs and fowls are afflicted with a disease known as cholera.

Osborne.—Cholera prevails extensively among fowls in this county, and the losses have been very heavy. A few sheep have been afflicted with common diseases, but the losses will not amount to much. The general condition of stock is above an average.

Pottawatomie.—The total loss occasioned by disease among farm animals in this county during the past year will perhaps not exceed $15,000. The cattle that die are mostly calves and yearlings, and the principal disease affecting these is black leg.

Sedgwick.—We have had the usual diseases among our farm animals in this county. Those common to swine have been the most general and destructive, though we have lost a good many horses and cattle.

Stafford.—A large number of horses have died in this county of a disease known as blind staggers. It is thought to be caused by feeding unsound and wormy corn.

Wyandotte.—The only losses of any consequence that have occurred during the past year among the farm animals in this county has been among hogs. The destruction of these animals has not been so great as in former years.

KENTUCKY.

Barren County.—No diseases of any consequence have prevailed among farm animals in this county except those common to swine. From long and close observation I am satisfied that diseases among these animals are caused almost entirely by lice and intestinal worms.

Breckenridge.—Hog cholera has prevailed to a limited extent in some localities in this county, as has a similar disease among fowls. Other classes of farm animals have been free from infectious and contagious disease.

Carroll.—The only loss of farm animals I have to report from this county has occurred among hogs. The loss among these animals has not been very heavy—perhaps not to exceed $5,000 for the year.

Clark.—Considerable losses have been sustained by the farmers in this county during the year among sheep and fowls. Diseases among swine have not been as widespread as usual, and the losses for the year have been comparatively light.

Clay.—Some cholera has prevailed among hogs in this county during the past year. The loss has not been very heavy, and only about one hundred head have died up to this time. Other classes of farm animals are healthy.

. *Clinton.*—Horses and cattle are healthy. There is some disease among sheep, but it does not seem contagious or very destructive. Hogs have had cholera, but the disease is not very prevalent at this time. Fowls have a disease called cholera, and many die.

Cumberland.—A few hogs occasionally die in this county of a disease somewhat resembling cholera, but they are so few that I can arrive at no accurate estimate of loss. Other classes of stock are healthy and in good condition.

Edmonson.—The only losses of a serious character sustained by the farmers in this county is from the disease known as cholera among hogs.

Harlan.—Some little disease has prevailed among horses, cattle, and hogs during the past year, but the losses to the farmers of the county up to this time have been very light.

Jackson.—While farm animals in this locality are in an average condition, still diseases prevail among horses, cattle, and swine. The losses among sheep and fowls have amounted to but little during the past season.

Jessamine.—Our county has lost a great many hogs by cholera, a disease for which we have no remedy. The best preventive I know of is to avoid in-and-in breeding. Frequent crossing of the stock, and salt and ashes given occasionally with the food, will be found beneficial. Jacob Thornton, one of my correspondents, says: "My plan is to prevent the disease by giving kitchen slops. I give them slop every day. At the age of eight months their weight is from 250 to 275 pounds net. Previous to the adoption of feeding kitchen slops I lost one-half of my hogs by cholera. I change my male hog each year. I have not had a diseased hog in four years, although the disease has been on all the farms around me. Copperas, lime, and ashes are all good preventives. I once slopped 1,200 head of hogs at a distillery, and lost eleven by cholera. Two distilleries within four hundred yards of mine lost from one-third to one-half by cholera. I bought my hogs in the same counties, but I used a bushel of assafœtida as a preventive. I put this in a sack, and laid it in the lead trough to the cooling tanks, and the result was a prevention of the disease. I found that if the hogs lived for six or eight weeks after being placed on this slop they were proof against the disease. If the disease was in the hog when put on slop, it soon developed. I lost no hogs after they became seasoned on this slop. I have had fresh hogs in adjoining pens die of cholera without affecting the older or seasoned animals After the disease is fully developed, I do not believe there is any remedy. A preventive is what we must rely on."

Knott.—Murrain among cattle is quite prevalent at certain seasons of the year, and is generally fatal. Hogs die at a fearful rate of cholera and blind staggers, as do also chickens of cholera. A good many horses are annually lost by bots.

Lewis.—Hog cholera has prevailed to some extent in this county, and the aggregate loss for the year will amount to $5,000.

Logan.—Horses and cattle are usually very healthy. I heard of one man who lost three cows by letting them eat corn-stalks that had been masticated by hogs. I hear of but little cholera among hogs this season. There have been some cases of quinsy. Usually a good many hogs die in this county of cholera, and a few with measles. Sheep are often affected with foot-rot, scab, &c. Fowls generally are subject to cholera.

Lincoln.—We have had no disease among either horses, cattle, or sheep; but about the usual number of hogs and chickens have died of the disease generally known as cholera. The condition of farm animals is about 20 per cent. below the average.

Livingston.—The only diseases of an infectious or contagious character prevailing among farm stock in this county are those affecting swine and fowls. All other classes of stock are healthy and in good condition.

McLean.—With the exception of hogs all classes of farm animals in this county are healthy. Every disease affecting swine is termed cholera. Occasionally cases of scab occur among sheep.

Martin.—The disease known as hog cholera has been very destructive to these animals in this county in former years. Recently, however, our county has not been visited by the disease to any considerable extent.

Monroe.—There has been some hog cholera in the county, but I hear of none prevailing at present. The losses have not been very heavy up to this time. Other classes of stock are healthy. This is a good county for sheep-raising, but no one seems to turn his attention to it.

Owsly.—We often suffer heavy losses in this county by the ravages of cholera among swine, but during the past season these animals have been unusually healthy.

Rowan.—This is a mountainous county and but little stock is raised here. What we have seems to be free from infectious and contagious diseases. A good many fowls are annually lost by diseases incident to them.

Russell.—Sheep frequently die here of rot. Probably 25 per cent. of the losses among these animals is caused by want of shelter and lack of attention. Hogs die rapidly of cholera and quinsy. Frequently as high as 75 per cent. of all these in certain neighborhoods die during some seasons of these diseases. Chicken cholera is also very destructive, and often kills as high as 75 and 90 per cent. of those attacked. We occasionally lose a horse from distemper, which is regarded as a contagious disease.

Spencer.—Hogs have had no disease this summer, and are in better condition than for several years past. What is the cause? The mortality among sheep was very heavy in the latter part of last winter.

Wayne.—We have had no contagious diseases among horses for several years past other than common distemper. Cattle generally are in excellent health. Cholera occasionally prevails among hogs, but it has not been very destructive this year. So far as my information extends cholera is the only disease prevailing among chickens.

LOUISIANA.

Bienville County.—We rarely have hores die of any disease except colic. A good many cattle die annually, but no one stops to inquire into the nature of the disease.

MAINE.

Cumberland County.—The losses of all classes of farm animals in this county for the past year, from the various diseases incident to them, will not amount to over $7,000 or $8,000.

Knox.—I have not heard of the prevalence of any diseases among either horses, cattle, hogs, or sheep in this county. A small percentage of fowls are affected and die of a disease which I take to be roupe.

Piscataquis.—Our farm stock is remarkably exempt from all kinds of diseases, the result, I think, of much better care in wintering farm animals than was observed in former years. The condition of stock at this date is more than an average. Fall feed was good, and they came to the barn in extra condition.

York.—I have been farming forty years without the loss of a single animal from infections or contagious diseases. I once lost a sheep from maggots in the head, and occasionally lose a fowl from breeding in-and-in. Abused horses, starved cattle, exposed sheep, and filthy hogs are often sick. Overfeeding and want of pure air and proper exercise kill many pet animals.

MARYLAND.

Howard County.—Diseases have prevailed to a considerable extent during the past year among both horses, cattle, and hogs, though the aggregate loss for the county will perhaps not exceed $5,000.

Montgomery.—My experience with horses and cattle is limited. My opinion is that the topography of a country being so diversified as to do with a healthy or unhealthy development of animal life than many are aware of. Fortunately this county, being an elevated and healthy locality, causes the prevention or introduction of diseases among our stock of more importance than the cure of the same. Since the epizootic epidemic the horses of the county have remained comparatively healthy. During the war glanders and other contagious diseases were quite prevalent. Since that period the health and improved condition of horses equals their former fine qualities.

I am of the opinion that several years ago a disease similar to pleuro-pneumonia was brought here, and that several cattle died of it. So far as I could ascertain at the time no remedies were of any avail after the disease had fully developed. Some cattle survived, probably from being in a previous healthy and robust condition and from the general healthy character of the locality. Ample room for exercise, with pure air and water, are nature's best remedies.

With the exception of cholera the diseases of swine are new to us. This disease, when fully developed, may possibly be cured. I once saved a hog that had been reduced to an almost dying condition, and had eaten nothing for four days. My remedy was wheat flour, salt, flaxseed, hog's lard, and tar, liquified by moderately warm water. The hog was drenched, as it could not swallow, and its life saved, but it never recovered its full health. It was fattened and slaughtered in the fall, but the pork was found to be very inferior. If the disease can be checked in the first stages the hog may be saved, but if diffused through the system I believe it would be economy to kill the animal, or at least to remove it from all contact with healthy animals. I consider the prevention of the disease more important than the cure. The hog must have room for exercise, pure air and water, dry and comfortable beds, and wholesome food. Hog beds require frequent changing or thorough cleansing. I have a small quantity of salt daily added to my hog slop, and by careful personal attention I have not had a single case of cholera in my herd for two years. Formerly the neglect of preventives and proper care and attention cost me many valuable animals.

MICHIGAN.

Calhoun County.—The only remarkable loss of stock has been in a few flocks of lambs in this county from goitre. No disease of a contagious character has affected any of our stock since the epizootic.

Cass.—The only disease of a contagious character prevailing among farm stock in this county is cholera among hogs. Even this disease prevails to a very limited extent this year.

Clinton.—Some losses occurred among lambs last spring, occasioned by goitre. In my opinion this disease is hereditary, as it can be traced to no known cause. It is a query in the minds of our farmers as to what can produce it. No disease exists among any class of farm animals at present.

Gladwin.—This being a new county, we have but little farm stock as yet. The only losses that I have heard of have occurred among horses. About ten or twelve have died during the past season, but from what cause I am not informed.

Houghton.—The only losses I have to chronicle as occurring from the death of farm animals by disease during the past year have occurred among cattle and hogs. These losses have been very light, and are hardly worthy of mention.

Otsego.—On account of the terrible ravages of a contagious disease which prevailed among swine in this county in 1878, but few efforts have since been made to raise hogs.

Presque Isle.—With the exception of cattle, all classes of farm stock have been quite healthy during the past year. A large number of calves have been affected, and a great many of them have died.

MINNESOTA.

Carver County.—No diseases of any importance are prevailing among farm animals in this county. Fowls are frequently carried off in large numbers by a disease called chicken cholera. They live from one to three days. The disease is contagious, and usually kills the whole flock.

Clay.—Epizootic in horses is common here, though in few cases proves fatal. What is commonly known as glanders is the only other infectious disease to which horses are subject in this county, and probably nine-tenths of the losses of these animals have been occasioned by this one malady. All other classes of animals are healthy.

Isanti.—Quite a large number of cattle have died in this county during the past year of a disease locally known as "big head." The swelling usually commences on the side of the head, or under the jaw, and continues until it reaches the size of a child's head, if not before slaughtered. A tumor forms, and finally bursts and becomes a running sore. This continues until the disease terminates fatally, which it generally does. No remedy has been discovered.

McLeod.—Some losses have been occasioned by black-leg among cattle. Otherwise the health of farm animals in this county has been very good.

Locquin.—Quite a number of cattle in this county have been affected with black-leg during the past year. Chickens are subject to the usual diseases.

Martin.—Among horses we have what is called distemper, from which a small per cent. of those attacked die. The same may be said of cattle and sheep. Young cattle (yearlings and two-year-olds) die occasionally of black-leg or dry murrain.

Pope.—Horses generally are in a healthy condition. Cattle suffer from black-leg, a disease which seems quite fatal. I am assured by our assessor that as many as fifty head died of the disease in this town during the past year. I lost two heifers myself by the malady. Hogs, sheep, and fowls are generally healthy.

Wadena.—Distemper, as it is usually called, has had a general run among horses during the past summer. But few animals have died, however, and they were comparatively worthless. Other classes of farm animals are in an average condition.

Watonwan.—Many horses have distemper, but few die. Black-leg is the only infectious disease among cattle. No animal attacked by it was ever known to recover.

MISSISSIPPI.

Amite County.—The losses occasioned by the various diseases incident to farm animals will aggregate from $10,000 to $15,000 in this county for the past year.

Calhoun.—A large proportion of colts and young horses in this county are subject to distemper; but as the disease usually prevails in a mild form, deaths do not often occur. The disease is similar, but not near so severe as the epizootic of a few years ago. Hogs are frequently almost all destroyed by cholera. Sheep are subject to rot, of which they die rapidly. Fowls have what we call cholera, and when it attacks a flock of poultry it generally kills about all.

Choctaw.—Staggers and charbon have prevailed among our horses in their worst and most fatal forms. No purely contagious diseases have visited our animals during the past season.

Covington.—The only contagious disease prevailing among any class of farm animals in this county is cholera among hogs. All other farm stock is healthy and in good condition.

Franklin.—The only losses that have occurred from disease during the past year have been among horses and swine. It is difficult to say whether the diseases among swine are contagious or not. There is one farm where the disease frequently prevails, but the affected hogs do not seem to communicate it to others, although they often run together.

Jefferson.—A few cases of glanders in horses, sore mouth among cattle, and cholera among hogs are the only cases of disease among farm animals to report from this county.

Lee.—Except for the hogs and fowls this is the healthiest locality for farm stock I have ever known. The hog and fowl plague, however, is truly discouraging. With fowls it is not now the cholera, but either a vegetable or an insect poison. I lost 300 head last year and 200 this year. Not one of my fowls confined in coops raised from the ground either sickened or died.

Marshall.—I estimate the losses of farm stock in this county for the last year, from all diseases to which domesticated animals are subject, at $24,000, as follows: Horses, $14,000; cattle, $2,400; hogs, $6,000; sheep, $1,500; fowls, $100. Mules are included under the head of horses.

Neshoba.—So far as infectious and contagious diseases are concerned among horses and cattle they seem to have entire immunity. Cholera is incident to hogs here always, and this accounts for the heavy loss among that class of animals.

Noxubee.—There have been but comparatively few horses and cattle lost in this county during the past year or two by diseases of any kind. Hogs suffer more than any other class of animals, and this generally from a disease commonly known as cholera. The losses among this class of animals during the past year will reach a total value of $15,000. A like disease prevails annually among fowls.

Oktibbeha.—No infectious or contagious diseases are prevailing among any classes of our farm animals except that among hogs. I find that where these animals are properly cared for they generally escape disease. Farm animals generally are better treated than formerly, and hence are not so liable to contract disease.

Prentiss.—Some cholera prevails among hogs in this county, but otherwise farm stock is in better health and condition than usual.

Tippah.—The aggregate losses among all classes of farm animals by disease, for the year just closing, will not exceed, in value, the sum of $10,000 for this entire county.

Tishomingo.—I have no losses of consequence to report except among hogs, a number of which (perhaps 500 head) have died of cholera during the past year.

Winston.—Horses die here from bots, grubs, glanders, colic, overfeeding on pease in the fall, distemper, &c. Chickens die of a disease commonly known as cholera, and hogs of an affection resembling measles. Sheep have foot-rot, of which some die.

Yalabusha.—No class of domesticated animals in this county is subject to contagious diseases except swine and poultry. Cholera is the disease from which they usually suffer. No satisfactory remedy is known here. It frequently occurs that nearly whole flocks and herds die when attacked.

Yazoo.—Distemper prevails among horses, and is often quite fatal. In the spring the buffalo gnats poison a great many horses and colts. Lung fever and murrain prevail among cattle, and the losses have been quite heavy. Hog cholera or pneumonia, accompanied by inflamed bowels, diarrhea, &c., prevails extensively among swine, as do roupe and catarrhal troubles among chickens.

MISSOURI.

Adair County.—No contagious disease exists among horses in this county, and but very few cattle, sheep, and hogs have suffered from such maladies during the year.

Andrew.—The season has been remarkably healthy for all kinds of stock excepting poultry, which suffered greatly from the effects of hot weather, and large numbers were lost. Some swine disease has been reported as existing in the northwestern section of the county. During the past decade a large number of barns have been built (formerly they were very rare), and this fact is supposed to greatly contribute to the health and general welfare of domesticated animals.

Ballinger.—With the exception of distemper, infectious and contagious diseases are very rare among our horses. Cattle are rarely diseased, and the annual loss is small. Hogs have the cholera and typhoid fever, and many of them die annually. No contagious diseases among sheep.

Barton.—Texas fever prevails among the cattle in this county occasionally, and some seasons it is quite fatal and destructive. The losses among cattle from this and other causes during the year will reach $9,000 or $10,000. Hogs have had the usual disease, but the loss has not been so heavy as among cattle.

Caldwell.—All classes of farm animals have been remarkably healthy during the past year except hogs, and the losses in this class have not been so heavy as in former years.

Chariton.—Very little disease prevails among domestic animals in this county except among hogs and fowls. During the past season hog cholera has prevailed to a less extent than for many years previous.

Douglas.—The losses among hogs in this county during the past year have not been very heavy—perhaps not to exceed $7,000 or $8,000 in value. Other farm animals are in excellent health and condition.

Franklin.—Disease among stock of all kinds is generally brought on by neglect and a want of proper care and attention. This applies to hogs as well as to other classes of farm animals, a number of which are annually lost in this county which might be saved by intelligent care.

Henry.—The only loss of farm animals in this county during the past year has been among hogs, resulting from a disease generally known as cholera.

Iron.—Along the line of the Texas Railroad cattle are frequently attacked by Texas fever, a disease which is very fatal to native stock. We had about 7,000 head of hogs in the county, four per cent. of which died of cholera.

Johnson.—The only infectious or contagious disease prevalent among the farm animals of this county is that of cholera among hogs. Young pigs and shoats suffer most. A number of horses and cattle have died from diseases incident to these classes of animals.

Lewis.—All kinds of stock in this county seem to be healthy except hogs. The disease known as cholera continues to prevail among these animals, and we have no remedy and, I am afraid, no preventives. But I think I can see that prudent men, who are watchful, and feed well and take good care of their stock, suffer least. We have the usual diseases among poultry.

Livingston.—Horses have suffered some from distemper. Cattle are subject to pneumonia, black tongue, and Texas fever. Hogs are suffering from cholera and pneumonia; sheep from hoof-rot, and fowls from cholera. Our losses during the year have been quite heavy, perhaps unusually so.

Macon.—No contagious diseases exist among the horses and cattle of this county. A great many hogs die every year; at least one-half of all that are pigged die of disease before they are in a marketable condition. A great many fowls are also lost by cholera.

Marion.—From an examination of the assessor's books, and from information obtained by consultation with persons from different sections of the county, I am enabled to give the total value of hogs lost in the county during the year at $4,000, and of fowls at $1.095. The disease causing this loss is commonly called cholera.

Mercer.—Horses, cattle, and sheep are remarkably healthy in this locality, but the same cannot be said of hogs. Diseases have prevailed to an alarming extent among them the past year, and the losses have been very heavy. Out of 100,000 head in the county, at least one-fifth of the number have died during the year.

Moniteau.—With the exception of cholera among hogs, we have had no contagious diseases among our farm animals for years past. Cholera is still prevalent among fowls.

New Madrid.—The losses from diseases among swine in this county during the past season have been very heavy—will aggregate, I think, up to this time, $9,000 to $10,000. Chicken cholera has also been very destructive to fowls.

Osage.—The only disease affecting farm animals in this county is that of cholera among swine. Several horses were lost, but not by contagious diseases.

Ozark.—A few horses have died in this county of a disease called blind staggers, and a number of cattle of murrain. A recent visit of the hog cholera played sad havoc among the swine. It destroyed at least 25 per cent. of all the hogs in the county.

Pemiscot.—No contagious diseases are prevailing among any kinds of farm stock in this county that I am aware of, except among hogs and chickens in a few localities. During the past season from seven hundred to a thousand head of hogs have been lost by this disease.

Perry.—Horses and cattle are very healthy in this county. Some hog cholera prevails. At times this disease is very destructive, but so far this year it has not prevailed extensively. Heretofore fowls were raised quite extensively here, but for two or three years past chicken cholera has prevailed so extensively among them that farmers' wives have become utterly discouraged, and now attempt to raise only enough for family use.

Platte. — The mortality among farm animals in this county has been fearfully heavy during the past year. Our losses of hogs alone will reach 50,000 head, worth, say, $200,000. In addition to this we have lost some 200 head of horses, 500 head of sheep, 500 head of cattle, and perhaps 100,000 chickens, making a total money loss for the year of about one-quarter of a million dollars. I have never known a hog attacked with the so-called cholera to entirely recover, and the same might be said concerning chickens when attacked by a disease similar in character.

Polk.—Our hogs have been afflicted to some extent with cholera during the past year, but the losses have not been so heavy as usual. The worst disease among sheep is that of scab, which prevails to a considerable extent among these animals.

Saint Genevieve.—Chicken cholera made its appearance in our county last year, and has since caused a heavy loss of fowls. No remedy has been found for either hog or chicken cholera, and about all attacked die.

Saint Francis.—This is a very healthy locality for all classes of farm animals except hogs, and a larger or smaller number of these animals die every year from cholera. The same may be said of chickens.

Saint Louis.—The prevailing disease among hogs and fowls in this county is that commonly known as cholera. There seems to be no remedy for this disease. Copperas fed to them, with a good supply of pure fresh water, is regarded as a good preventive. Sheep are troubled to some extent with scab.

Shelby.—Hog cholera is the only disease of any consequence prevailing among farm animals in this county. The losses have been quite heavy.

Stoddard.—Horses here are afflicted with distemper, cattle with black tongue, and hogs and fowls with cholera.

Texas.—The only disease prevailing among farm animals in this county is cholera among hogs. This disease is not so general this season as in former years.

Vernon.—This county has lost quite a large amount of stock of all classes during the past year by various maladies. No class of animals has been entirely exempt.

Washington.—We have no disease among domesticated animals in this county, except what is known as cholera among hogs. Chickens are also subject to a similar disease. We have not had as much hog cholera as usual this year, but the disease has been very extensive and destructive among fowls.

Warren.—I have no losses of farm animals to report, caused by disease, except among hogs. The so-called hog cholera has not been as general as in previous years, hence the losses have been lighter. Chicken cholera has prevailed to some extent.

NEBRASKA.

Boone County.—Diseases common to horses and cattle have prevailed to a considerable extent in this county during the past year. Ten or a dozen horses have been lost, and perhaps one hundred head of cattle.

Clay.—The only disease that affects horses seriously is distemper. I estimate the losses during the past year at $30,000. Some cattle have died of disease, and a good many sheep. Perhaps the losses among these two classes will amount to from $7,000 to $8,000.

Dakota.—Distemper prevails among horses, and black leg among young cattle and calves. Some years we lose a good many cattle from running on corn-stalk pastures. Large numbers of hogs and fowls annually die of cholera.

Fillmore.—The disease mostly affecting horses in this county is distemper, from which several deaths have resulted. An occasional case of glanders also occurs. Cholera proves very destructive among fowls.

Hall.—Every year we have more or less of a disease among cattle called black leg, and some seasons the losses from this malady are quite large. The so-called chicken cholera is prevailing in several sections of the county. I myself have lost 200 out of a flock of 300 fowls within the last year. About 5 per cent. of the horses in the county have died during the year from other than contagious diseases.

Knox.—Black leg is about the only disease that has prevailed among cattle in this locality, and it has been confined almost exclusively to calves and yearlings. It does not appear every year. But few hogs are raised in the county.

Nemaha.—So little disease is known among stock in this county that it is scarcely worth mentioning. For one or two years we had a considerable amount of "hog disease," but hog-growers soon became convinced that breeding to *fine points*—to color and hair, &c.—had sacrificed the constitution of the animal. A change of breeding—good substantial boars to native or common sows—gave them better constitutions and less disease. Another reform was also resorted to: they ceased ringing noses and allowed hogs to root as nature had intended they should. These reforms in breeding and treatment, with good shade and water, and disease is so rare as to not warrant mention.

Otoe.—I have to report a loss of hogs by disease during the past year of $20,000. The malady which proves so fatal to these animals is generally known as cholera. It is supposed to be caused in most part by a too exclusive corn diet, and the too frequent practice of feeding them in their filthy quarters. The disease is regarded as preventable but not curable.

Pawnee.—Diseases among cattle and hogs have been very destructive here the past season. I estimate the losses among cattle at 1,200 head, and hogs at 2,500. Large numbers of fowls have also been lost.

Pierce.—A disease prevails among cattle here every year which none of our farmers seem to know anything about. It appears in the fall shortly after the cattle have been turned on corn-stalk pasture. The animals die within from ten to twelve hours after the disease sets in.

Saunders.—An infectious disease has appeared among horses here. Twelve have been attacked, one has died, and two more will probably die. Four out of six are on one farm, and are regarded as well. One of those that died, and another one that will probably die, were fine stallions. The disease has been variously called gleet, catarrh, and glanders. Young cattle occasionally die of black leg. In the late fall and early winter cattle die of dry murrain, or what is usually called "smut poison." Swine plague still prevails. The largest percentage of deaths occur among pigs and shoats. Foot rot prevails among sheep.

Sarpy.—The disease prevailing among swine has taken a new turn this year. Very few pigs over six months old have been attacked. The animals become blind, and their tongues become black, resembling burnt meat more than anything else. They seem to be unable to either drink or eat, their tongues are so dry and hard. In many cases the sows cease to give milk when the pigs are only two or three weeks old. The black leg has not made its appearance among our cattle this fall. It is not an annual visitant, as we frequently escape it for two or three years in succession. I think if saltpeter was used, say once in two weeks, with salt, it might prove a preventive. Having used it with success last year I will continue it in the future, especially for young stock.

NEW HAMPSHIRE.

Sullivan County.—A number of horses and sheep have been lost in this county during the past year, but not by diseases regarded as contagious. Sheep have died from foot rot, worms in the head, and catarrhal diseases.

NEW JERSEY.

Camden County.—Strangles or distemper attacks about all the horses raised here and a good many that are brought into the county from elsewhere. A new disease made its appearance among horses during the latter part of September. It seemed to be contagious, as it went through a stable, some animals, however, suffering but slightly. The first symptom was a dullness of the animal and great lack of elasticity in its movements. This was followed by swelling of the legs, as in farcy, attended with great pain in attempts to walk, with considerable discharge of a thick mucus from the nostrils. This, in some cases, would last for weeks. Animals turned out to pasture recovered most rapidly. Applications of warm water to the legs appeared to be of benefit in reducing the inflammation. I heard of no deaths from the disease.

Cape May.—No infectious or contagious diseases have prevailed among farm animals in this county during the past season. I regard your report on diseases of domesti-cated animals as a valuable work, and am glad you have taken hold of the matter. The annual losses of farm animals by diseases incident to them are great, and they can only be combated and overcome when fully understood. Your report will go a good way in this direction.

Salem.—I hear of the prevalence of no diseases among domesticated animals in this county. There has been a heavy loss in fowls, as in previous years, from the disease known as cholera. From the most accurate information I can get I place our losses for the year at $15,000.

NEW YORK.

Franklin County.—Thirty years ago heaves was quite common among our horses, and also a disease then called glanders, but now a horse is seldom attacked with either. The cause of this change is unknown. Some attributed it to the then mode of thrash-ing grain, which required nearly all winter, the continued dust causing the heaves, which many thought ultimately terminated in glanders.

I am a firm believer in cause and effect throughout all nature. In those sections of the country where the hog fever has prevailed most, if they had not violated natural laws their hogs would not have had the fever. It is much easier to prevent disease than to cure it. One physician that will prevent the contraction of disease is worth three that can cure the same. If a person lives long on the finest brands of flour he is quite likely to contract dyspepsia, or some other disease, and, after resorting to dif-ferent medicines from which he finds no relief from his sufferings, perhaps the last resort will be coarser brands of food and a speedy return to health. Can we not so care for and feed our animals as almost to defy all manner of diseases?

Fulton.—Losses among classes of farm animals by disease will not aggregate $2,000 in this county for the past year.

Montgomery.—Losses by death and damage by abortion of cows during the year in this county will approximate $3,000. Whether "abortion" is a contagious disease is yet an unsettled question among farmers and professionals. The stock of the county the past year has been free from epidemic, endemic, and ordinary inflammatory dis-eases.

Onondaga.—A form of epizootic distemper has killed many horses in this county. Some have been saved by calling in veterinary surgeons. Cattle have been more or less affected by a similar disease. Sheep suffer with foot rot, scab, and grub in the head. The two former diseases yield readily to thorough caustic treatment; the latter usually results in death. We have but few cases of hog cholera. Many of these an-imals die in pens for lack of proper alteratives. Sulphur, soda, or rotten wood would cure if given in time.

Ontario.—The usual ailments have prevailed among the farm animals of this county during the past year, but no infectious or contagious diseases have been reported.

Putnam.—During the past summer and autumn pleuro-pneumonia, or lung plague of cattle, broke out in three distinct localities or centers, equidistant about ten miles, in consequence of which the towns of Kent, Patterson, and Carmel were placed under quarantine by General Patrick. The appropriation made for the suppression of the disease not being adequate, the State authorities were unable to inaugurate measures for completely stamping out the malady. Recognizing the importance of immediate and decisive action, the board of supervisors of the county assembled in special session, and by a unanimous vote pledged the credit of the county to indemnify the owners of all cattle killed by the order of General Patrick in pursuance of law, the same to be paid in case the State failed to make the necessary appropriation. The general, act-ing under the advice of Professor Law, caused the following herds to be destroyed, viz: herds of G. W. Patrick, town of Patterson, 33 head, seven of which had died; D. C. Elwell, 4 head; E. Leonard, 5 head; Freeman Sprague, town of Kent, 40 head, eleven of which had died; Alvah Hyatt, town of Carmel, 50, two of which had died. A herd of sixty head adjoining Mr. Sprague, of Kent, has been pronounced affected, but have not yet been condemned by Dr. Law. While the efforts put forth to stamp out this terrible scourge in this county are highly commendable and we trust will prove successful, yet we cannot but think that the danger to the whole country is so great that the earnest attention of the general government should be given to the

subject. The course marked out by Dr. Law, and as set forth in your Special Report No. 12, would seem to be the proper one.

Saint Lawrence.—Within the past few days I have heard of five cases of horses afflicted with a malignant throat disease, of which two of them died. The first symptom of the disease is observed in a refusal of the animal to eat grain or drink water. Unless relieved at an early stage of the disease the animal soon dies. I hear of no other ailments among farm animals.

Suffolk.—An unknown type of distemper prevails among horses in this county, but it does not affect these animals very seriously. They have a slight cough and a limited discharge of mucus from the nostrils. A few cases of glanders have occurred, and quite a large number of deaths have resulted from lockjaw, colic, and other acute diseases. There have been a few cases of lung plague among cattle brought in from other counties in the State, but the disease seems to have been so thoroughly stamped out by the State commissioners that at present I can hear of no case in the county. Sheep, where properly cared for, are healthy, but I hear of some flocks in which scab prevails. In limited localities there has been some loss from swine plague. The general plan in this county is to keep swine confined in inclosures, which affords some immunity from a rapid spread of the contagious disease incident to them. A good many fowls are annually lost from various diseases.

NORTH CAROLINA.

Caldwell County.—The usual diseases have prevailed among our farm animals during the past year, such as cholera among hogs, distemper among colts, and epizootic among horses.

Catawba.—But little disease of any kind exists among farm animals in this county, except among hogs. The disease affecting this class of animals is called cholera. In some portions of the county, and especially where they are permitted to run at large, the disease is very general and destructive.

Chowan.—We have had more lung fever this year among our horses than any other ailments. The animal often lives until the lungs are entirely gone, and continues to eat until death stops it. Our hogs have been afflicted with the usual diseases, and the losses will amount to three or four thousand dollars during the past season.

Gates.—With the exception of diseases among swine, all classes of farm animals in this county are and have been in good health.

Greene.—No diseases have recently prevailed among domesticated animals in this county, except among hogs and fowls. The disease is known only as cholera, and proves very destructive.

Haywood.—Hog and chicken cholera frequently prevails in this county, but neither have been very prevalent the past season. Horses are subject to distemper. All other classes of farm animals are healthy.

Hyde.—The diseases most common among horses are staggers and pneumonia. The animals attacked rarely recover. Cholera is still prevailing among hogs and fowls. Other classes of domesticated animals seem to be healthy.

Iredell.—Stock of all kinds are only raised in this county sufficient for home use. No special disease prevails among any class except that of hogs. This disease, which is also fatal to fowls, is generally known as cholera.

Lenoir.—A large number of hogs and fowls are annually lost in this county by a disease known as cholera. The disease, as it relates to hogs, has not been so extensive as in former years.

Lincoln.—The usual diseases have prevailed among farm stock in this county during the past year, but they have been of rather a mild form.

Macon.—I hear of no deaths from infectious or contagious diseases except among hogs and fowls. A few of the former and a large number of the latter have died from a disease generally known as cholera. There has been a very serious loss of cattle from milk sickness, a disease which seems to be increasing in our county.

Mitchell.—Hogs suffer with a disease which seems to be universally known as cholera. The symptoms vary, and what at one time proves efficacious, at others does no good whatever. There is no disease here among other classes of live stock.

Montgomery.—The heaviest losses among farm animals in this county during the past year have been among horses and cattle. The total value of losses of all classes of farm animals will reach about $10,000.

Pamlico.—The loss of horses is confined principally to colts, 25 per cent. of which die before reaching maturity. We rarely lose any cattle by disease, but about 20 per cent. of them are annually destroyed by storms and high tides. At least 25 per cent. of our sheep are annually destroyed by dogs. The loss of hogs and chickens sometimes reaches higher figures than these from the disease known as cholera.

Person.—Hogs are about the only farm animals affected by contagious diseases in this county. The losses have not been as heavy as usual during the past season.

Polk.—Cattle have been affected with distemper, and a few deaths are reported.

During part of the year the cholera raged fearfully among the hogs, but it is not now prevailing so extensively. Fowls are suffering greatly at this time with what we call chicken cholera.

Randolph.—The usual diseases have prevailed among farm animals in this county during the year, but the losses have been light.

Rowan.—Farm animals generally are in very good health and condition. We have lost some cattle, sheep, and hogs from diseases common to each class. Hogs have suffered from the disease so widely known as cholera.

Sampson.—There is no epidemic or contagious diseases prevalent among either horses, cattle, or sheep in this county. In many sections of the county, however, hogs and fowls are rapidly dying from cholera. About four-fifths of all the hogs attacked die, and those that recover had better die, as they are never of any value.

Tyrrel.—We have had the usual amount of sickness and loss from hog and chicken cholera during the past season. The general condition of farm animals, however, is very good as compared with previous years.

Washington.—Aside from hog cholera there has been but little ailment among domestic animals in this county during the past year. I should think the losses among swine would amount to $16,000 or $18,000. A great many flocks of fowls have also been terribly decimated by a disease known as cholera.

Wilkes.—Distemper killed a large number of cattle in this county during the past season. The value of those lost by this disease will amount to over $2,000. Cholera also destroyed a large number of hogs and fowls. No disease has prevailed among horses or sheep during the year.

Wilson.—I estimate our losses among hogs for the past year at $1,500. The disease has not been so severe as usual during the past season.

OHIO.

Ashland County.—The only epidemic or endemic diseases that ever prevail among the farm animals of this county are cholera among hogs and foot rot among sheep. A disease known as cholera sometimes attacks our fowls, but during the past year our stock has been singularly exempt from all maladies.

Fayette.—The only disease prevailing among farm stock in this county worthy of mention is cholera among hogs. The disease has been more than usually destructive during the past season. From careful inquiry I am led to estimate the losses in the county for the past year at $24,423. Cholera among fowls is also very destructive, but I have been unable to secure any reliable statistics as to losses.

Gallia.—A considerable number of horses, cattle, and swine have been lost by disease in this county during the past season. The aggregate will reach $12,000 or $15,000.

Geauga.—No infectious or contagious diseases have affected stock in this county the past year. Occasionally an animal dies, say from two to four of each class in a township. Horses generally die from bad management and over driving. A few sheep die every year, but I think more are lost from exposure and want of care than from any particular disease.

Guernsey.—The losses among all classes of farm stock in this county for the current year will amount to from $10,000 to $15,000. Hogs and chickens have suffered to some extent with cholera.

Holmes.—No disease of any kind prevails among farm animals in this county save a few cases of common distemper among horses, and some cholera among fowls.

Henry.—Some losses have occurred among horses by distemper and colic. Hogs die of what is called cholera, but I think the disease is similar to pneumonia or lung plague of cattle. The losses have been quite heavy the past year.

Logan.—The following are the losses among the various classes of farm animals in this county by disease for the past year, viz: Horses, $7,179; cattle, $3,659; hogs, $5,906; sheep, $1,922; fowls, $2,265.

Mercer.—Farm animals are more healthy at this time than for several years past. There is less complaint of hog and chicken cholera than any time during the past five years. This, I think, mainly attributable to the precautionary measures resorted to by all judicious farmers in the administration of antimony, copperas, &c., to their hogs as preventives of disease, and in not allowing them to remain too long in one place. Those who feed their fowls plenty of poke-root have no diseases among them.

Ottawa.—Upon inquiry, I cannot learn of the prevalence of disease among any class of farm animals. Even in the district of Bay Township, where the cholera was so fatal to swine last year, they have been entirely free from it this season, for which thanks be unto a kind Providence.

Preble.—A contagious disease termed distemper has prevailed to some extent among horses in this county. It is not very fatal, as not one in a hundred of those attacked have died. Quite a number of milking cows died in one locality of the county during the hot weather of October of a disease known in this county as dry murrain.

Shelby.—There is but very little cholera among our hogs this year. Other animals seem to be in very good health. I give you the following remedy for bloody murrain: Take one tablespoonful of alum and one of composition powder; pour a small quantity of boiling water over this, and after it has cooled inject with a syringe. The disease is not deep seated, and this injection will reach and cure it in almost every case. The alum acts as an astringent, while the composition powder is healing.

Trumbull.—The following losses have been sustained by the people of this county during the past year: Horses, $4,903; cattle, $10,004; hogs, $1,041; sheep, $4,206; total, $20,154. These losses were occasioned by the various diseases incident to farm animals.

Wayne.—The usual number of farm animals have been lost in this county during the past year, but I hear of no disease prevailing as an epidemic.

OREGON.

Columbia County.—Our swine are of the long-nosed kind. They run in the woods, eat worms, snails, and roots, drink pure water from mountain streams, and sleep under big cedar trees. During the six years I have resided here I have never known a hog to die of disease.

Douglas.—I can hear of the prevalence of no contagious diseases among either horses, cattle, or hogs in this county. Sheep are more or less affected with scab. I estimate that 25 per cent. of all the sheep in the county are annually affected by this disease, and that three-tenths of those affected die, not directly, but by continual scratching they rub off all their wool, and when a cold storm comes they perish for want of warmth.

Linn.—The heaviest losses that have occurred among farm animals in this county during the year have been among horses and sheep. The diseases among horses are mostly distempers. Hogs are occasionally affected by leaches, and sheep quite generally with scab and leaches.

Marion.—A great many horses die here of a disease we call staggers or yellow water. It is very fatal, and an animal attacked with it rarely recovers. Sheep have leach and scab, and a great many of them die. About ten per cent. of the fowls in the county annually die of cholera.

PENNSYLVANIA.

Beaver County.—A mild form of distemper has prevailed during the past year among our horses, but few have died from its effects. We annually lose from forty to fifty head of cows from puerperal fever. Foot rot and pales among sheep are very troublesome. Something like dysentery has played sad havoc among spring lambs of the long-wool flocks this fall. So far it has proved incurable. It attacks all classes in common. Parched corn seems to afford some relief. A large number of fowls die annually from cholera. One of my neighbors lost three hundred this fall.

Indiana.—We have no infectious or contagious diseases among our farm animals. There are a few sporadic cases of disease and death, but the loss is not large. Farmers in different localities suffer every year to a greater or less extent in the loss of chickens by cholera. Black leg prevails to some extent among young cattle, some distemper and lung fever among horses, and occasionally sheep are affected with scab and grub in the head.

Lehigh.—I estimate the entire loss of farm animals in this county for the past year, from diseases incident to the various classes of domesticated animals, at $39,729.50.

Perry.—Horses and cattle are healthy. Hog cholera prevails in some localities. A similar disease exists among fowls, which, during some years, carries off entire flocks. The few sheep we raise are healthy.

Sullivan.—Pneumonia prevails among horses and cholera among hogs. I find among all classes of farm animals that improved breeds are more generally liable to disease than the native stock.

Warren.—An occasional case of glanders among horses, or a disease similar to farcy, is the only contagious disease I have to report among any class of farm animals in this county.

Wayne.—Farm animals are very healthy in this locality at present. Some few cases of hog cholera are reported, but the disease is not prevailing to any great extent.

SOUTH CAROLINA.

Colleton County.—A few cases of blind staggers and unknown diseases have proved fatal to a good many of our horses during the past year. Hogs continue to die rapidly of cholera and measles. I have tried strychnine, in small doses, with good success in this disease. Fowls also die of a disease called cholera. By peppering them often I keep my fowls remarkably healthy.

Fairfield.—Horses and mules in this county die generally from bad treatment. Cattle are free from diseases—scarcely ever hear of a death. Hogs are frequently affected with a disease generally known as cholera, and this is regarded as epidemic and contagious. Fowls also die frequently with what we call chicken cholera. When it gets among a flock it generally kills all.

Georgetown.—We have epizootics among our farm animals, but, except among fowls, our losses are almost nothing. Fowls are destroyed by a variety of diseases in great numbers. The so-called chicken cholera, during the last spring months, swept through this whole county with great fatality, leaving in some townships scarcely a duck, turkey, or chicken. No remedy for the disease is known here.

Marion.—Horses to the value of $8,000 or $10,000 have died during the past year in this county. Staggers and a sort of malarial infection seem to be the principal causes. Perhaps better drainage would remedy this. Hogs and fowls die rapidly of cholera. Fresh green pastures to graze upon, and the administration of sulphur, tar, ashes, and fine coal are all claimed as preventives.

Oconee.—There has been a heavy loss of hogs and sheep in this county during the past year. Hogs suffer from cholera, and it is a rare thing for one that is attacked to recover. There have been some losses among cattle also.

Pickens.—The usual diseases incident to domesticated animals have prevailed here during the past year, but they have been less fatal than in previous years. The loss of swine has been greater than all other classes combined.

TENNESSEE.

Benton County.—Horses are frequently affected with blind staggers, and cattle with murrain and mad itch. Hogs are affected with cholera and quinsy, and chickens with cholera. Sheep are sometimes seriously affected with rot and staggers.

Campbell.—The losses during the past year in this county have been quite heavy among horses, mules, and swine, from the various diseases which annually afflict these animals.

Cannon.—Horses and mules are healthy. No contagious diseases are prevailing among cattle, but from 2 to 3 per cent. of our entire crop are annually lost for lack of care and attention. Our losses the past year will no doubt reach $20,000. Cholera is very destructive to swine. We have no preventive or remedy for this malady.

Carroll.—Big head, big jaw, swinney and big shoulder, colic, and bots cause some losses among horses and mules in this county every year. Glanders prevailed to some extent in the spring, but it was confined to limited localities. About one-third of the hogs in the county annually die of cholera. Fowls are subject to cholera, and nearly all die that are attacked by the disease.

Carter.—The losses among stock have been heavy in this county the past year. The ruling disease among horses and cattle is distemper. We have a new disease among our cattle this season—at least it is new to us. It is termed black leg, and kills very suddenly. Hogs die of cholera, sheep of distemper, and chickens of cholera.

Dickson.—A disease which the farmers call cholera is proving very fatal among hogs and fowls in this county. It is a contagious disease, and readily spreads from one locality to another. Some horses, mules, and cattle have died, but not from infectious or contagious diseases.

Dyer.—We have had little or no hog cholera this season, and our hogs are doing remarkably well. Horses are occasionally affected with distemper, and sheep frequently die of a similar disease. Many chickens die every year of cholera.

Fentress.—The losses in farm stock in this county from the various diseases to which they are subject, have this year been a little less than the average of former years.

Hancock.—Hogs have been unusually healthy during the past year; consequently there is an abundance of these animals in the county. I do not think the losses will exceed $5,000 for the past season. Our poultry crop has been enormous, notwithstanding some disease has prevailed among fowls.

Hardeman.—About 2 per cent. of all the horses in this county have been lost during the past year. This is more the result of bad treatment and neglect than of disease. A number of cattle and hogs have also been lost by the usual diseases.

Henderson.—The greatest loss of farm animals in this county has occurred among hogs. These animals were remarkably healthy up to within a few weeks past. The disease from which they are now suffering proves very fatal, but fortunately it is yet confined to but a few localities. It has been found that those affected are very lousy, and an application for the destruction of these vermin has to some extent prevented the spread of the disease.

Jefferson.—Cholera, as it affects hogs and fowls, has prevailed in a milder form during the past year than usual. Our losses, however, impose a heavy tax upon us. A good many horses have died by local diseases, and a large number of sheep for want of proper care and attention.

Lake.—The total value of horses, hogs, and fowls lost in this county during the past year by disease will reach $12,000 or $14,000. Hogs and fowls have died of the disease generally known as cholera. No infectious or contagious disease has prevailed during the year among our cattle or sheep.

Lawrence.—Although our climate is high and dry, our hogs are subject to cholera and our cattle to dry murrain. These diseases are generally fatal. Fowls also are subject to cholera.

Marion.—Our horses are subject to colic and occasionally to blind staggers. Cattle in some localities have the murrain, and hogs die only of cholera. Fowls are also subject to cholera. Sheep occasionally suffer with rot. Stock of all kinds seem to be in a better and healthier condition than usual at this season of the year.

Morgan.—No disease among cattle except hollow horn. Hogs are seriously affected with disease in different forms, but always called cholera. Fowls are visited with a fatal disease also usually called cholera.

Overton.—Fowls have suffered terribly from cholera during the past year. Our losses among farm animals have been comparatively light, but I think I am placing the estimate very low when I state that our losses among fowls will reach fully $6,000 for the year.

Robertson.—I hear of a limited amount of hog cholera prevailing in some localities of the county. The short corn crop will prove more damaging to our stock interests than anything else.

TEXAS.

Austin County.—Three or four thousand dollars will cover all the losses of farm animals in this county from disease during the past year.

Bowie.—A good many cattle and hogs were lost last winter and spring for want of care and nourishing food. The extreme winter destroyed the grass.

Brazos.—Glanders in horses is about the only contagious disease we have to contend with. Cholera destroys a large per cent. of our hogs and chickens every year.

Burnet.—All kinds of stock are remarkably free from contagious diseases except hogs, which suffer occasionally from cholera. Remedies are hardly ever attempted, as none have ever been known to prove effective.

Cameron.—The losses of all classes of farm animals in this county by disease will not exceed $10,000 for the past year. Owing to the free range, information as to sickness is very uncertain, as in most cases the owner himself will not know whether he has lost any animals or not.

Camp.—A good many horses have been lost by disease in this county during the past year, but more die every year from neglect and want of shelter than from disease. Several cattle have died of bloody murrain. Cholera still prevails among hogs and fowls, and is generally very fatal. No specific for this disease seems to have been discovered.

Chambers.—From the best information I am able to procure, I am led to believe that the losses of horses from disease in this county for the last year will amount to $1,500, and of cattle at least $20,000. Other farm animals have been comparatively healthy, except, perhaps, sheep. These animals are never entirely clear of disease of some kind, which prevail at certain seasons of the year more malignantly than at others.

Clay.—Stock is much healthier this season than last year. We attribute this to extremely dry weather, which has measurably prevented the decay of vegetable matter and kept the bulk of the food in sound condition.

Collin.—Horses and mules have suffered with a mild type of distemper during the past year. Sheep have also had scab, but it, too, has been in a mild form, and the losses have been comparatively small. Hogs have been entirely free from disease during the year.

Comal.—Some five hundred head of horses, and, perhaps, one hundred head of cattle, have been lost in this county during the past season. The losses mostly occurred from want of water, as the season was very dry.

Eastland.—A few calves and yearlings died in this county last spring of a disease called black leg. Some hogs have died, but I do not believe there is a case of the so-called cholera in the county at this time.

Gillespie.—Some eight or ten head of calves recently died here of black tongue. The carcasses were burned. Horses, cattle, and sheep are in a healthy condition.

Gonzales.—The cattle in this county are healthy, and the only losses that occur are occasioned by starvation in winter. The only disease which horses have suffered from for some years is distemper. Cholera prevails among hogs in a few localities. This disease is not so general as in former seasons. Sheep are afflicted with scab, but not many are lost by this malady. They also suffer from starvation during the winter months. Fowls, like hogs, are subject to cholera, and sometimes the losses are quite heavy.

Hamilton.—Distemper is the principal contagious disease affecting horses in this county. It is generally confined to colts and yearlings. Scab is quite prevalent among sheep, but its fatality is principally confined to old and feeble animals. Owing to the long drought, horses, cattle, and sheep will enter the winter in bad condition.

17 C D

Kerr.—A good many calves and yearlings are now affected with a disease known as black leg. Those in the highest condition are the ones generally attacked. Scab among sheep causes about nine-tenths of the losses among this class of domesticated animals. Chicken cholera has prevailed to an alarming extent during the past dry season.

Lamar.—A great many diseases have prevailed among farm animals in this county during the past season, and the losses have been heavier than usual. I estimate the losses among the various classes as follows: Horses, $16,665; cattle, $10,000; hogs. $80,000; fowls, $10,000.

Lee.—Large numbers of sheep are raised in this county. Nearly all are more or less affected with scab. But comparatively few animals die of this disease, as it is not very fatal, but it greatly weakens them, and in seasons of short range and inclement weather many of them perish.

Marion.—The only contagious and infectious diseases that prevail among farm animals in this county is cholera among hogs and fowls, and a kind of rot among sheep. But few attacked of either ever recover.

Maverick.—Scab prevails to a considerable extent among sheep, and a great many animals are annually lost by the disease. Chicken cholera is the most fatal and destructive disease we have to contend with. It affects almost all classes of fowls. Very often, within the short space of two or three days, flocks composed of a hundred or more fowls will be entirely destroyed.

Medina.—Various kinds of distemper prevail among horses in this county, and several animals have been lost. Cattle have a disease which affects their eyes; hogs are afflicted with cholera; sheep with scab; and chickens with a disease generally known as cholera. The losses among sheep have been greater than among other classes of animals.

Morris.—Glanders, staggers, and bots occasion heavy losses among horses in this county. We have no remedy for any of these diseases.

Parker.—Cattle generally die for want of proper food, and horses from various diseases. A good many hogs and chickens are annually lost by a disease known as cholera. Sheep die of scab. Cattle subsist on native grasses, and during winter some few die from poverty, but much the greater number are lost by dry murrain, a disease supposed to be caused by an insufficient supply of stock water.

Plano.—Very few farm animals are lost in this county by contagious and infectious diseases. The losses are generally heavier in sheep than in any other class of animals.

Red River.—Horses and cattle are generally healthy, as they are mostly reared on grasses and shift for themselves. It is true a great many die, but this is generally from neglect. Blind staggers is the worst disease we have to contend with—at least more losses occur from this than from any other disease among horses. The most fatal disease among hogs is cholera, and the annual losses occasioned by this disease are very heavy. The same can be said of cholera among fowls.

Rush.—Horses and cattle have had no contagious diseases. A good many sheep are annually lost by grub in the head. Hogs are subject to a disease called cholera, although the symptoms appear to be those of a fever. A nostrums seem to be of no avail. Chickens also have cholera, and a great many of them die.

Shelby.—All stock in this county is remarkably healthy this season, except hogs. They occasionally die in large numbers of the disease known as cholera.

Trinity.—A few hogs have died in this vicinity from a disease said to be superinduced by eating cotton seeds. Other classes of farm animals have remained healthy.

Tarrant.—The only farm animals affected by contagious diseases in this locality are hogs, which have the cholera. Chickens are also subject to this disease.

Uvalde.—Horses and cattle here are mainly affected with some kind of slow fever, known generally as Spanish fever, with an occasional case of blind staggers. Staggers is most common and generally fatal among horses. Hogs and fowls are affected with that fatal disease known as cholera. The most common disease among sheep is a skin disease, though last winter many flocks suffered with a malady similar to the fever prevalent among horses and cattle.

Walker.—A very destructive disease among horses, called "button farcy," prevails in the lower portion of our county. Murrain is prevalent among cattle, and but few attacked recover. A number of Durhams have died of a disease called mad itch, which terminates fatally within too short a time after discovery to attempt alleviation of the animal, which seems in great distress. A great many hogs die of a disease which is known here only as cholera. Sheep suffer greatly from screw-worm—the Texas pest.

Williamson.—The usual amount of sickness has prevailed among the domesticated animals of this county during the past season. As grass is very poor and stock will go into winter in bad condition, it is hard to tell how great the losses may be during the next few months.

Zapata.—No diseases are prevalent among cattle, but a disease similar to scab among sheep is extensively prevailing among horses, and seems to be very fatal.

Sheep are affected with inflammation of the bladder. The urine is bloody. About all die that are attacked.

VIRGINIA.

Caroline County.—In some localities in this county chicken cholera has proved so destructive that all efforts to raise fowls have been generally abandoned. Other classes of farm stock are healthy and doing well.

Dinwiddie.—Hogs here are more or less subject to cholera. By the free use of poke-root boiled and mixed with meal this disease is often cured. A free use of salt and a dose of spirits of turpentine once in two weeks is thought by us to be a sure preventive. The chicken cholera is our greatest foe. As no remedy has as yet been found for this disease, our farmers are afraid to go into the poultry business on a large scale. Horses, cattle, and sheep are unaffected by contagious diseases, and seem to be in good condition.

Floyd.—In some localities in this county a few cattle have died of black leg and murrain. Cholera prevails among hogs, in some localities quite extensively. Farm animals generally are in better condition than usual.

Gloucester.—The only infectious or contagious disease prevalent in this county at this time is cholera among chickens. I think that feeding fowls a little carbolic acid mixed in corn-meal dough once in a while is both a preventive and a cure for this disease. I have used it six years with great success, during which time I have taken no extra precautions as to cleanliness of roosts, &c.

Goochland.—The only class of farm animals in this county that I can hear of that are suffering with disease is that of swine. They have a disease generally known as cholera, which proves quite fatal. It is a rare thing for an animal to recover from it. Fowls suffer with the same disease, and it seems equally fatal to them.

Halifax.—Distemper is the only disease prevalent among horses in this county. The fatal disease referred to in special report No. 12, page 253,' under the head of "A strange disease among cattle," reported as existing in North Carolina, is prevailing among the cattle here. The disease is identical, according to the description given. Cholera or swine plague prevails more or less every year. It is quite prevalent and very fatal this year.

Madison.—Hogs have died of cholera to some extent, but the disease has not been very general. Sheep, particularly lambs, have been affected to a greater or less extent this year than for many previous years; consequently more losses have occurred. Fowls have suffered severely, turkeys and chickens particularly. I have lost fully one-half of my own flock of chickens.

Pulaski.—Farm animals in this county are in a healthy condition, Some losses annually occur among cattle from a disease commonly known as black leg. In the winter season sheep are troubled with catarrh, from which some are annually lost.

Rockbridge.—Some few cases of black leg have occurred among cattle, but the losses have been inconsiderable. Hogs and fowls have suffered greatly from cholera. I estimate the losses of the former at $10,000, and of the latter at $9,375.

Smyth.—Hog cholera has prevailed to some extent in this county, but has neither been so general nor so fatal this season as in some previous years.

Surry.—Horses are afflicted with distemper, and hogs and fowls with cholera. Cattle and sheep are healthy. The general condition of farm animals is above an average.

Sussex.—The disease commonly called cholera kills almost every hog and chicken that it attacks in this locality. We have suffered but little from the ravages of other diseases among our farm animals.

Warwick.—Some hogs have died during the past season of diseases common to this class of animals. All other kinds of stock are in good health and condition.

WEST VIRGINIA.

Brooke County.—A slight distemper or cough has prevailed among horses and cattle here for some months past. As yet it has not assumed a fatal type. Hogs and sheep are healthy, and cholera has almost disappeared from among fowls.

Cabell.—We have had no contagious diseases among either horses, cattle, or sheep. Some hog cholera has prevailed, and in almost every case it proves fatal. Chicken cholera has also prevailed to some extent, and it kills in nearly every case.

Calhoun.—Distemper prevails among horses in this county, murrain among hogs, and rot among sheep.

Gilmer—The losses among farm animals the past season have been unusually light. Some little cholera among chickens and rot among sheep are about the only diseases reported as prevalent.

Hampshire.—Glanders are occasionally quite fatal to horses in this county. Some animals seem to have been constitutionally affected by previous epizooty. We have hog cholera and a mangy scrofula, often fatal to these animals. Horned cattle are generally healthy.

Harrison.—Horses in this county are comparatively exempt from disease. At certain seasons of the year they are subject to distemper, but I do not think it prevails at present. We occasionally lose young cattle by a disease known as black leg. Cholera prevails among fowls at this time.

Jefferson.—The losses among farm animals in this county for the past year, from all diseases, will aggregate in value from $7,000 to $8,000. No contagious diseases have prevailed except among fowls.

Marion.—The losses among all classes of farm animals in this county during the past year, from the various diseases and ailments to which they are subjected, will aggregate from $7,000 to $8,000. Cholera is very destructive to hogs occasionally, and to fowls every year.

Monroe.—There is not now, nor has there been in a good many years, any other disease among cattle than that known as black leg, which carries off a good many calves each year and a few yearlings. At least 15 per cent. of all the hogs in the county are annually lost by the disease known as cholera.

Morgan.—This is a remarkably healthy county for all classes of domesticated animals except hogs and fowls. The losses for the past year have been comparatively light.

Nicholas.—There has been some distemper among horses in this county, and a few deaths among cattle caused by hollow horn. In the southern part of the county hogs are affected with disease, but the losses are not so heavy as in former years.

Ritchie.—The only contagious disease prevailing among horses here is a distemper, which is confined almost exclusively to colts under one year of age. It has proved quite fatal, as it has caused a loss of about forty head. Hog cholera prevails in a very malignant form. Nearly all die that are attacked. Sheep are affected with grub in the head. This disease is confined mostly to lambs, and nearly all die that are affected. Chicken cholera prevails extensively, and proves fatal in almost every case.

Taylor.—Hogs and fowls in this county are afflicted with cholera, and sheep with foot rot. The loss of hogs has been great, but I cannot give anything like a true estimate of their value.

WISCONSIN.

Brown County.—The only disease not mentioned last year, which prevails to some extent this year, is one located in the teats of cows, which causes them to give bloody milk. In several cases it has been noticed in two or three cows in a herd of ten or a dozen, and in several cases where one cow only would be attacked in a herd of two or three.

Dodge.—A few horses have died the past season of a disease resembling epizootic distemper, and some, perhaps, from improper feeding. The loss has been from one to two per cent. of the whole number. Fifteen or twenty head of cattle died last winter of a disease resembling bloat. They died suddenly and before help could be administered. Some were believed to have died for want of proper attention and care.

Lafayette.—Diseases among horses, cattle, and sheep in this county are quite rare. Hog cholera has prevailed for two or three years, but is not so extensive at present as formerly. I give the value of the losses as follows: Horses, $1,500; cattle, $3,000; hogs, $16,000; sheep, $1,000; chickens, $3,750.

Ozaukee.—Farm stock has been remarkably healthy during the past year. Horses have had the distemper, but in rather a mild form. A good many chickens have been lost by a disease which first caused their combs to turn black. After lingering a few days they would die in spite of all remedial measures.

Portage.—The only infectious or contagious disease prevailing among farm animals in this county is a disease among hogs. Last winter and spring a few were stricken with a sort of paralysis of the hind parts, and about fifty head died. This was not 5 per cent. of the whole number attacked.

Racine.—The prevailing disease among horses is distemper. Black leg among cattle prevails to a limited extent, and is very fatal. Sheep are affected with foot rot, but the disease does not seem very fatal.

Richland.—Distemper is prevailing to a limited extent among horses in this county, and hollow horn and garget among cattle. There is also some cholera among hogs, but the disease is not very general. A few sheep have been lost by grub in the head.

Rock.—I am unable to get anything like accurate statistics in regard to losses by disease among farm animals in this county. I think, however, that the losses among all classes, from the various diseases to which they are subject, will not amount to over $10,000 for the year.

Trempeleau.—The only infectious disease prevailing among our stock at present is an epizootic distemper among horses. This malady carries off a few horses every year.

Washington.—The only infectious disease prevailing among horses in this county is distemper. Until recently we have heard of no disease among swine, but for the last three or four weeks the so-called cholera has been raging fearfully in the western tier of towns, especially in the town of Hartford.

NUMBER OF DOMESTICATED ANIMALS IN EUROPE.

The Statisque Internationale de l'Agriculture, edited and published by the general statistical service of France, issued in 1876, contains an official statement showing the number of domesticated animals in Europe in 1873, the year in which the enumeration seems to have been made. The table is given below, and to this is added the number of farm animals in the United States in 1878, as shown by returns made to this department.

Table showing the number of domesticated animals in Europe.

Countries.	Horses.	Mules and asses.	Cattle.	Sheep.	Swine.
Great Britain	2,101,100		6,002,100	29,495,900	2,519,300
Ireland	532,100		4,142,400	4,482,000	1,042,244
Denmark	316,570		1,238,398	1,842,481	442,421
Norway	149,167		953,036	1,705,394	96,186
Sweden	438,090		2,026,330	1,636,201	382,811
Russia	16,160,000		22,770,000	46,432,000	9,800,000
Finland	254,820		997,960	921,745	190,326
Austria	1,367,023	42,976	7,425,212	5,026.398	2,551,473
Hungary	2,158,819	33,746	5,279,193	15,076,997	4,443,279
Switzerland	105,792		992,895	445,400	304,191
Germany:					
Prussia	2,278,724	9,708	8,612,150	19,624,758	4,278,531
Bavaria	351,669	228	3,066,263	1,342,190	872,098
Saxony	115,792	112	647,972	206,833	301,369
Würtemburg	96,970	199	946,228	577,290	287,350
German Dukedoms	133,122	674	1,114,178	544,611	621,067
Holland	253,393	3,466	1,469,937	898,715	611,004
Belgium	283,163	11,849	1,242,445	586,097	632,301
France	2,742,708	705,943	11,721,459	25,035,114	5,755,656
Portugal	79,716	188,640	520,474	2,706,777	776,868
Spain	680,373	2,319,846	2,967,303	22,468,969	4,351,736
Italy	477,906	718,222	3,489,125	6,984,049	1,553,582
Greece and Ionia	69,787	93,688	109,904	1,200,000	55,776
Roumania	426,859	6,734	1,842,786	4,786,317	836,944
Total for Europe	31,573,663	4,136,031	89,678,248	194,026,236	42,686,493
To which is appended United States (returns for 1878)..	10,938,700	1,713,100	33,234,500	38,123,800	34,766,100
Balance in favor of Europe..	20,634,963	2,422,931	56,443,748	155,902,436	7,920,393

INDEX.

	Page.
Alabama, Southern cattle fever in	103
Anthrax or charbon, virus of	130
Arkansas, diseases of farm animals in	233
Southern cattle fever in	103
Bacillus anthracis	131
B. malariæ	211
B. suis, period of incubation	31
Bacteria, changes of	61
late researches in	131, 133
California, diseases of farm animals in	235
Carbolic acid as a disinfectant	34
a preventive in swine diseases	66
in Southern cattle fever	135
Catarrh in sheep	214
Cattle in Europe	231
Charbon, a blood poison inserted by horse-flies	85
a disease of the bacteremia	209
a gangrenous erysipelas	87
germs from dead animals must be destroyed	210
inoculation of	209
lesions in the primary digestive organs	209
or anthrax, virus of	130
remedies for	83
Chicken cholera, caused by microscopic organisms	212
circular of French Government	214
contagious	212
experiments with	211
inoculation of guinea-pigs and fowls with its virus	212
pathology	213
Coal oil as a preventive of swine plague	92
Concealing animal diseases by farmers	37
Colorado, sanitary regulations in	216
Connecticut, diseases of farm animals in	235
pleuro-pneumonia in	166
Contagious lung-plague of cattle	142
Criticism on Dr. Detmer's curative methods	90
Cow died of charbon and swine plague combined	84
Dakota, diseases of farm animals in	235
Dead animals not to be thrown into streams	65
Delaware, diseases of farm animals in	235
pleuro-pneumonia in	176
Detmers, Dr. H. J., second report on swine plague	13
supplemental report	36
Diagnosis and pathology in Detmer's report very clear	92
Diseased animals, marketing of	36
Diseases of farm animals in Alabama	232
Arkansas	233
California	235
Connecticut	235
Dakota	235
Delaware	235
Florida	235
Georgia	236
Illinois	239
Indiana	241
Iowa	242

Page.
Diseases of farm animals in Kansas.. 244
　　　　　　　　　Kentucky 245
　　　　　　　　　Louisiana 246
　　　　　　　　　Maine... 246
　　　　　　　　　Maryland 247
　　　　　　　　　Michigan.. 247
　　　　　　　　　Minnesota : 247
　　　　　　　　　Mississippi 248
　　　　　　　　　Missouri.. 249
　　　　　　　　　Nebraska.. 251
　　　　　　　　　New Hampshire 251
　　　　　　　　　New Jersey 251
　　　　　　　　　New York 252
　　　　　　　　　North Carolina 253
　　　　　　　　　Ohio 254
　　　　　　　　　Oregon.. 255
　　　　　　　　　Pennsylvania...................................... 255
　　　　　　　　　South Carolina 255
　　　　　　　　　Tennessee 256
　　　　　　　　　Texas 257
　　　　　　　　　Virginia 259
　　　　　　　　　West Virginia 259
　　　　　　　　　Wisconsin... 260
Dogs susceptible to swine-plague infection............................. 25

Effect of eating flesh of cattle affected by Southern cattle fever 117
Excrement of horse-flies will decompose blood............................ 88
Experiments in swine-plague 18, 46, 56, 76
　　　　　　with chicken cholera...................................... 211
Extermination the only cure for swine plague 32

Facts and experiments in preventing and treating swine plague............... 46
Farcy in horses..:.. 202
Florida, diseases of farm animals in.................................... 235
　　　　swine plague in 94

Georgia, diseases of farm animals in 236
　　　　sanitary regulations in....................................... 217
　　　　Southern cattle fever in...................................... 102
Germany, regulations against rinderpest 197
Glanders, both contagious and infectious............................... 202
　　　　confounded with catarrh, ozena, strangles, and epizootic influenza.... 206
　　　　method of propagation... 203
　　　　post-mortem examinations...................................... 205
　　　　recoveries very improbable.................................... 207
　　　　reproduced by inoculation..................................... 205
　　　　spontaneous development rare.................................. 203
　　　　symptoms... 204
Grass-fed hogs 92

Hepatization of cattle by inoculation with swine-plague virus.............. 24
Horse-flies, description of.. 85
Horses affected by Southern cattle fever............................... 117
　　　　glanders and farcy in 202
　　　　in Europe.. 261
Human beings liable to swine infection 45

Infected animals should be killed and buried or cremated 64
Infection by cohabitation 68
　　　　in swine, facts illustrating the spread of..................... 40
Illinois, diseases of farm animals in 239
　　　　sanitary regulations in 218
　　　　Southern cattle fever in 106
Indiana, diseases of farm animals in.................................. 241
　　　　Southern cattle fever in...................................... 106
Iowa, diseases of farm animals in..................................... 242
　　　　sanitary regulations in....................................... 220

Page.

Kansas, diseases of farm animals in 244
    sanitary regulations in 221
Kentucky, diseases of farm animals in 245
    sanitary regulations in 222

Louisiana, diseases of farm animals in 246
Lyman, Dr. C. P., report on contagious pleuro-pneumonia 163

Maine, diseases of farm animals in 246
Malarial effects upon swine 94
    fever, inoculation of 210
Maryland, diseases of farm animals in 247
    pleuro-pneumonia in 176
    sanitary regulations in 223
Massachusetts, pleuro-pneumonia in 144
Michigan, diseases of farm animals in 247
    sanitary regulations in 223
Minnesota, diseases of farm animals in 247
Mississippi, diseases of farm animals in 248
    Southern cattle fever in 103
Missouri, diseases of farm animals in 249
    sanitary regulations in 224
Mules and asses in Europe 261
Murrain, a synonym of Southern cattle fever 103

Nebraska, diseases of farm animals in 251
    sanitary regulations in 226
New Hampshire, diseases of farm animals in 251
    pleuro-pneumonia in 163
New Jersey, diseases of farm animals in 251
    pleuro-pneumonia in 144, 161, 173
New York, diseases of farm animals in 252
    pleuro-pneumonia in 144, 153, 167
    sanitary regulations in 227
North Carolina, diseases of farm animals in 253
    sanitary regulations in 227

Ohio, diseases of farm animals in 254
    sanitary regulations in 227
Oregon, diseases of farm animals in 255

Pasteur's researches in bacteria 132
Payment for hogs destroyed 32
Pennsylvania, diseases of farm animals in 255
    pleuro-pneumonia in 170
    sanitary regulations in 227
Pigs in the uterus affected with the disease of the mother 86
Pleuro-pneumonia, a case of 21
    a second attack very rare 148
    compared with dangerous parasitic diseases 181
        glanders, canine madness, &c 181
        swine plague 181
        Texas fever 181
        yellow fever 180
    fully as infectious as in the Old World 146
    herds quarantined 158
    imparted by infection 152
    in Connecticut 166
        Delaware 176
        Maryland 176
        Massachusetts 144
        New Jersey 144, 161, 173
        New Hampshire 163
        New York 144, 167
        Pennsylvania 155, 170
    latency in the animal system 148
    losses from 144, 179
    methods of communication 146

P.ge.

Pleuro-pneumonia, most virulent in summer............................... 149
    must be stamped out by the national government.....154, 179, 182
    post mortem appearances................................. 150
    progress southward from New York...................... 144
    report of Dr. C. P. Lyman 163
    review of Dr. Law's work 143
    symptoms.. 148
    synonyms.. 143
Potash a disinfectant and destroyer of *bacilli* 96
Putrid measles, symptoms and course of............................... 84

Rabbits receive and impart swine plague............................... 25
Railroads should be forbidden to receive diseased animals............. 33
Rats receive and impart swine plague.......................11, 70, 71, 72
Rendering-tanks tend to spread swine plague.......................... 65
Review of Special Report No. 12...................................... 89
Rinderpest, ante-mortem phenomena.................................... 188
    causes.. 186
    contagious principle not discovered................... 186
    history... 185
    measures for prevention and extinction................ 196
    pathology... 192
    proportion of fatal cases............................. 192
    repressive regulations in Germany..................... 197
    symptoms.. 189

Salmon, Dr. D. E., report on Southern cattle fever.................... 98
Sanitary regulations in Colorado.................................... 216
    Georgia... 217
    Illinois 218
    Iowa 220
    Kansas 221
    Kentucky.. 222
    Maryland.. 223
    Michigan 223
    Missouri 224
    Nebraska 226
    New York.. 227
    North Carolina.. 227
    Ohio.. 227
    Pennsylvania 227
    Tennessee... 228
    Texas... 229
    Virginia.. 229
    West Virginia... 230
    Wisconsin... 230
Sheep in Europe... 261
    catarrh in.. 214
    receive and impart swine plague...............19, 68, 69, 70
Schizomycetes always come from outside the animal 64
    classification of..................................... 60
    no proof of their chemical action 63
    sometimes easy, sometimes difficult to destroy........ 63
    synonym of *Bacillus suis*............................ 60
South Carolina, diseases of farm animals in 255
Southern cattle distemper 142
Southern cattle fever, acclimation 137
    an epizootic.. 126
    bearing of late researches............................ 134
    beyond permanently infected districts................. 104
    boundary of infected districts........................ 104
    carbolic acid in 135
    caused by a vegetable parasite 139
    chemistry... 122
    climatic origin 112
    definition 98
    differs from enzootics 126
    disinfectants 136
    do Southern cattle suffer from it?.................... 113

Page.
Southern cattle fever, early accounts 98
       effect of weather and temperature....................... 110
       exists where intermittent fevers are unknown 125
       germs show increased resistance to cold................. 100
       great outbreak of 1883................................. 106
       has some characters of yellow fever 129
       histology 120
       horses affected 117
       how communicated...........................115, 116, 129
       interpretation of morbid changes 122
       infection not chemical, but organic....................128, 130
       in Alabama 103
            Arkansas 103
            Georgia 102
            Illinois.. 106
            Indiana 106
            Mississippi 103
            Texas 103
            the West 104
            Virginia... 100
       its extremely fatal character........................... 116
       legal extension of infected districts 99
       medical treatment..................................... 135
       nature of the contagion...........................123, 128
       not confined to its original localities 125
       peculiarities of incubation.............101, 107, 108, 128
       points of investigation needed 140
       post-mortem appearances 118
       preventive measures................................... 138
       retention of virus in the animal system 114
       suppression within an infected district 139
       synonyms...98, 117
       the disease permanent in infected districts............96, 101
       will it be permanent in the North?..................... 111
Swine in Europe 261
       methods of keeping................................... 34
       diseases less virulent in winter 7
Swine-plague, carbolic acid as a preventive 66
       coal oil as a preventive............................... 92
       combined with charbon................................ 85
       definition of.. 13
       dogs receive the infection 25
       experiments in...................................9, 18, 76
       facts and experiments in treating and preventing............. 46
       illustrating the spread of 40
       guinea-pigs infected and infecting cattle 74
       hygienic treatment 66
       infected animals must be at once removed.............. 66
       infection preserved by old straw-stacks 9
       probably disseminated by rats 72
       received and imparted by rabbits, rats, and sheep..... 10,
                                11, 25, 20, 71, 72
       by the human system 45
       retarded by drought and heavy rains.................. 39
       through food, water, and wounds 9
       in Florida... 94
       Henderson County, Illinois............................ 38
       inoculation by cultivated virus........................ 73
       from a pig infected by a sheep..................... 70
       in the Southwest...................................... 84
       lungs affected in summer and intestines in winter 14
       may be stamped out in winter 66
       measures of prevention................................ 32
       medical treatment..................................... 67
       microscopic examinations............................59, 84
       morbid changes 14, 31
       no case of spontaneous development.................... 66
       no immunity from second attack 8, 20
       not attributable to entozoa 7, 14

Page.

Swine plague, pig inoculated from sheep 69
      post-mortem examinations.............................. 16, 94
      prognosis 13
      propagated by *bacilli* 8, 20, 25
      results of examinations 64
      spread by rendering-tanks.............................. 65
            transportation of diseased hogs and carcases 65
      symptoms.. 13
      treatment.......................................35, 67, 96
      useless preventives 66
      very often spread through criminal carelessness 65
      vitality of the infectious principle 25
      white blood-cells first destroyed....................... 91
Synonyms of Southern cattle fever 98

Tennessee, diseases of farm animals in........................... 256
      sanitary regulations in 228
Texas, diseases of farm animals in 257
      sanitary regulations in 229
      Southern cattle fever in 103

Virginia, diseases of farm animals in........................... 259
      sanitary regulations in................................. 229
      Southern cattle fever in 100

West Virginia, diseases of farm animals in 259
      sanitary regulations in 230
Wisconsin, diseases of farm animals in 260
      sanitary regulations in.................................. 230
Worthless prescriptions for animal diseases 36